uild Your

wn Small

Power

System

About the Authors

Kevin Shea (Calverton, New York) founded and operates RE Power, Inc., a greenroots small-scale biodiesel manufacturing plant. He also designed, built, and preserves a not-yet-recognized national landmark appropriately called the Long Island Green Dome. It is the nation's largest and first LEED-qualified, Energy Star residential geodesic dome home made of renewable and reused material, and equipped with a solar array, a wind turbine,  and an ad hoc geothermal system. Mr. Shea is also designing and building the world's first "smart" wildlife reserve in the dry tropical forests of Ostional, Nicaragua, integrating old and new technology (including a wind turbine) for a unique experience that will allow people to have fun while saving the world. He is also crowd-sourcing alternate reality (pervasive) games for a new world myth. For more on all this and other information about Mr. Shea, visit www.google.com/profiles/kshea19.

Brian Clark Howard (Washington, DC) is an environmental journalist with a decade of experience in websites, magazines, books, and other media. He serves as a web editor at National Geographic.com, and before that worked for TheDailyGreen. com, part of Hearst Digital Media. Mr. Howard was previously managing editor of *E/The Environmental Magazine*, the oldest, largest independent environmental magazine in the United States. He has written for TheAtlantic.com, MailOnline.com, *Plenty*, *The Green Guide*, *Miller-McCune Magazine*, *Popular Mechanics* online, *Men's Health*, *Mother Nature Network*, *Oceana*, *AlterNet*, *Connecticut Magazine*, and elsewhere.

Mr. Howard co-authored two books for McGraw-Hill in 2010: *Green Lighting* and *Geothermal HVAC*. He also wrote the chapter on saving energy for the 2009 book *Whole Green Catalog* and the chapter on green power and green lighting for the 2005 book *Green Living*, which he also co-edited. Mr. Howard has bachelor's degrees in biology and geology from Indiana University and a master's degree in journalism from Columbia University. He was a finalist for the Reuters/IUCN Environmental Media Awards and appears on numerous radio and TV programs.

Build Your Own Small Wind Power System

Kevin Shea

Brian Clark Howard

New York   Chicago   San Francisco
Lisbon   London   Madrid   Mexico City
Milan   New Delhi   San Juan
Seoul   Singapore   Sydney   Toronto

The *McGraw·Hill* Companies

McGraw-Hill books are available at special quantity discounts to use as premiums and sales promotions, or for use in corporate training programs. To contact a representative, please e-mail us at bulksales@mcgraw-hill.com.

Build Your Own Small Wind Power System

2 3 4 5 6 7 8 9 0   DOC/DOC   1 7 6 5 4 3

ISBN 978-0-07-176157-4
MHID 0-07-176157-8

 The pages within this book were printed on acid-free paper containing 100% postconsumer fiber.

Sponsoring Editor
Judy Bass

Editorial Supervisor
Stephen M. Smith

Production Supervisor
Richard C. Ruzycka

Acquisitions Coordinator
Bridget L. Thoreson

Project Manager
Patricia Wallenburg, TypeWriting

Copy Editor
Lisa McCoy

Proofreader
Claire Splan

Indexer
Judy Davis

Art Director, Cover
Jeff Weeks

Composition
TypeWriting

This book is dedicated to everyone who is working hard
to make small wind energy part of the solution
for a cleaner planet, for today and future generations.

Contents

Preface

As a homeowner with an interest in saving money with less conventional means, I am proud to play a role in the early adoption of renewable energy, and to share my experiences with others.

I planned on installing my first wind turbine on my property after the challenging experiences of building my home, which I call the Long Island Green Dome. It is a large geodesic dome located in Calverton, New York, and it required unique prep work. This involved erecting numerous 20- × 10-foot canopy tents and securing them in the ground with posts. Well, the ground posts didn't hold well to the wind. When we returned one day, we found two tents missing. They took flight

Kevin Shea's unique home in Calverton, New York, which he dubbed the Long Island Green Dome. One of the largest residential wood domes in the United States, it is powered by solar panels and a small wind turbine. *Kevin Shea.*

and tore apart. One imbedded itself on the dome frame, and the other was in pieces, threatening to go over the angry neighbor's property.

After that, we tied the canopy to heavy railroad ties. But some of the workers were afraid to go on the top of the frame during windy days for fear of blowing off. Debris from the site was flying over to the neighbors' properties, making for more angry neighbors. Recently applied but not secured sheathing was constantly blown off the frame. Installing storm tarps on the structure required many hands and ropes, and I'll never forget the day I was lifted off the ground while holding the edge of a tarp that rapidly filled with the wind. I remember when I first had the doors on, the wind came with such force that it blew one of the heavy, 4 × 8 metal and glass doors out, breaking the control arm right off. It was that year that I began to plan to grab that energy for my own.

On windy and sunny days, Kevin's home produces an average of 8 kWh (kilowatt-hours). Since he gets paid 9 to 11 cents per kWh from his utility, he makes about $5.76 every eight hours. *Kevin Shea.*

I didn't have a lot of money, and didn't want to spend a lot of money. So, I spent time looking for something that was powerful enough, yet easy enough for a layman like me. It was the Skystream 3.7 by Southwest Windpower. I liked the idea that it had the inverter in the machine, that it looked sleek, didn't have guy wires that would take up space in my yard, and could be a bit hidden from the angry neighbors. I applied for my permit, and found out there was no code except an old rule stating that any structure must not exceed 35 feet. I said, "No problem," because the tower is only 33 feet.

The town official said, "Oh, no. You have to measure from the average grade of your property to the top of the blade as it is facing up." That was disappointing. I knew the six-foot blades would throw it over the limit. Skystream offered that size because it met many local code requirements and could still produce energy

Building the dome home was challenging, in part because of the property's high winds. It was this blustery experience that first gave Kevin the idea to explore small wind power—although it's also true that gusty days can be misleading at some sites. *Kevin Shea.*

(although as we'll soon explain, higher towers are almost always much better for energy production, and are ideally in the 100- to 200-foot range).

So, initially, the only option was to apply for a variance, which meant opening up the decision of building the turbine with my angry neighbors. Although that's a process we now recommend for a number of reasons, at the time, I decided instead on installing it exactly to the letter of the code. So, I dug the foundation deeper than necessary, bringing it lower than the average grade. When the town came to inspect it, the inspector would see that my tower was installed in a small "crater" garden, no taller than 35 feet from the average grade of my property. I didn't know exactly

The geodesic dome takes shape. Kevin did much of the work himself, although he recruited friends and hired some assistance. A lot of folks do the same for small wind projects. *Kevin Shea.*

Kevin originally called his house "Shea Stadium," after the New York Mets' arena. But after the pro baseball team built Citi Field, Kevin renamed his home the Long Island Green Dome. *Kevin Shea.*

what performance loss I would receive from this drop, but I was still ready to call victory.

People drop by and tell me that I am doing a good thing. That makes me feel great. And my power authority gives me a check (I combined wind and solar to exceed my needs) as per our net-metering agreement (more on this later). I am as close to being my own power plant as one can be on Long Island.

People frequently ask me a question that I didn't readily have a quick answer to ... until now: "Hey, Kevin, it's windy here in _____ (fill in the blank). So why can't we install some 'windmill' right on _____ (fill in the blank) to get some free electricity?"

That persistent question, and knowing the limited knowledge most people have of wind power, led me to propose this book on local, small-scale wind power. Fortunately, the answer is that we can, and should! And here is how!

Kevin Shea

Acknowledgments

Kevin Shea: Thank you to my coauthor Brian Clark Howard, who has offered me help to make sure this book is seamless and more digestible. His photos, interviews, and case studies bring this book alive. In addition, he was my Jiminy Cricket to the Pinocchio in me, a wooden author hoping to be a real author. His e-mail comments on my work as "awesome" were at times comforting, and other times (when I felt the chapter was weak) nerve-wracking, but he was also succinct and direct with the constructive criticism. I would also like to thank Seth Leitman for having enough confidence in me to bring me on board to this worthy project. It is as if he gave me a desk with my name engraved on it and an office window, turned my chair to face the laptop, and said, "Start writing! We trust it will be great." Special thanks to Judy Bass and the McGraw-Hill team for their insight to see value in this book (and e-book), and providing us the support to get it done.

I would like to thank the many seasoned innovators that brought the wind turbine to us. Without it, there would be a large hole in the world. I couldn't imagine a world where there is renewable energy from the sun, water, earth, but not from the wind. And also, I would like to thank those who took that call to adventure and returned to not only show us a turbine, but taught us the right way to do it. Mick Sagrillo, Ian Woofenden, Hugh Piggott, Paul Gipe, and the Southwest Windpower team (Ross Taylor, Therese, Russell Dixon Thayer, etc.) provided me additional training on wind power technology. In addition, I give a single standing ovation to all of the wind advocacy organizations, such as American Wind Energy Association (AWEA), which stand taller than a turbine tower and provide a clearinghouse of time-tested info on installation, performance, and safety.

I would like to take the time to personally thank Carl Vogel, author and friend extraordinaire, who always moves me with his acts of kindness, humble attitude, and keen intellect. He somehow convinced Seth that I can offer you, the reader, something worthy, encouraged me, and even consoled me when I was venting. I would also like to thank the young energy brat pack, President Jamie Minnick and volunteers of Eastern Energy Systems, a clean energy company whose team welcomed me in as the unofficial mayor of the company (since my position at the up-and-coming small company is still undefined). I was permitted to be a sponge as I learned on the job about assessing, installing, and selling wind turbines. Also, admiration and thanks to my lovely Luna (translates to moon in Spanish), Karla Cruz, who permitted me all unreasonable hours to work on this book, breaking sometimes only for a kiss, a hug, a loving look, and a large mug of cafe con leche. Also thanks to my mom, Mary Ellen Maggio, for not telling my good brothers that I was writing the book until it was printed, so they couldn't rub my face in it.

And lastly, but not least, I thank YOU. You bought this book/e-book. And you are getting as far as this section. Books that sell you wizards, wands, and wishes-that-could-come-true are usually the big sellers, so your interest in something that offers renewable power, revenue, and real-world solutions is encouraging to me—even if you give us mixed reviews in Amazon. Now, read the rest.

Brian Clark Howard: Thank you to my coauthor Kevin Shea, who has worked tirelessly on this book over many months. His hard work and dedication has been inspiring. I would also like to thank Judy Bass and the McGraw-Hill team for their professionalism and unwavering support. Hearty thanks go to Patty Wallenburg, who was essential in bringing this project to completion. Thank you to Seth Leitman, who honored Kevin and me with the invite to work on this book.

I would like to thank my mentors, Doug Moss and Jim Motavalli, who taught me so much about the possibilities of going green, and life, at *E/The Environmental Magazine*. I'd also like to thank my brilliant colleagues at National Geographic, especially David Braun, and at The Daily Green, especially Dan Shapley and Gloria Dawson, who inspire me every day. Thank you to Remy Chevalier (remyc.com), who has taken many hours to explain complex topics to me and who has steered me to many invaluable sources. I'd also like to thank all my friends in the green blogosphere and throughout the green movement. There are too many names to list here, which perhaps is a testament to how collaborative, supportive, and creative this space is. Every day I am honored to be a part of it.

I'd also like to thank my family, who taught me to respect and appreciate the natural world and to strive to leave everything better than how I found it. Thanks to my parents, Allan and Diana, and my sisters, Amy and Lisa. Thank you also to my wonderful, beautiful girlfriend, Gloria, who supports and challenges me.

Introduction

Welcome to a book about installing a small wind turbine system. Before we try to entice you with the world of wind through stories, mathematics, physics, price tags, laws, and the wrenches thrown into your machine that complicate things, let us say that with this book we hope to give you a good foundation for installing, or having someone install, a wind turbine. We want you to reach the turbine's top. This book is for you to not only understand wind from abstract text and photos, but will get you in the field, exploring, finding, and harnessing your own wind.

Kevin's "Ship Bedroom," one of several themed spaces inside the Long Island Green Dome. Others include the Paleolithic Bathroom (complete with cave wall shower), the Atlantis Bathroom, and the Noah's Ark Bedroom. *Kevin Shea.*

Although no book is a substitute for hands-on training, we hope to get you started on the process.

Will Wind Work for You?

So long as you have *property with sustainable wind above nine miles per hour (six meters per second)*, yes. It may be possible for wind to work for you with a bit lower average wind speeds, possibly with new technology, but we'll get to that, and show you how to understand the risks.

This guide is largely for those who own a plot of land, home, farm, or building and are interested in renewable, home-grown energy, as well as investing in removing or lowering electrical bills. We also include information for energy professionals, such as contractors, developers, energy auditors, installers, and other trades people. And for those who have none of this but want to see it happen in their community, wind power will work for you, too.

We have been acquainting ourselves with a sample of the people who are interested in small wind turbines; the plant nursery owner on Long Island who has an annual energy bill equal to the price of a mansion; the farmer and community that would like to earn from wind; the wealthy couple who drive a "luxury" Toyota Prius, have a soapstone counter, and polished fly-ash concrete ballroom floor and want to do more (and show it off); the middle-income homeowner who strives to

Kevin Shea with his Skystream 3.7 by Southwest Windpower. People often notice it from the road and stop to admire it, or ask if they could get a small wind generator of their own. *Kevin Shea.*

eliminate his energy bill; the hunters, Harry and Harriet, in their DC-powered weekend cabin; the people living without electricity or running water in a developing nation. We hope to address their needs and wants, and we look forward to acquainting ourselves with many others.

Periodic inspection and maintenance, as well as occasional repair, of a wind turbine are essential to the long-term success of a wind energy system. If you are a "put it up and forget about it" kind of person and can't afford to hire someone to perform an annual inspection and maintenance, we recommend installing a photovoltaic (PV) system.

Please note that this book has a strong do it yourself (DIY) component, so you or at least one of your team members should start procuring or cleaning your tools.

Is Wind Energy a Passing Fad?

No. While experimentation of energy from wave power, algae, and even human poop is promising, wind power is the fastest-growing energy resource in the world. And the amount of energy carried by the Earth's winds is much greater than current world energy consumption. The most comprehensive study of the issue, done in 2005,[1] found the potential of wind power on land and near-shore to be 72 terawatts, equivalent to 54,000 MTOE (million tons of oil equivalent) per year, or over five times the world's current energy use in all forms. In the United States, installed wind capacity now totals 35,603 megawatts—enough to power 10 million homes—and experts predict it could ultimately supply 20 percent of the nation's electricity without the need of energy storage. In the shorter term, the U.S. Department of

Wind power is the fastest growing energy resource in the world. *Kevin Shea.*

Energy (DOE) has set a goal of sourcing 5 percent of the country's electricity needs with wind power by 2025.

In 2010, work began on the world's largest wind farm, in the Mojave Desert, which is predicted to serve electrical demand for more than 275,000 homes. Wind power is also gaining ground in many other places, particularly in Europe (led by Germany, Spain, and Denmark), but also in developing countries, such as Nicaragua and Costa Rica. Denmark already gets about a quarter of its energy from the wind, and this is expected to rise in the near future.

Are Big Utility-Based Wind Turbines the Best and Only Solution?

Most wind power is now—and will continue to be—generated by large-scale wind farms composed of a series of interconnected towers, standing over 300 feet tall and weighing nearly 200 tons each. Large-scale wind is an exciting and emerging clean energy source, but it is located on the far side of your own electric meter.

Many consumers can now readily buy wind power, but at a cost: the power still has to be delivered to our homes and businesses. In fact, what usually happens is that people can only buy wind "credits," which support wind farms but don't guarantee that any electrons pumped to you were generated by those turbines. Once a wind farm is plugged into the grid, the power it produces goes into the general pool, which gets moved around by operators at blinding speed.

There's currently no way to distinguish wind-generated grid power from coal-generated power, or from any other source for that matter. Still, in principle, buying

For the foreseeable future at least, most wind energy will continue to be provided by utility-scale wind farms, which benefit from economies of scale and ability to access the strongest breezes. But the small wind industry still has a valuable role to play in a cleaner future. *Kevin Shea.*

wind credits can result in others recognizing you and your company as "green," and it does help support the growing industry. These credits are often also called "green tags" or "renewable energy certificates" (RECs), and they are an important and growing field, but they are frankly not the subject of this book.

Think Locally, Act Locally, Reap Locally, Affect Globally

Instead, this book focuses on the untapped potential for wind power much closer to home, literally. Local, home-scale wind power stands between your home and the meter. So right off the bat, you don't pay any delivery charge for those electrons, and you don't result in power lost via transmission. And because small-scale wind lowers your home's demand for power from the grid, it works beautifully with other renewable and conservation efforts.

Depending on how expensive the prevailing grid rate for electricity is in your area, and on whether your state or national government offers incentives for local wind power, you may discover that you can get financial support for a wind energy system. You may be very pleasantly surprised how quickly you see an annual return on your wind investment. This book will walk you through that analysis. If you happen to be building somewhere off the grid entirely, such as a rural cabin or on an isolated island, then small-scale wind can readily become a cost-effective part of your energy solution.

To be clear, small wind power turbines are electric generators that use wind to produce clean, emissions-free power for individual homes, farms, and small

Guests enjoy the swings inside the Long Island Green Dome. Renewable energy and efficiency projects can be a lot of work, but they can also be a lot of fun, and they can be a way to meet neighbors and like-minded folks. *Kevin Shea.*

businesses. The United States leads the world in the production of small wind turbines, although Asia and other countries are rapidly catching up. A small turbine is typically defined as having a rated capacity of 100 kilowatts or less—enough for providing for ten or fewer average U.S. homes. The turbine tower should be a minimum of 30 feet taller than nearby trees or buildings (although taller is usually much better, and the most productive home systems often approach 200 feet in height). It should be far enough from your neighbors in case an unlikely collapse occurs (follow our instructions and make sure you work with well-engineered designs, and you shouldn't go wrong).

The homegrown energy potential is enormous for local, smaller-scale wind power. The fastest-growing market segment within small-wind power is among grid-connected, residential-scale systems with a capacity of one to ten kilowatts (kW). This book focuses on how to install this kind of residential-scale system so you can watch your meter run backwards, and perhaps get paid by your power authority, when the wind is blowing.

Thoughtful utility companies like wind power, whether industrial or small-scale, as a cost-effective way to diversify their energy portfolio and avoid the price

Kevin wired up his wind and solar systems for easy monitoring and maintenance. For homeowners concerned with rapidly rising utility bills, a home wind power installation can immediately lower monthly electricity costs. *Kevin Shea.*

shocks common with coal, oil, and natural gas. For those concerned with reducing their greenhouse gas footprint to mitigate climate disruption, a single residential-scale (less than 10 kW) wind turbine displaces the carbon dioxide produced by more than 1.5 average cars. (If you are a truck or Nascar driver, get a 10 kW.)

For homeowners concerned with rapidly rising utility bills, a home wind power installation can immediately lower monthly electricity costs. For the "green jobs" angle, U.S. manufacturers have engineered one-third of small wind turbines worldwide. (Although there is rising competition from Europe and Asia.) If a farmer in Haiti, Botswana, or anywhere else wants to invest in small wind power to help them raise their standard of living, hopefully this book can help.

All that being said, we want to stress that installing and/or operating a small wind turbine is neither easy, cheap, nor without real risks to life and limbs. We asked Ian Woofenden, author of the excellent guide *Wind Power for Dummies* (For Dummies, 2009), for some advice. "Small wind is one of the hardest renewable energy technologies to implement," Woofenden told us via phone while he took a break from one of his many long bike trips.

"I probably talk ten times as many people out of it as I talk into it," Woofenden added. "It isn't right for everyone, and most people have unrealistic expectations about it [we'll get into that shortly]. Small wind power is not easy or cheap, and the smaller it is the worse it is."

We don't want to scare you off, but we want you to be aware of the real risks and limitations, and we hope this book helps guide you through what should be an enjoyable process.

How This Book Is Organized

The major remaining obstacles to a small wind power revolution are educating consumers and making public policies more wind-friendly. This book will address both these areas, drawing on examples from around the world. It will thoroughly explain the fundamentals of the technology, with facts and figures, and demonstrate why it is such a fast-growing industry. We will walk the reader through the "how to" basics of getting their own system up and running, including how to evaluate their site for wind power potential (and it's not about licking your finger), getting permits (pull hair here), and (gulp) financing it.

This book will take you through the step-by-step procedures to install your own turbine system, from purchasing options to maintaining the system. For those who can't build one right now, we also show the reader how to help promote more wind-friendly public policies, from local permitting and zoning to working with local utilities. It will serve as a valuable toolbox to help you get things done and make a difference. The final chapters in this book will analyze the breakthroughs in wind technology that are just around the corner, as well as provide a list of resources that readers can plug into to stay abreast of developments.

Many green technologies work well together, especially solar and wind energy. Kevin's unique home includes both, as well as a "green roof," which helps reduce harmful runoff and naturally cools indoor spaces. *Kevin Shea.*

Assumptions

- You are interested in effective wind electric systems.
- You want to know if a wind turbine is feasible for your home, farm, office, or development.
- You want to know how to achieve a "wind-win" scenario with neighbors and your town's code.
- You want to separate myth from truth about wind and wind turbines.
- Your goal is to support or experiment with clean technology.
- You are curious if you can actually install your own wind turbine.
- You want to begin developing the knowledge base or skill set to install a small wind turbine.

Feel free to skip anything you don't want to read. Each chapter is a stand-alone. And there will be cross-references among the chapters, so if you encounter a gap in your personal knowledge base, you will be able to quickly catch up.

Our Responsibilities

Facilitate the understanding of the core knowledge you need to manage the entire lifecycle of a small wind power system.

Important Warning

We did our best to be comprehensive, but no book can tell you everything you need to know about installing a complete wind system. That involves a wide range of technical skills, including wiring and electrical work, tower engineering, climbing, pouring concrete, hoisting heavy equipment, and much more. No book is a substitute for hands-on experience and training, and this is certainly true when it comes to small wind systems. There are numerous hazards in this industry that can lead to serious injury and even death, and it sometimes happens. Risks include falls, falling objects, electrocution, and blade strikes, to name a few.

If you're serious about getting your own wind system, we strongly recommend going with an experienced and qualified installer, or getting trained beforehand if you are going to do some of the work yourself. It is possible to build your own wind system with your own two hands, but it is difficult, and you will need more than this book. You will, at a minimum, need a mentor to show you how to complete some of the trickier tasks required. This book can't show you everything you'll need, but it is a good place to start.

The small wind industry does involve a range of technical skills, as well as risks. This book will help you understand some of the challenges and hazards, but no book is a substitute for hands-on training. *Courtesy of Rosalie Bay Resort.*

A small wind system is a considerable investment and takes some work to maintain, but it can also be very rewarding. *Kevin Shea.*

Conventions Used in This Book

We would like to share a word about the form of this book. The beginning of each chapter will have an inspiring or provocative wind quote from a VIP:

> *I can't change the direction of the wind, but I can adjust my sails to always reach my destination.*
>
> —JIMMY DEAN

This is followed by an overview box, which will inform readers what questions we are going to answer. Here is the model:

Overview
- What is the science behind wind energy?
- Why is wind the solution for me?
- Which systems will work for me?
- Is this going to be expensive?

In addition, we sprinkle subject-specific sidebars throughout the book.

 Hazard   *Installing a wind turbine is an inherently risky activity. Some of the potential risks include falls, being struck by falling objects, electrical shocks or even death. We have tried to place this symbol near instructions that carry significant risk, to underscore that extra vigilance may be warranted.*

 Power Up!   *Some eye-popping information or tidbit that may require consent from your guardian to read.*

 Tech Stuff   *A deeper, more technical look at wind science, often involving math equations, applied logic, and statistics. If you aren't technically-minded, you may want to skim over these sections.*

INTERCONNECT

Inspiring stories that bring the small-scale wind turbine community network across the globe to your home. So get out the welcome mat.

Resources

Resources that fully explore and explain this topic so thoroughly that you shouldn't just take our word for it.

QR Codes

In some cases, we want to provide additional information that can't fit into this book, so we provide an encoded URL or other content in a matrix barcode. Aim a mobile phone with a camera or smartphone with the proper scan software at the square, called a QR (quick response) code, click, and, voilá, up pops a link to websites of content, manufacturers, coupons, videos, and maps.

Scan me!

Kevin built his dream home on Long Island, and he's now busy building an innovative wildlife reserve in Nicaragua. What will you build? *Kevin Shea.*

After You Read the Book, Then What?

We plan to support this book on our own websites, and we will be doing follow-up podcasts, videos, blog posts, and so on on the topic. Our long-term goal is to build and keep alive a networked audience through social media. If you have no idea what we are talking about, just know that we are not done with YOU after you finish reading this book. You will be able to come back to us for more. Now that is a guidebook with a good return that keeps coming.

Now That We Have Your Attention

If you want to take part in one of the great undertakings of our time, get involved now.

In the process of your daily agenda, make a difference. After reading this book, you will know many of the core concepts and terms, advantages and limitations of small wind power, but we need to get your commitment in the beginning before you get totally enthralled with your project.

So here are five easy ways you can help *now*:

1. If you are acquainted with the legislators who represent you at the local, regional, and/or national levels, talk to them.

   Tell them you would like to see the promotion of safe, effective, and efficient small wind energy systems to reduce the on-site consumption of utility-supplied electricity and improve our economic and environmental "situation." And ask them what they're doing to
   - Accelerate the massive amounts of renewable energy that are needed
   - Give everyone an equal opportunity to participate in the renewable energy revolution
   - Make renewable energy feed-in tariffs available to everyone (see Chapter 6)

2. If you live in the don't-know-my-legislators category, please submit the statement and questions above to your legislators' e-mail at this website:
   - United States: www.congress.org/congressorg/directory/congdir.tt
   - Canada: www2.parl.gc.ca
   - European Union: www.europarl.europa.eu/parliament/public/near You.do?
   - List of National Parliaments: www.ipu.org

3. For those who use social media tools (e.g., Facebook, GoogleMe, YouTube, Flickr), please take action at the Repower America Website (shared by the Alliance for Climate Protection and the Climate Protection Action Fund) at: repoweramerica.org/take-action.

4. To register your support for electricity feed laws, also known as feed-in tariffs (FITS), and the renewable energy payments that make them work, go to the website for the North American Alliance for Renewable Energy at: www.allianceforrenewableenergy.org.

5. Become a member of any of the following wind energy associations listed on this website: www.ecobusinesslinks.com/wind_energy_association.htm.

Do not turn the next page until you have completed at least one act. Note: Your e-mail is *not* being placed on some third-party mailing list.

Did you complete one of the five items? If so, please continue to read.

First Wind: Introduction to Small Wind Power

I can't change the direction of the wind, but I can adjust my sails to always reach my destination.

—JIMMY DEAN

Overview

- What is wind energy?
- Where did wind turbines come from?
- What are the components of a wind energy system?
- What are the main drivers for wind energy?

Regardless if this is your first project or your umpteenth project, you are charged up about harnessing your own wind (Figure 1-1). So let's breeze through this first chapter.

Topics will include a brief overview of wind power history, advances in large-scale wind worldwide, and how these developments drive up the potential for local, small-scale wind. The chapter will also give a brief explanation of how the wind turbine system works. Last, we will outline the significant environmental benefits of wind power, as well as potential impacts, including addressing common myths and areas for concern.

A Brief History of Wind Energy

You might already have a good understanding of wind energy, but let us start with the basics.

FIGURE 1-1   A 10 kW small wind turbine on lattice tower at Windy Acres Farm in Calverton, New York. The farm is a popular tourist stopover, and so helps educate people about wind energy. *Kevin Shea.*

Wind is simply air in motion caused by uneven heating of the Earth's surface by the sun. The Earth absorbs and releases the heat at different rates because it is made of different types of materials. This produces cold and warm air masses that shift continuously. As warm air rises, cool air moves in. Wind is caused by differences in pressure. Where these differences develop, air is accelerated from higher to lower pressure.

Wind has been converted into useful forms of energy for millennia (Figure 1-2). Long ago, wind was used to build or "feed" a fire, spread seed, and dry cloths. Historians believe wind has been tapped to move sailboats for at least 5,500 years. In the 17th century B.C.E., the powerful Babylonian emperor Hammurabi recorded plans to use wind power for ambitious irrigation projects. (That's the same ruler responsible for Hammurabi's Code, one of the earliest written systems of law.)

It seems that people understood wind energy as a public good even back then. Over time, and with more innovation, wind began to be converted into mechanical energy through windmills, which turned grinding stones for making flour and drove pumps for moving water.

Many historians think the first windmills were invented by the Persians around 500 C.E. (Figure 1-3). These devices were made out of bundles of reeds and wooden frames, and were mounted on vertical shafts (taking the shape of what we now call vertical axis wind turbines, or VAWTs). They were housed in brick or clay walls, with an opening that allowed the wind to enter, and were used to grind grain and pump water. In the first century C.E., the Greek mathematician and engineer Heron

FIGURE 1-2 Wind has been used for millennia to fan fires, spread seed, and dry clothes, as well as for artistic purposes, such as this re-creation of traditional German wind art in Texas. *Brian Clark Howard.*

(aka Hero) of Alexandria invented a wind-powered musical organ. Similarly, by the fourth century, devotees in Tibet were cleansing their karma via wind-powered prayer wheels. (For good measure, they also had stream- and candle heat-powered prayer wheels.)

Many historians say the world's first truly practical windmills were invented in Sistan—a region now on the border of Iran and Afghanistan—sometime around the seventh or ninth centuries C.E. These were also vertical axis models, with rectangular-shaped blades and six to twelve sails of reeds or cloth. These early windmills were used to grind flour, process sugarcane, and pump water. Over time, the designs spread throughout the Middle East and Central Asia. They were also sometimes adapted to pump seawater to make salt.

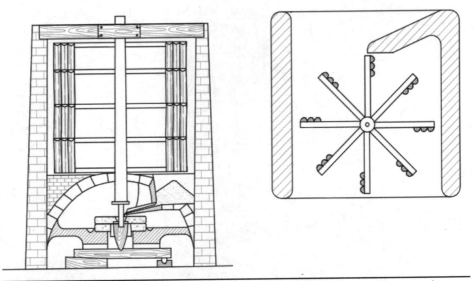

FIGURE 1-3   Many historians think the first windmills were invented by the Persians around 500 C.E. Their devices gradually spread throughout the Middle East, and were mentioned in early Islamic texts. *Kaboldy/Wikimedia Commons.*

The earliest definitive record of a windmill in northern Europe is a report from Yorkshire, England, in 1185. This example had a horizontal axle, predating today's horizontal axis wind turbines (HAWTs), and it was used for grinding grain and pumping water. Beer a la wind!

Some scholars have theorized that the horizontal design was adapted after contact with earlier vertical designs, perhaps as a result of the Crusades, but medieval technology scholar Lynn White, Jr., has argued that the Europeans independently invented their windmills (Figure 1-4). According to White, it wasn't so much a leap of technology as much as an evolution, given that waterwheels were already in use there. Wind-driven versions offered a number of key benefits, such as the ability to work even when the rivers are frozen. Plus, there are only so many babbling brooks to go around, especially when the best ones are already claimed by the nobility.

In the 14th century, the people perhaps most associated with windmills today, the Dutch, began using the machines to pump water out of their dammed watersheds, expanding their amount of arable farmland, and even permitting them to live below sea level. Many beautiful historic examples of the windmills remain standing today.

American history scholars believe the first windmill built on U.S. soil may have been erected in Jamestown, Virginia, in 1621. After that, a number were built across the continent as colonists expanded. Between 1850 and 1900, an estimated six million small windmills were installed on American farms to drive irrigation pumps. They were made by such firms as Star, Eclipse, Fairbanks-Morse, and Aeromotor.

FIGURE 1-4 During the late Medieval and Renaissance periods, horizontal axis wind turbines sprouted across Europe for pumping water and grinding grain. Can't you imagine Don Quixote tilting at these windmills in Spain? *Lourdes Cardenal/Wikimedia Commons.*

In 1887, Professor James Blyth of Anderson's College (now University Strathclyde) in Glasgow, Scotland, built the world's first *wind turbine*—meaning a device that harnessed the power of the winds to actually generate electricity. Blyth's machine was 33 feet high and had cloth sails, and it powered lights in his vacation cottage. Allegedly, when he offered to share his excess electricity with the towns-people, they declined, calling it "the work of the devil." Blyth would build another wind turbine for a local hospital, but his designs never caught on. Damn him!

In 1888, inventor and engineer Charles F. Brush built a massive experimental wind turbine in the backyard of his Cleveland, Ohio, mansion (Figure 1-5). Atop a 60-foot, 40-ton wrought iron tower, Brush's team placed a 56-foot-diameter horizontal axis turbine with a rotor composed of 144 blades—which provided a sail surface of 1,800 square feet (an average area of a U.S. home)—and a tail that was 60 feet long and 20 feet wide.

According to Green Energy Ohio, the 20-foot shaft inside the tower turned pulleys and belts, rotating this dynamo up to 500 revolutions per minute. It was extremely inefficient, in part because it was such a solid mass and because it turned slowly. And despite its size, the turbine could only produce a maximum of 12 kilowatts. It could power your New York home today, but the wind had to be strong enough to compensate for the steel shield of blades.

Brush's wind turbine reportedly lasted for 20 years, although it was obsolete within just a few, since Cleveland soon received grid power, which was more reliable and easier to work with.

FIGURE 1-5   In 1888, Charles F. Brush built a 12-kilowatt wind turbine in his Cleveland, Ohio, backyard. *Wikimedia Commons*.

In the 1890s, Danish scientist Poul la Cour also built wind turbines, and the technology really caught on in Denmark, which is still a global leader in the field. By 1908, Denmark had at least 72 electricity-producing wind turbines, ranging in output from 5 kW to 25 kW. The largest of these were mounted on 79-foot towers, with four-bladed, 75-foot-diameter rotors. In 1957, another Dane, Johannes Juul, reached an important innovation with a 79-foot-diameter horizontal axis wind turbine at Gedser. This design had three blades and faced upwind, and it established the most dominant design in the industry.

The 1920s, 1930s, and 1940s saw a "golden age" of wind turbines in America, as millions of units were installed across the country (Figure 1-6). Some were used on bridges and for other isolated structures, but most were bought for farms and ranches, where they provided lighting and ran machinery, typically with batteries. Manufacturers included Jacobs Wind, Wincharger, Miller Airlite, Universal Aeroelectric, Paris-Dunn, Airline, and Winpower. Most designs generated a few hundred watts to several kilowatts. The most popular model was a two-bladed, horizontal design from Wincharger, which produced up to 200 watts.

FIGURE 1-6   In the early 20th century, millions of small wind turbines were installed around the world, before widespread rural electrification. Many of these machines still spin today, like this one in Idaho, and some are still used for remote power and pumping. *Brian Clark Howard.*

During this period, some turbines were used in Africa and in other developing areas, and some were even used on Admiral Byrd's expedition to Antarctica. At the same time, the Dunlite Corporation sold many small turbines across rural Australia.

The rural electrification programs that started in the 1930s ended this golden age of wind turbine use, since getting grid power meant that it was no longer convenient or economical to keep the small blades spinning. As a result, fewer wind turbines were built for several decades. True, some very small turbines have long been used on boats (Figure 1-7)—in fact, German U-boats even had them in World War II—but there wasn't a lot of demand.

However, in the 1970s, rising environmental consciousness, the Arab oil embargo, and the back-to-the-land movement combined to feed new demand. At first, many people sought out vintage wind turbines—Jacobs Wind models were especially prized. Gradually, a small industry developed to recondition and service these machines, and over time, companies began to design and build new equipment. Some people also tried making their own wind turbines out of locally available scrap parts, such as hand-cut wooden blades, used car alternators, and old bicycle chains. These designs can work, as triumphantly proved by young Malawian tinkerer William Kamkwamba, the subject and coauthor of the best-selling book *The Boy Who Harnessed the Wind* (William Morrow, 2009). However, they are a lot of work, are often noisy, and usually don't produce much juice.

FIGURE 1-7   Small wind turbines have been used on boats for a long time, as proven by this 1902 image of the grounded ship *Chance* off New Zealand. *David De Maus/ Wikimedia Commons.*

Today, consumers have considerable choice when it comes to small wind turbines. They can shop for full-service wind solutions, or order kits and parts online and do the whole thing themselves. The industry is still relatively immature, especially compared to commercial wind, and many designs are new and without track records or real-world data to back them up. But that is changing, and new certification programs are coming online.

A Brief History of Commercial-Scale Wind Energy

On the utility scale, not much developed during the early decades of the 20th century. In 1931, the Soviets built the WIME-3D turbine near Yalta. This 100-foot-diameter, three-blade rotor was mounted on a 100-foot steel lattice tower. It produced up to 100 kW, meaning it would technically be classified as small wind by most experts today, despite its big size.

In 1941, the world's first megawatt-scale turbine was built, on the top of Grandpa's Knob Summit in Castleton, Vermont (Figure 1-8). The massive, 1.25 megawatt (MW) turbine was connected into the local grid, but it only lasted for 1,100 hours, since a known structural weakness resulted in failure. The builders, Palmer Cosslett Putnam and the S. Morgan Smith Company, reportedly couldn't make reinforcements due to wartime material shortages. It would be about 40 years before anything was tried to that scale.

In the 1970s and 1980s, the National Aeronautics and Space Administration (NASA) conducted extensive research on utility-scale turbines. Efforts were coordinated out of Lewis Research Center (now John H. Glenn Research Center) in Sandusky, Ohio, not too far from Brush's pioneering machine. The wind projects

FIGURE 1-8 The world's first megawatt-scale wind turbine was built on blustery "Grandpa's Knob" near Rutland, Vermont, in 1941. Rated at 1.25 megawatts, the project only lasted a brief time, since war shortages prevented repairs. *DOE/Wikimedia Commons*.

were funded by the National Science Foundation and the Department of Energy (DOE), with support from a public that was hungry for renewable, homegrown energy sources.

Several designs were tested at sites in Ohio, North Carolina, Washington state, and elsewhere (Figure 1-9). Major contractors included General Electric, Boeing, Westinghouse, and United Technologies. The work eventually led to many key innovations, including steel tube towers, variable-speed generators, composite blade materials, partial-span pitch control, and structural components. In 1987, the massive Mod-5B turbine was built in Hawaii, with a rotor of 328 feet and a rated power of 3.2 megawatts. The blades were segmented so they could be transported more easily.

Also in the 1980s, California passed tax rebates for wind power, and this kicked off the world's first commercial wind farms, the most famous of which is located in Altamont Pass. Those turbines are small (100 kW) and inefficient compared to today's commercial operations, but they set the stage for an exciting new industry, and they still spin today.

By the way, have you ever noticed which way wind turbines spin? Most old-fashioned windmills had their blades rotate counterclockwise, and that's how many wind turbines worked, too. However, in the 1970s, the Danes started making all their designs work clockwise, and soon it caught on as an industry standard.

Vertical axis turbines haven't seen as much development over the years, although NASA has tested them, and there has been a recent resurgence in them, despite some controversies (Figure 1-10). The Darrieus wind turbine, a key vertical style, was patented in 1931 by French aeronautical engineer Georges Jean Marie Darrieus. (More on VAWTs in Chapter 8.)

FIGURE 1-9   In the early 1980s, NASA, the DOE, and Boeing built a 7.5-megawatt cluster of three big wind turbines in Goodnoe Hills, Washington. The experimental wind farm was supported by the Carter administration and a public shocked by the Arab oil embargo. *NASA/DOE/Wikimedia Commons.*

Today, wind energy is primarily extracted to generate electricity, although some windmills are still used to pump water, primarily on remote ranches. Although small-scale wind is now overshadowed by commercial projects in terms of investment, political support, and public awareness, the small wind industry continues to show strong growth and considerable promise. Small wind turbines don't make sense for everyone or for all locations, but they definitely have a role to play in a cleaner future and in smart energy economics.

It's worth remembering that wind is a renewable energy source, because the wind will blow as long as the sun shines. There is no controversial belief regarding some "peak wind" theory, as there is with fossil fuels. Wind will be here for at least 20 millennia more, and we hope that human beings like you will be innovative enough to reap the benefits.

Recent Growth of the Global Wind Market

You are standing on the highest peak in the American Northeast, the 6,288-foot Mount Washington in New Hampshire. It's a forbidding landscape of wind-swept

FIGURE 1-10   A micro-VAWT mounted above solar panels on a light pole in a Brooklyn parking lot. VAWTs have enjoyed a resurgence recently, though they remain controversial. *Brian Clark Howard.*

rock, home to some of the planet's fiercest winds. All of a sudden, you are blown over by a terrific gust. You are swooped into the air, out of control, and at the mercy of winds with the amazing speed of 231 mph (103.3 m/s), the fastest ever recorded with an anemometer (outside of a tropical cyclone).[1] While you are up in the air, let's ride the trade winds across the globe to see what advances are being made in the wind industry worldwide.

Figure 1-11 shows a graph of world-installed wind power capacity for 1996 through 2008. Take this chart with you, because it can give you a glimpse of small wind turbine growth to expect on your journey.

Circling the United States, the wind rushes through numerous large-scale onshore (i.e., on land) wind farms. You are blown past the Altamont Pass Wind Farm in California, a bold step in the infancy of the market (Figure 1-12). The 5,400-turbine collection of residential-size turbines initially installed in the early 1980s provided lessons for future development. It has long been cited by conservationists as a primary reason why wind turbines can sometimes do more harm than good, as it ended up killing thousands of birds, including eagles and other raptors (more on this later this chapter). Note that 80 megawatts of turbines are to be upgraded by the end of 2011 to remedy this problem.

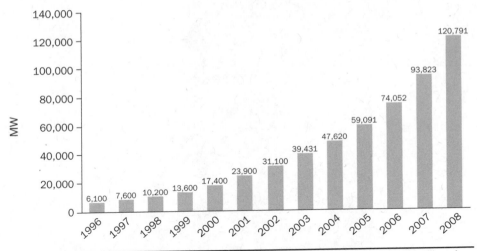

Figure 1-12 California's sprawling Altamont Pass is home to one of the world's most famous wind farms, which was started in the early 1980s. Shown are Enertech models that are considered vintage today. *Xah lee/Wikimedia Commons.*

On the other side of the coast, a massive offshore wind farm network is being planned from Maine to North Carolina. In between are 100,000 small wind turbines now operating throughout the country, providing about 100 megawatts of generating capacity. This includes boats, homes, farms, and businesses with spinning flags of wind power.

Swoop up to Canada, home to more than half of the small wind manufacturers worldwide, with models in the 30-50 kW range and to three-fourths of manufacturers with models in the 50-100 kW range. Small and utility-size wind turbines power over 1 million Canadian homes, with estimated wind energy potential for 17 million homes.[2] Despite its recent economic troubles and extended cold temperatures and harsh weather, Canada's small wind industry has increased by 55 percent over the past two years, according to a new market study conducted for the Canadian Wind Energy Association (CanWEA).

According to CanWEA's manager for small wind policy, Emilie Moorhouse, the country has an estimated 12 MW of installed small wind capacity. Almost 90 percent of that is due to systems smaller than 1 kW. A large percentage are off-grid, in very isolated and far northern areas.[3]

Taking the trade winds, we pass over sparsely populated areas of Greenland, the world's largest island, with harsh climatic conditions and little infrastructure due to cost constraints. Wave a salute to the task forces looking for ways to harvest one of the world's best onshore wind resources, due to its high average wind speed of 15 mph (7m/s).

At the gateway of Europe, you swoosh through history. You might see remnants of windmills that once dominated the landscape of Holland, also home of the Zeeland small wind turbine test site (Figure 1-13). Also, tip your cap to Denmark up north, which has long been home to pioneers of wind turbine development. Then there is Germany, Spain, and numerous other countries in the European Union that have seen small wind turbines planted like seeds in a field after the government agreed to pay feed-in tariffs for clean electricity wherever the site permits (see Chapter 6 for more on financing).

In the United Kingdom, which recently added a feed-in tariff for small wind, the growing trend is small, distributed, grid-connected projects (Figure 1-14). The United Kingdom has historically been the second-largest market after the United States, representing 20 to 25 percent of global demand.[4] The United Kingdom added an estimated 10,000 small wind turbines since 2005, for a total of over 15,000.[5] Installed capacity now exceeds 20 MW.[6] There is rising interest in many other countries in the European Union, particularly Italy, where small turbines are seen as offering "made in Italy" potential.[7]

At closer look, you can see a unique turbine application that is just beginning to enter the market around the world, in urban regions. Soaring from the forest of skyscrapers, flats, and municipal parks like Peter Pan on his search for Wendy, you can feel the sudden drop in wind energy as it wraps through the concrete maze.

Figure 1-13   The Zeeland small wind turbine test site in Holland, where researchers evaluate different designs. *Jeroen Haringman/www.solarwebsite.nl.*

Figure 1-14   The United Kingdom recently added a feed-in tariff for small wind, which is helping spur the industry. Pictured is an Ampair 6000 wind turbines in Berkshire. *Ampair.*

FIGURE 1-15   Placing wind turbines on rooftops and in urban areas is unlikely to produce much energy, unless the winds are very strong. Pictured is an experimental Honeywell wind turbine on the roof of the Solarium, a green-themed apartment building in Queens, New York. *Brian Clark Howard.*

This environment is not generally conducive to effective use of large or even small wind turbines. However, micro-turbines, dubbed urban wind turbines (UWT), are in development (Figure 1-15). If new field-test certification standards arrive for UWTs, it may bring a breath of fresh air.

Currently, it is wise to be skeptical of those who say any turbines can be placed on rooftops, in urban areas, or in places with a lot of turbulence. Such obstacles take much of the power out of the wind, and wind speed is a crucial factor in the amount of energy that can be produced, as we'll soon explain. However, most experts think there is potential in this area, if the right factors align and the right technology comes along, for the right price.

Pushing forward, the wind currents accelerate as you approach the western coast of India. The estimated potential for wind-energy production in India is 40,000 MW at 100 feet above ground level. In rural India, small and micro-turbines are lighting up homes of remote villages and powering hundreds of telecommunications towers.[8]

We will conclude your world tour with one of the biggest growth markets: China, which reportedly added about 50,000 small turbines in 2009, for a total of some 400,000 in place by the end of that year.[9] There may be as many as 100,000 small wind turbines in use by nomadic herdsmen in northwestern China. These small turbines—so compact they can be carted on horseback from one encampment to another—are among the few sources of power available on the Asian Great Plains.[10]

"For a young married couple in inner Mongolia, their most revered gift is a wind turbine," Trudy Forsyth told us via phone from the National Renewable Energy Laboratory (NREL)'s National Wind Technology Center in Colorado. Forsyth is a senior project leader who specializes in testing and certification of small wind turbines, and serves as NREL's liaison to the American Wind Energy Association (AWEA). "They can take it with them, where they can set it up, and get some lights and TV," said Forsyth. "Wind turbines are more rugged than solar panels, and it's windy there."

Even today, three-fourths of all small wind turbines built are destined for stand-alone power systems at remote sites, far from the nearest village. Others serve mountaintop telecommunications sites where utility power could seldom be justified.

Although there are other countries to visit, there is only so much space in this book. But take a look at the interactive map from the Global Wind Energy Council on www.gwec.net/index.php?id=9 to get a bigger picture.

We hope you get the idea that there is huge and growing global demand for emissions-free wind power, which can be installed quickly, in many places in the world. Over the past ten years, global wind power capacity has continued to grow at an average cumulative rate of over 30 percent [11] (Figures 1-16 and 1-17). Although

Evolution of Commercial Wind Technology

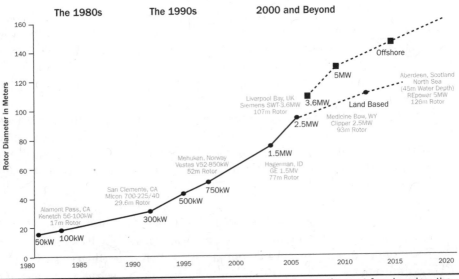

FIGURE 1-16   Commercial wind technology has advanced rapidly over the past few decades; the industry is much more mature than the small wind power sector. *Wind Powering America/NREL.*

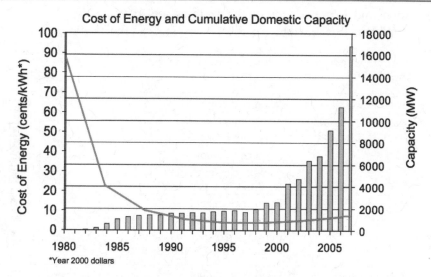

Capacity & Cost Trends

Cost of Energy and Cumulative Domestic Capacity

*Year 2000 dollars

Increased Turbine Size - R&D Advances - Manufacturing Improvements

FIGURE 1-17 As the cost of producing energy with large wind turbines has dropped dramatically, production has skyrocketed. *Wind Powering America/NREL*.

commercial-scale projects have led the way, the small wind industry has seen strong growth as well, despite the global recession. In general, small wind is still a niche energy player, and it does have some of the limitations of being an immature market, such as difficulty in finding replacement parts and access to skilled labor.[12]

Still, improving technology has given rise to a more cost-effective small wind turbine over the last 15 years, moving closer to the cost of conventional energy sources. For years, small wind turbines have been used to generate power to charge batteries. Technical developments have increased the power rating, efficiency, and reliability of these turbines. Current wind turbines can also now be used either on or off the grid to power homes and businesses. In fact, the fastest-growing segment of the small wind turbine market is that of residential-scale, grid-connected units, between 1 and 10 kW.

Due to advanced technology, wind turbine blades have become lighter, yet more durable and efficient in extracting wind energy. Similarly, the rotor speeds have been slowed down to decrease noise, and vibration isolators have been introduced, particularly for roof-mount models. Self-protecting technologies like the ones developed by WindTide in Ireland help the wind turbine protect itself in case of high wind speeds, including with active pitch controls (explained in Chapter 8) that maximize energy capture at high speeds without permitting the turbine rotor to get the brunt of the heavy gusts.

Some small wind turbines are also available with wireless connectivity, enabling owners to control the system from a distant location, while others come with nifty electronic data displays. Such technological advances have made these wind turbines more like an appliance, enabling it to be part of the household dinner conversation, and a desirable gift on the wedding registry.[13]

Want to fly there for a closer look now…. Scan the locations.

Resources

Wind Power Facility	What and Where	Quick Access
Altamont Pass Wind Farm (U.S.)	Article on progress: bit.ly/altamount Fly to geolocation: maps.google.com/maps?q= 37.739499,+-121.656643	
Half-Hollow Nursery 100kW	Fly to geolocation: Laurel, New York maps.google.com/maps?q= 40.969019,+-72.57088	
Various homes and facilities (Canada)	Fly to geolocation: Canada www.canwea.ca/swe/casestudies .php?id=69+	
Zeeland Wind Test Site	Fly to geolocation: Buys Ballot Street 8-A Schoondijke 4507 DA, Netherlands 0117 406045 maps.google.com/maps?q= 51.358035,+-3.55696	
Inner Mongolian Wind Farm, Bailingmiao, China	Fly to geolocation: Inner Mongolia Wind Farm Bailingmiao, China www.panoramio.com/photo/21179085	
The Green Building	Fly to geolocation: 19 New Wakefield St, Manchester, M1 5NP bit.ly/thegreenbuilding	
Bahrain World Trade Centre	Fly to geolocation: en.wikipedia.org/wiki/Bahrain_World _Trade_Centre Map maps.google.com/maps?q= 26.24000,+50.58131	

INTERCONNECT

"The Boy Who Harnessed the Wind" Shows the Importance of Scale

A small, landlocked country in southern Africa, Malawi is among the world's poorest, and only about 2 percent of its people have electricity. But one young Malawian tinkerer has proven to the world that anyone can improve their lot in life through the power of the wind.

William Kamkwamba's story has become internationally known, thanks to a TED fellowship, a bestselling book (*The Boy Who Harnessed the Wind*) and a globetrotting schedule of speaking and TV engagements (Figure 1-18). As Kamkwamba explains in his inspiring, gripping memoir, he was first exposed to the concept of wind energy in 2002 when he was 14 years old, while looking through a textbook at his rural town's tiny library. With Malawi suffering through a heartbreaking famine, Kamkwamba's farmer parents couldn't afford his school fees, and there was only so much work he could do in the parched earth, with no way to irrigate or afford fertilizers.

Impressed by the graceful, intrinsically simple shape of a wind turbine, Kamkwamba wondered if he could build one to pump water and bring lights to his house so he could study when the sun went down. He spent months pouring through tattered engineering texts and scavenging through junkyards and scrap heaps. He even took a part-time job to earn the few dollars he needed to purchase a bicycle dynamo.

While Kamkwamba endured harsh torment by his peers and confused stares from adults, he pressed on. Near his family's modest home, he raised a 16-foot tower from blue-gum poles he cut himself. At the top of the tower, he and a friend cobbled together a 12-watt wind generator out of used bicycle parts, plastic pipes, tractor fans, and wood.

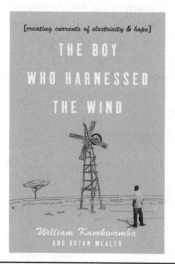

Figure 1-18   William Kamkwamba of Malawi inspired millions by building a wind turbine out of junk.

In one of the book's most dramatic moments, Kamkwamba amazed a crowd of his critics when he first wired a light bulb to his spinning machine … and it glowed dimly.

Soon, Kamkwamba was powering several lights and a radio in his home, and he charged cell phones of grateful passersby for free. He added car batteries for some energy storage, and he improvised a simple overcurrent protection feature (from a nail and speaker magnets) and a 600-watt inverter to convert the machine's wild alternating current (AC). The blades turned a pulley, which turned an old bicycle wheel, which powered the dynamo. According to his blog, Kamkwamba and friends recently added better deep cycle batteries, and they hope to add solar panels.

Early on, the young boy's impressive contraption attracted attention from the national authorities and the news media, and before long, he was offered a scholarship to a prestigious school in South Africa (Kamkwamba has since been accepted to Dartmouth University in the United States).

Kamkwamba also built several more wind generators, including some that pump water for irrigation and one that serves a local school in his native Kasungu District. The largest of his machines stands at 37 feet.

Kamkwamba is exceedingly humble, and he is already hard at work helping others in his community. His Moving Windmills Project has partnered with the U.S.-based buildOn .org nonprofit to rebuild his old school, which is in serious disrepair. Kamkwamba also hopes to empower others in the developing world to generate their own energy.

It's easy to see why William Kamkwamba's story resonates around the world. It's a touching example of how a bright, determined individual can lift himself out of poverty through hard work. It's also a good reminder that, to be a success, a wind turbine doesn't have to be big or made from shiny, expensive parts.

Although a 12-watt turbine isn't going to make much of a dent in the typical American home's energy budget (the annual average is 10,000 kilowatt-hours), that says more about our consumption patterns than it does about the limits of do-it-yourself (DIY) projects. For a family like the Kamkwambas, having the ability to read at night, pump well water, or charge a cell phone without walking into town and paying a fee is a significant improvement in quality of life. Sometimes, we have to rethink our notion of scale and remember that what doesn't work in one situation may be just the thing for another.

Wind Energy System Components

Here's a gentle breeze through a complete wind energy system.

Turbine

A wind generator, or turbine, has two functions:

- Collect the wind (blades)
- Convert wind energy into mechanical then electrical energy (generator)

FIGURE 1-19   Assembling a wind energy system is a challenging task that requires teamwork. Scotsman Hugh Piggott leads workshops on both sides of the pond that have taught thousands of people to do it themselves. *Hugh Piggott.*

Turbines may come in different shapes and sizes, but in general, the *blades* and the *hub*, the center part that turns the shaft, combine to make up what is called the *rotor*—what we see spin (Figure 1-19). The spinning of the shaft (or stator in some cases) results in generating an electrical charge within the alternator. To maximize the energy harvested, a turbine needs a device that keeps it facing the wind, either mechanical—as in a tail—or electronic—such as the electronic orientation yaw drive.

Power Up!   *Blade size and robustness of the alternator, rotor, and tail to weather through years of high-velocity winds will be one of your major factors in deciding which turbine to purchase. The next factor will be the generator's conversion efficiency from wind's kinetic energy to electricity, and the durability of the generator's components to long-term exposure to air turbulence and high revolutions per minute (rpms) (see Chapter 8).*

Tower

The tower puts the turbine up in the location of the "fuel"—the smooth, strong winds that provide the best energy. Wind turbines should be sited at least 30 feet higher than anything within 250 feet (76 m).

Three common types of towers are *tilt-up*, *fixed-guyed* (the ones with the supporting wires), and *freestanding*, or the one where the structure is wider at the base than it is at the top and has a significant foundation. Towers must be specifically engineered for the lateral thrust and weight of the turbine, and should be adequately grounded to protect your equipment against lightning damage.

Transmission Wires

Transmission wires are usually made of copper or aluminum, and they deliver the electricity from the turbine to your conversion assembly (see "Balance of System") and ultimately to your home and the utility grid. Proper sizing of wire to handle the load, grounding patterns, trenching, and safety measures are discussed in detail during installation.

Balance of System (BOS)

In addition to the primary hardware of turbine, tower, and wires, you'll need other parts to make your investment work. These other components are often collectively referred to as the "balance of system."

Charge Controller

In a battery-based system, this smart little box keeps your batteries properly fed and safe for the long term. As the name suggests, a controller governs the charge produced by the wind turbine. It blocks reverse current and prevents battery overcharge. That's important, because overcharged batteries need more maintenance and wear out quickly.

Nearly all installations require a controller to divert excess electricity if too much is being produced at once, due to high winds. Many controllers also provide intelligent charge regulation, array output optimization, automatic battery equalization, and even built-in datalogging. Chapter 10 on BOS will cover selection and installation of charge controllers.

Battery Bank (Storage)

If your system is off-grid, you'll need a battery bank—a group of batteries wired together—to store energy so you can have electricity when it's not windy. A wind turbine can also be hooked up to both batteries and the grid. Produce excess energy that your batteries can't hold, and you can feed it into the grid, hopefully for a credit. When insufficient electricity is generated or the batteries are drained, electricity drawn from the grid can make up the shortfall. Yet battery banks can provide emergency backup during blackouts. ·

Only *deep-cycle batteries* are recommended for wind-electric systems, although William Kamkwamba built his first small turbine in Africa with regular off-the-shelf batteries, and he still had a little juice for a few lights. For wind systems, *lead-acid batteries* are the most common battery type chosen due to their low cost and wide availability. Sure, lithium-ion batteries would work great, but they aren't cheap.

System Meter

System meters can measure and display several different aspects of your wind system, including how much power you are producing at the moment, the charge status of your batteries, and how much electricity your home is using. Many of today's inverters have remote access via personal computer and mobile devices. "Hold on, let me check my status on my smart phone."

Dump Load (aka Diversion Load, Load Resistor)

A *load*—some device(s) drawing energy—must be kept on a turbine at all times to prevent it from spinning too fast and overheating internal parts or getting damaged. That load could be the grid or a battery bank. However, if the batteries are fully charged, they can't accept more juice without damage. In that case, the controller diverts excess electricity to a dump load, which is usually an air or water heating element that "burns off" energy as heat. If you can put the heat to good use, even better!

(Technically, some controllers do not require dump loads if they operate using pulse width modulation shunt techniques to prevent the battery bank from overcharging).

Disconnects and Overcurrent Protection

You'll need to have a direct current (DC) disconnect one-switch breaker between the batteries and inverter (see the next paragraph) to allow for easy maintenance and to protect the wiring against electrical fires. In many countries, an additional AC disconnect must be located outside the house and near the panel. It typically is a large, one-switch breaker mounted in a metal enclosure. It can provide an extra defense to protect line workers against electrocution from your electricity during a utility grid repair.

Inverter

An inverter converts direct current (DC) to alternating current (AC), which is used by appliances, lights, gadgets, and other devices in most households and facilities. There are two basic types of inverters:

- **Battery-based inverters**   These inverters require large deep-cycle batteries to operate.
- **Batteryless, or grid-tied, inverters**   These inverters are connected to the electrical grid and provide no backup. When the utility grid is down, you can't make any electricity, because it shuts down the turbine. This is a safety precaution to prevent the accidental electrocution of electric line workers during grid repair in case of an outage.

Back-up Generator

Sizing a system to cover a worst-case scenario, like no wind for weeks, can result in a very large, very expensive, and inefficient battery bank. Therefore, a backup, fuel-powered generator may be necessary, and in fact is almost always necessary in off-the-grid settings.

Petroleum- and diesel-fueled generators produce AC electricity, though a battery charger (either stand-alone or incorporated into an inverter) can convert that to DC energy, which is stored in batteries or can be hooked directly up to the …

AC Breaker Panel

The AC breaker box is the point at which a building's electrical wiring meets with the power authority. It contains a number of (hopefully labeled) circuit breakers that route electricity to the various rooms throughout a structure. These breakers allow electricity to be disconnected for servicing, and also protect the building's wiring against electrical fires.

In addition, for their use, utilities usually require an AC disconnect (mentioned earlier) between the inverter and the grid. These are usually located near the utility kWH meter.

Kilowatt-Hour Meter

A bidirectional kilowatt-hour meter (kWH meter) simultaneously tracks how much electricity is being used (measured in kilowatt-hours, kWH) and how much is being generated. It measures power coming from the grid and going to the grid.

Utility Interconnection Equipment

Along with the meter, transformers and any relay switches are commonly provided by the power utility after a net-metering agreement has been signed to ensure that the proper voltage and frequency are provided once the inverter is connected to the grid.

With new technology, you might find that many of the components listed are integrated into various other components. For example, it is common to see an inverter that includes charge controller, monitoring, AC/DC disconnects, and even dump load capability. This makes your shopping quicker and reduces compatibility problems and complications that can arise with separate components. Some products, like Southwest Windpower's Skystream 3.7 and 600, have integrated nearly all the components within the nacelle, complete with wireless communication to your computer.

Table 1-1 shows a summary of the three common wind system configurations. The grayed areas are components that are found in all systems. We are aware that some of you might want to design a 24-volt system to power your boat, cabin, or one or more direct current appliances, like a DC submersible water pump (we explain electricity in more detail in Chapter 3); however, even the strictest of wind power users will eventually want to use a power tool or appliance that requires AC power. When you are at a point of purchase, print and fill out this form. Note: Any point void of a check box is an indication that the item is not required. The fuel generator is considered an optional item if you are grid-tied, though you may want to consider it in a region with frequent blackouts. In Nicaragua, there could readily be two days without electricity in a given week, and overdraining your batteries shortens their life.

TABLE 1-1   Core Options for a Small Wind System

Wind System Configurations			
Components	Grid-tied	Off-grid	Grid-tied battery backup
Wind turbine	X	X	X
Tower/foundation	X	X	X
Transmission wiring	X	X	X
Inverter	X	X	X
Brake	X	X	X
DC disconnect switch		X	X
Charge controller		X	X
Dump load		X	X
Utility interconnect equipment	X		X
Battery bank		X	X
Back-up generator		X	X

Direct Drivers for Wind Energy

Let's start putting wind into your sails, whether you have already committed to building a turbine or are just browsing. Let's review why the rest of the world thinks small wind energy makes sense.

In fact, the growth of the wind market is being driven by a number of factors, including economic incentives, practicality, values, care for the environment, and the impressive improvements in the technology itself.

To Hugh Piggott, a Scot who has trained thousands of people around the world to make their own small wind turbines through his company Scoraig Wind Electric, and who writes and speaks extensively on the topic, the industry is about what he calls "human scale." In an interview, Piggott told us he thinks small wind "works on a personal level."

He said, "It makes people feel that they're part of the energy generation process, and it gives them a sense of how much energy they use and what it takes to make. We live increasingly disassociated from the stuff we use, but more localized production of food and energy is a healthy thing."

Piggott told us he doesn't think small wind is likely to make a big impact on energy trends any time soon. And although not everyone agrees with that assessment, recent history would certainly suggest that he's right, or at the very least that the industry would have to ramp up vastly more than they are currently to really make a dent in fossil fuel use.

"Compared to offshore wind farms, small wind is incredibly inefficient," Piggott rightly points out, but that doesn't mean he sees no value for his work. To him, small wind turbines have a lot of utility, as long as they are smartly sited and installed so they can generate a meaningful amount of electricity. "If you just want to see something spin, buy a whirlygig. They're a lot cheaper," Piggott quipped.

"People generally have huge misconceptions about wind energy, and a lot of the time they get deflated," Piggott added. "Most people assume that they can put something on their roof that will generate most of their electricity in an urban environment. That's really, really a long way from the truth," he said.

Author Ian Woofenden told us, "We need to be real about the wind resource, about the fact that energy efficiency is often a better buy, about load analysis, and about what machines can actually do."

We hope to do just that in this book, and to help you see the situations when small wind does make sense, on a number of levels, including economics. We created this brief list to help you understand some of these factors. Feel free to place a check mark on anything that piqued your interest. In fact, you can send it in to your local wind energy association.[14]

Economic Benefits of Small Wind Power

What can wind turbines provide? This list may surprise you.

Free Fuel

Wind is a by-product of solar insolation of the Earth, and is modified and enhanced by terrestrial topography, large bodies of water, and ecological and anthropomorphic activity.[15] But there is no exchange of common currency for the production and transportation of wind.

Return on Investment (ROI)

ROI is a tricky concept, and it's something most human beings are bad at calculating in their daily lives. Is it better to drive a half-hour out of the way to avoid a $2 toll? Is it worth it to study one more hour for the big test? Is grad school a good decision? Unless you are an investment banker, chances are good that you don't spend a lot of time estimating your return on investment for very many decisions. Yet most of us opt to consider the question when it comes to renewable energy projects.

The good news is that, with proper understanding and design of a wind power system, including site assessment, personal needs assessment, a reliable product, and government incentives, owners of that system should be able to reap financial benefits that exceed current investment rates over the long term (e.g., certificates of deposit and bonds).[16] Granted, it's vital that all those factors we just mentioned are in place. Areas with stronger incentive structures and higher electric rates clearly get an advantage when it comes to ROI, and you've got to have a system that is high quality and able to access a good wind resource. Small wind installer and expert Mick Sagrillo recently calculated the ROI on a high-profile small wind project to be more than 4,000 years, largely because the owners grossly overestimated the strength of the breezes on top of their building.

"Many projects are being installed as a greenwash thing: they're not producing any useful energy, and they cost money that could have been spent on insulation and other more effective measures," Piggott told us.

Something else to think about is maintenance costs. If you can get a service plan, or at least know that your installer can back you up for regular checkups and repairs, that can make a big difference in ROI. As Piggott told us, "Your machine may generate a few hundred dollars a year, and half of that could be blown on a service call if the only technicians available are far away."

When asked when he thought small wind did have a reasonable ROI, Piggott told us "when you live in a pretty windy place." He also pointed out that it tends to work really well off the grid, when costs of extending the lines would be prohibitively expensive, or in developing countries, where the only power available is through car batteries that have to be charged at stations, for a fee, then bicycled in. "A lot more people are trying to develop local skill base and manufacturing capabilities, in the developing and developed world, so costs should come down," Piggott explained.

When asked to estimate a payback period on a "typical" small wind energy system, Woofenden responded, accurately, that there is no such thing as a typical wind generator. But when pressed, he said, "Depends on the resource, utility rate, incentives, reliability, but I would say between 10 and 20 years is a reasonable range."

Woofenden, who maintains several wind turbines for his off-the-grid home on an island off Washington state, also has a grassroots passion for the technology (Figure 1-20). "Most things we buy have no return at all, because almost everything has a negative return," Woofenden told us. "I think the payback return question is sort of out of line. I tell potential buyers who ask me about it to cite one thing they bought that they thought about ROI. We subject renewables to that question but not our other choices."

At the other end of the spectrum, wind author Paul Gipe told us that ROI is a critical factor, and that small wind won't really take off until it gets more favorable. In Gipe's view, robust feed-in tariffs are the best way to get there. In any case, we'll get into the money behind the decision in Chapter 6.

Financial Hedge Against Rising Energy Prices

As soon as a grid-tied wind generator (with net metering) is switched on and producing energy for the grid, your cost per kilowatt-hour has been immediately reduced, and the benefits increase as price volatility or inflation raises consumer

FIGURE 1-20   Workshop participants build their own small wind energy system from scratch, thanks to the tutelage of Dan Fink. That's a lot of work, but it can be tremendously rewarding. *Dan Fink.*

electric rates. This is worth keeping in mind, because energy prices for many people have been rising sharply over the past few decades, far outpacing inflation. There are a number of reasons for this, including demand outpacing supply, increasing restrictions on new energy projects like transmission lines and power plants, and geopolitical instability. Most experts expect energy rates to continue to climb. Meanwhile, the cost of wind will always be free.

Further, since every kilowatt-hour of electricity produced by wind helps reduce demand for fossil fuels, it can actually help keep costs of those legacy fuels down.

Access to Financial Incentives for Developers and Consumers

It's no secret that the global small wind energy market, like all renewable markets today, is largely driven by government support policies and regulations. (A lesser-known truth is that conventional energy businesses also receive massive tax benefits and incentives, and part of the reason why energy is so "cheap" is because it is already subsidized.) In the United States, the wind market is supported by federal tax credits and several state-level renewable portfolio standards (RPS), which mandate that a certain amount of energy be provided with renewables.

In December 2008, the European Union (EU) agreed to a new Renewable Energy Directive for a binding 20 percent renewable energy target by 2020. Germany and Spain have had feed-in tariff schemes for several years, and the United Kingdom recently added one for small wind. In the Asia Pacific, government support is also driving wind energy growth. China's wind-installed capacity has doubled every year since 2006. The Renewable Energy Law, along with other policy measures, is driving the Chinese market even further. In addition, many states in India also have feed-in tariff schemes and RPS in place.[17]

Becoming Competitive with Conventional Technologies

With today's rising coal and gas prices, and the chance of increased environmental regulations on those dirty industries, new wind farms already compete favorably when it comes to economics. With small-scale projects in remote sites or on boats, or for pumping irrigation systems, a wind turbine can be more affordable then connecting to the grid or installing a back-up generator system.

Job Creation and Regional Economic Development

New wind energy will bring significant economic benefits to the participating states, and not just construction and maintenance jobs. Over time, wind component manufacturing should follow expansion. Wind can make a positive impact in terms of employment and revitalization of rural and declining areas.

In fact, CanWEA's Emilie Moorhouse told us that one of the motivations behind Nova Scotia's recently passed small wind feed-in tariff was to support the major small wind manufacturer in the province, Seaforth Energy. "They'll have this double benefit, of environmental and social benefits, plus jobs," Moorhouse told us via phone from her office in Montreal.

More Americans already work in the wind industry (85,000) than the coal industry (81,000), according to 2009 data from the American Wind Energy Association and the U.S. Department of Energy. Expect the gap to widen in the coming years.

Values and Small Wind

Small wind power also has a number of important benefits that aren't strictly financial, or that aren't easy to put a price tag on, at least not with current methods of accounting (Figure 1-21). Let's look closer.

Environmental Benefits

Speaking of environmental benefits, wind power, of course, is based on capturing the energy from natural forces, so it has few of the polluting effects associated with "conventional" fuels. Wind energy contributes to the reduction of emissions of various harmful air pollutants (carbon dioxide, nitrogen oxide, sulfur dioxide), and has already played a role in improving regional air quality. Wind power supports

FIGURE 1-21 Small wind turbine designer, installer, and trainer Hugh Piggott told us he thinks his machines "work on a human scale" to teach people about energy use and production. *Hugh Piggott*.

efforts to meet emission caps in a cost-effective manner.18,19 In addition, wind energy uses virtually no water, which, in an increasingly water-stressed world, is a major environmental consideration.

Independence

For many people, it is about energy independence from their power authority. Businesses that pay additional fees for electric demand in excess of their allotment plan, or others who usually couldn't see options other than accepting the rapid inflation of electrical costs every year, are now considering making their own mini-power plant. For some, it is about "stickin' it to the man," by turning the tables on the power authority, and compelling them to pay for the surplus energy produced from small wind turbines through a net-metering or feed-in tariff agreement.

Image Enhancement

Project directors of everything from department stores to manufacturing plants, nurseries, colleges, office parks, hospitals, hotels, municipalities, and much more understand the intrinsic value of appearing "green" to the ever-alert consumer. Wal-Mart and Home Depot, for example, have added very small wind turbines to light poles in selected parking lots. Such projects aren't always the best idea when it comes to the actual wind resource available, but under the right conditions, they might be a wind-win (Figure 1-22).

FIGURE 1-22   Small wind turbines, like solar panels, can be powerful symbols of environmental commitment. Pictured are AeroVironment AVX turbines on the roof of a new office building at Brooklyn Navy Yard. *Brian Clark Howard.*

In Queens, New York, Rick Rosa, the sales manager of the new Solarium "green" apartment building, told us that response has been "phenomenal" to the building's rooftop WindTronics wind turbine. "It has been a guiding light for the neighborhood," Rosa told us. "It's a symbol of us being a green building. It brings people over here, and makes them ask about what we did," he added.

In the case of the Solarium, the turbine is no window dressing, since the building is the first green-certified complex in the borough. That was achieved through energy and water efficiency, use of recycled materials, recycling of debris, and other factors. According to Rosa, the turbine helps makes the facility's common charges low.

Interestingly, Rosa told us that the main question people ask about the rooftop turbine is "how much power does it generate?", suggesting that consumers do care that green changes aren't just cosmetic, and that they make a real impact. (The turbine choice and location in this case are not without controversy, as we'll soon show.)

Consumer Choice

Over the last few years, small wind turbine technology has improved in quality and versatility. What was once a strictly marine or rural, off-grid solution has evolved to allow renewable energy to enter a wider consumer market. More options are available, including roof-mount turbines, more aesthetic shapes, noise and vibration reduction improvements, integrated sensors, and turbines sized to the customer's needs, from 150 watts up to 100 kW. As we'll show you, many of these newer designs are largely untested, and some are controversial, and it still holds that rooftops are rarely good places for turbines. But in general, choice is a good thing, and we'll walk you through the pros and cons of each major offering in Chapter 8.

Self-Reliance

Wind power technology has evolved and has become more standardized, reducing the learning curve and the risks of investment. This has opened the doors to that significant group who are insistent on relying on their own judgment, knowledge, and abilities in order to survive economic downturns with local resources and not a lot of additional help. Tightening up the bootstraps, they move forward, confident that wind energy is a feasible means to survive, sometimes with *most* of the energy and material for survival being provided locally. This is certainly not an easy pursuit, but it can be immensely rewarding.

Do-It-Yourself

Although we don't suggest you try to build your own wind system from scratch without at least some hands-on training or apprenticing, people have successfully done it, and there are bestselling books and websites to prove their feats. There are a number of plans and kits available, as well as instruction books and classes.

More premade small wind systems are designed to be relatively easy to install, but they are still definitely harder than putting together IKEA furniture, and most manufacturers strongly recommend professional guidance. Still, with more certified systems, integration of components, time-tested procedures, and equipment configurations, the guesswork is being reduced.

High Visibility

In history, success has often been celebrated by building prominent structures. When religion was dominant, many towns erected enormous pyramids, temples, and cathedrals. Then, government capitol buildings started to tower over the dwellings of the citizens. Finally, financial services companies came to be housed in glass spires that reached for the sky, seemingly defining the character of Western society. Well, wind turbines placed up on a high pedestal can communicate values for one's business, farm, or community that could supersede any provocative poster or flowery talk.

The Joy and Satisfaction of It

When it comes down to it, many of the folks we've talked to who own small wind generators, or who work in the industry, admit to being in love with watching those blades turn, almost as much as they like seeing their electric meters run backwards. As Hugh Piggott pointed out to us, just wanting to see something turn isn't reason enough to invest in a wind system—there are much cheaper pinwheels for that. But there is definitely something to the direct experience of making your own power based on the breezes. It's worth remembering that as much as we like to think of ourselves as rational beings, we make very few decisions based solely on number crunching. If we did, we'd all be driving stripped-down Honda Civics, wearing Dockers, and waiting until movies are released on basic cable.

To one of us, our connection to the wind runs even a bit deeper. Kevin still finds great joy in storms, even the ones hinting that the end of the world is near. He recalls the days of fighting the wind-driven rain to place a roof on his home. He remembers the wind chill that froze the marrow in his bones. Well, even four years later he still smells sweet revenge and wears an evilish grin while watching his spinning turbine as the neighbors' homes shudder with the gusts and the arcing trees.

Practicality

There are a number of related practical reasons to consider wind power.

Diversity and Reliability of Electricity Supply

So long as we have sunshine, land, and water, wind is highly reliable (although highly intermittent). That is, with enough well-designed systems, the output at any given time should only vary gradually due to varying wind speeds or storms. True, storms can require temporary powering down to protect equipment; however, with weather forecasting, that can be coordinated with complementary energy sources to ensure uninterrupted service. Currently, some utilities tend to malign wind power as "unreliable," but that seems to be changing as the industry matures, technology improves, and we continue to diversify sources.

Taking a step back, many analysts sensibly point out that it would behoove us as a society to diversify our energy sources. Currently, we are highly dependent on fossil fuels, as well as nuclear power plants that are aging and that are politically very difficult to replace in many countries. The more we can ramp up a diverse set of renewables, the better off we'll be to weather periods of uncertainty. That includes large and small wind, as well as solar electric and thermal technologies, geothermal, wave and tidal power, biomass, trash-to-energy, and other strategies.

Political Security

Despite a recent blip in the trend as a result of recession, global demand for energy has been increasing sharply. Significant investment in new power generation capacity, grid infrastructure, and geopolitical strategy is going to be required to continue to provide current standards of living in the developed world and to improve the lives of those in the developing world. Yet, supplies of fossil fuels are finite, and petroleum prices are volatile.

In contrast, wind is a massive indigenous power source, available virtually everywhere in the world. As mentioned earlier, wind energy can mitigate petroleum price volatility, reducing the friction among nation-states over scarcity. For example, wind generator installations currently being deployed on military bases are not only being seen as practical, but as one way we can contribute to reducing dependency on foreign oil.

Natural Synergism with Solar Technology

Wind power is directly related to the power of the sun, but the winds continue their ferocity even when the sun's rays are on the other side of the globe. Wind energy is thus an excellent complement to solar power when the sun has set or during a storm (Figure 1-23). In addition, the two emerging forms of renewable energy have a complementary relationship over the course of a year. In most temperate zones, the amount of energy that can be harvested by solar panels is less in the cooler months, when the sun is lower in the sky. In contrast, the winds tend to blow the strongest in late fall to early spring.[20]

FIGURE 1-23 Wind and solar power often work well together, as at this home site in Colorado. *Warren Gretz/DOE/NREL.*

Diversity of Applications, Including Remote and Off-Grid

Small wind turbines, alone or as part of a hybrid system, can power homes, businesses, farms, ranches, mines, logging camps, boats, and many other things. Wind energy often works very well for remote applications, such as water pumping, ice making, and telecommunications sites. Community small wind projects have been launched for schools, tribes, municipal utilities, and rural electric cooperatives.

CanWEA's Emilie Moorhouse told us that Canada has been working to develop incentives for hybrid wind-diesel projects for the northern reaches of the country, where some 300 communities live totally remote from the grid. Currently, a five-year grant program is targeting 20 northern communities and six mines. "These communities are dependent on diesel, which is dirty and expensive," Moorhouse told us. "In a sense, wind is the only renewable technology that's applicable to the far north. Solar: there's no sun up there in winter. Hydro: the rivers are frozen in winter. There is no biomass," she explained.

Moorhouse said wind-diesel projects have already proven themselves in at least 14 communities in Alaska, and she said Canadians are working to build the skilled workforce to deploy and support these hybrid systems. "In the north, there is icing and maintenance issues, so it's better to have several turbines rather than just one big one," Moorhouse added.

Resiliency

In the case of commercial wind farms, power is generated by large numbers of generators, so individual failure of one machine does not have large impacts on the total production. This feature of wind has been referred to as *resiliency*. Even at the home scale, hybrid renewable energy systems permit one form to compensate for the other system when either the wind is calm or the skies are dark.

In a larger sense, adding more wind turbines is part of a smart strategy of diversifying.

Summary

As we have shown, wind energy has a long and colorful history, stretching back hundreds of years. For millennia, people have felt the breeze on their cheeks, watched leaves blow across the path, and asked if that mighty power could be harvested for our benefit. The good news is that it can, and you can choose to be a part of it. In fact, the small wind industry has shown impressive growth in recent years, despite the global recession (Figure 1-24).

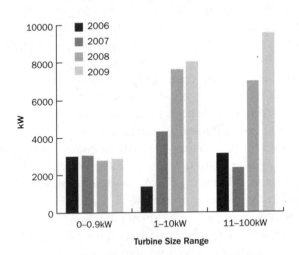

U.S. SMALL TURBINE MARKET GROWTH BY SEGMENT

(In general, systems 0–0.9kW correspond to the off-grid market, systems 1–10kW correspond to the residential market, and systems 11–100kW correspond to the commerical/light industrial market.)

FIGURE **1-24A**   The small wind turbine industry has shown strong growth over the last decade. *AWEA.*

2009 GLOBAL SALES

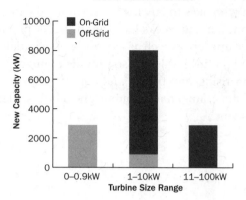

GROWTH OF U.S. SMALL WIND MARKET

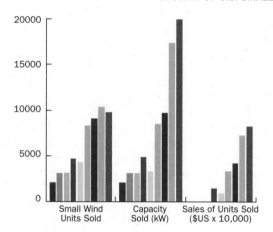

Year	Units	kW	Sales $US
■ 2001	2,100	2,100	(not avail.)
■ 2002	3,100	3,100	(not avail.)
■ 2003	3,200	3,200	(not avail.)
■ 2004	4,671	4,878	$1,489
■ 2005	4,324	3,285	$990
■ 2006	8,329	8,565	$3,320
■ 2007	9,092	9,737	$4,197
■ 2008	10,386	17,374	$7,266
■ 2009	9,800	20,300	$8,240

Manufacturers report that while interest in small wind has increased through the economic recession, consumers are delaying purchasing decisions until financing becomes more available and affordable. Therefore, as the economy recovers, many manufacturers expect a surge in sales.

2009 Global Sales

	Units	kW
Off-Grid	15,500	7,600
On-Grid	5,200	34,400
Total	20,700	42,000

FIGURES 1-24B, 1-24c   The small wind turbine industry has shown strong growth over the last decade. *AWEA.*

A small wind system is not cheap or easy to install and maintain. If you are the kind of person who likes to buy the latest shiny new toy, only to leave it in a closet and forget about it a few weeks later, wind power probably isn't for you. A wind system requires upkeep, as well as a considerable initial investment.

However, wind turbines can also provide many benefits, such as clean power, insulation from rising fuel prices, and considerable satisfaction. To those with a good wind resource, amendable zoning, and a little patience, a wind system can be a smart investment.

Second Wind: Challenges and Impacts of Small Wind Energy

Most people never run far enough on their first wind to find out they've got a second.

—WILLIAM JAMES

Overview

- What are the main drawbacks to small wind power?
- What are the real impacts of noise and appearance?
- Can small wind turbines disrupt radar or cell phones or throw ice?
- Can small wind turbines be harmful to birds and bats?

Not to deflate the helium out of your Magenn wind turbine balloon (see Chapter 12 on new and future technology), but no energy source is perfect, even if some enthusiasts can't see beyond their own nacelles.

Even if you are a small wind advocate, you may want to read about these common myths and shortfalls. For one thing, opponents to your wind turbine might appear at your zone appeal or during construction, with a hair-raising study about wind turbines that you might not be prepared to process.

Without advocates like you to clear the air, local ordinances could require an endless list of study reports and certifications, similar to what is required for utility turbines. To give you an idea, here is what is currently required for utility-sized turbines, although thankfully not for small turbines ... yet:

- Soil survey of the entire area
- Tower engineering

- Tower foundation engineering
- Structural certification for the tower
- Bird, bat, and noise studies
- Ice "throw" engineering and calculations
- Blade throw calculations
- Telecommunication interference testing
- Environmental impact
- Study average wind speed monitoring power
- Output reporting and economic viability analysis
- Agricultural impact statement
- Architectural projections of what the turbine will look like on the property
- Documentation of nearby sensitive environmental areas
- Wind access agreements with neighbors
- An abandonment plan
- Determination of the area affected by shadow flicker
- Stray voltage testing
- HAZMAT (hazardous material) reports
- Well testing
- Electrical wiring insulation testing
- Site reclamation plan
- Bonding

Obviously, every neighborhood is not so restrictive. There are many areas in the world where wind turbines are as welcome as a hot apple pie at the windowsill. However, some of these requirements came about due to previous experiences with older wind or related technology. Most of the reported negatives of wind energy relate primarily to industrial-scale turbines, and historically, many experts and researchers have essentially ignored small wind (Figure 2-1), assuming that it

FIGURE 2-1   How's that for a breath of fresh air? Whisper 0.9 kW small wind turbine in Spanish Fork, Utah. *Windward Engineering/DOE/NREL.*

lacks a significant size or quantity to even warrant research. Still, some work has been done, and many questions continue to be raised, sometimes by zoning boards and others with decision-making powers. At the very least, we think it's important to give some of these issues a thought when you envision your dream system so it can be as truly sustainable as it can be and meet all your expectations.

We cover these challenges in detail, but also include a concise chart that you can rip out of your own personal book (not from the library book!), or better yet scan and bring wherever you go. If that isn't good enough, we provide some footnotes and web links. Remember, information is power.

Energy Storage and Transmission

I want it, and I want it now!

If you want reliable electricity, wind will always be here, and you can count on it long after you pass. That makes it a reliable energy resource. Yet, wind does not blow 24/7/365 in most locations. If it did, we would certainly like to see the people who live there and their culture. They would probably be applying ChapStick to their faces and using some clever transport utilizing the prevailing winds, not unlike sailing ships. And they might all commute to work in a similar direction. For the rest of us, wind currents are typically strongest in the late fall, winter, and early spring, but generally calmer during the summer, unless you are in a monsoon or hurricane zone. And even during the day, wind speeds vary based on your location and terrain.

So, this reliable, renewable energy has an Achilles heel. Unlike Odysseus, you can't just open up a woven bag to release the wind force. One of the biggest advantages of coal-burning power facilities is *dispatchability*. Dispatchability, or maneuverability, is the ability of a given power source to increase and/or decrease output quickly on demand. Maneuverability is one of several ways grid operators match output (supply) to system demand. Need more energy? Insert more coal. Need less energy? Burn less coal. Unlike conventional fuels, we cannot slow down or speed up the wind. And shutting down a wind turbine is not the answer. Disengaging the rotor from existing wind force is technically not the same as controlling dispatch— it is more like letting the fuel line leak.

The increasing role of the wind's variable renewable sources has prompted concerns about grid reliability and raised the question of how much wind can be placed into the grid before vast energy storage is needed. The question becomes an economic issue. Currently, the grid can accommodate a substantial increase without energy storage, so long as we share the resources and loads over larger areas, perhaps over continents. Beyond this level, the impacts and costs are less clear, but currently, storing electricity is more expensive than dispatching when needed.

For the most part, practical options of storing energy at a time when demand is low are still in the exploratory phase. They consist of a number of strategies, including batteries, pumping water, conversion to heat, and other ideas.

Batteries are certainly one possibility (Figure 2-2), but they require maintenance, and they are currently way too expensive to use on the scale of the grid. Some analysts have predicted that electric and plug-in hybrid cars may soon be able to leverage their on-board batteries for energy storage. The idea, often called *vehicle-to-grid* (V2G), is that much of the excess energy that is generated by wind turbines at night, when the winds are blowing but people are asleep and not using energy, would be "stored" in the batteries of cars that are undergoing charging. Then, during the day, for those people who aren't using their cars—because they're at work, say, or home sick—they could sell back power to the grid at a time when demand is high, the wind isn't blowing as hard, and electric rates are higher. The concept of wind-powered cars also means that clean power would be directly offsetting use of foreign oil, which is something that has broad political appeal.

The catch is that we are, at best, several years away from V2G, in part because our grid isn't yet "smart" enough to have that fine a level of two-way control. Only a relative handful of smart meters have thus far been installed, and they aren't cheap. Further, there are very few electric and plug-in hybrid cars on the road, although that is expected to change over the next few years.

Finally, current battery technology isn't quite up to the task. At a recent media event, Andrew Tang, who heads up smart grid development for the utility Pacific Gas & Electric, said that he doesn't think it will be worth it for consumers to "rent out" their batteries for some time. That's because each time power is put in and then taken out of a battery, that cycle decreases its lifespan, and that's a significant concern when car battery packs cost several thousand dollars. According to Tang, the pennies that a V2G participant would be credited from a utility wouldn't cover

FIGURE 2-2 Batteries can be used to store renewable energy, but they are expensive, require maintenance, and are not completely efficient. These batteries in Mexico got corroded since they sit in an equipment room that hits 122 degrees Fahrenheit. *Charles Newcomber/DOE/NREL.*

the wear and tear on their battery—at least until batteries improve and electricity rates go up.

Another grid energy storage method that is being researched is to use off-peak or renewable-generated electricity to compress air, which could then be stored in an old mine or some other kind of geological feature. Then, when electricity demand is high, the compressed air can be heated and expanded to generate electricity. Similarly, some power plants use their excess energy at night to pump water uphill. Then, when they need extra juice during the day, they release the water, and convert the energy into electricity via water turbines, just like in the Hoover Dam.

In a more complicated process, utilities could make hydrogen at night with their excess energy, then run that through a fuel cell during the day to provide extra power, or use it to run fuel-cell cars (Figure 2-3). It would be a clean process, especially if the initial energy were generated by the wind. Fuel cells are currently too expensive to work on such a scale, but may be coming.

Another option is to convert the surplus electricity to public or residential heating services. The Bonneville Power Administration (BPA) has recruited homeowners in Washington to install special devices on their water heaters that communicate with the electrical grid. They tell the heaters to turn on or off, based on grid conditions and the amount of renewable energy that's available.

The benefit of your grid-tied small wind turbine is that energy storage is much less of a problem, because small generators are *distributed,* meaning they produce power right where it is being used. However, if your system is off the grid, then you currently have few options to store wind energy for later use: namely, batteries,

Figure 2-3   This Honda prototype car runs on hydrogen, thanks to an onboard fuel cell. If the hydrogen is produced by electricity generated by a wind turbine, that makes for a completely clean loop. *Brian Clark Howard.*

and trust us, you aren't likely to be able to afford, or want, that many of them. In a few years, you may be able to invest in a home fuel cell. (They're selling pretty well in Japan now, thanks to big government subsidies.) Ideally, you will be able to find some uses for any extra power you generate, such as running an electric saw, power washer, or floor cleaner. So, we guess you should have your wind dictate when you should get off the couch.

If your system is tied into it, then the grid can effectively act as your storage bank. In most places, where there is some type of net metering agreement, you'll get credited for energy you put into the grid, and that's likely the best deal you'll find. Even if there is a massive ramping up in the number of wind turbines, the grid should be well equipped to handle it, since it will be so distributed.

Finally, one more potential problem with new utility wind farms is that they often require new expensive transmission lines to pump the power they produce into the grid. Installing high-voltage lines is a major undertaking that disrupts the environment and makes for angry neighbors. However, with small wind, everything is at a much smaller scale.

Environmental Impacts

Everything we do has some kind of impact on the environment, and wind power is no exception. For the most part, wind turbines are very eco-friendly, when you consider their whole lifecycle and how much clean power they can produce. Still, there are a few things to be aware of.

Noise

Just like all moving things, wind turbines make some noise (Figure 2-4). It's definitely an important issue to consider, and you may notice that turbine manufacturers tend to heavily tout the alleged quietness of their respective machines. Brian Levine, cofounder and vice president for business development of Michigan-based manufacturer WindTronics, told us that, according to his experience, noise tops the list of consumer concerns (followed by effects on animals, vibration, and tower height). Levine's core product, a Honeywell-licensed small turbine, is said to be exceptionally quiet, although some wind experts say the technology is still unproven.

So, the question is, are small wind turbines loud enough to be considered annoying, and might they have physiological effects? There have been some reports of complaints about noise from small wind turbines, ranging from perceived annoyance to alleged illness.

However, an expert panel in 2009 prepared by AWEA and CanWEA concluded the following:

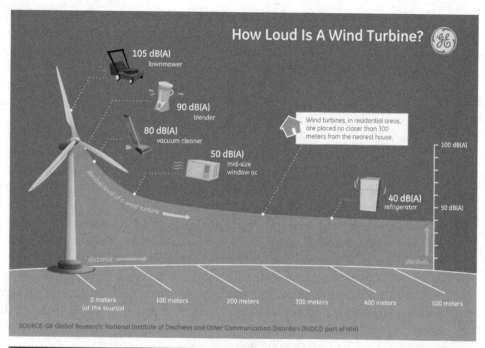

FIGURE 2-4   Wind turbines have moving parts, so they make some noise, but they are usually comparable to many other mechanical devices we are regularly in contact with. *GE Reports/Designed by Group SJR.*

- There is no evidence that the audible or subaudible sounds emitted by wind turbines have any direct health effects.
- The ground-borne vibrations from wind turbines are too weak to be detected by, or to affect, humans.
- The sounds emitted by wind turbines are not unique.[1]

"Wonder why I can't hear my turbine while watching football on TV?"

Let's take a closer look. Sound volume is measured in decibels, dB(A), with a range from 0 dB(A) (imperceivable) to over 150 db(A) (damaging over the long term). The human ear can distinguish sound differences of about 3 dB(A), or greater than ambient noise. But don't be fooled: 3 dB(A) is not 3 percent, it is a 50 percent increase in the sound level. Nearby trees on a breezy day will measure about 55 to 60 dB(A) on a decibel meter. The average ambient noise in a house is about 50 dB(A). A conversation is about 40 to 50 db(A).

The good news is that most of today's residential wind turbines produce the same level of sound as a conversation or background noise—40 to 50 decibels—over most of their operating range.

The second component of sound to consider is frequency. That's because even though the volume of sound coming from a wind generator may be the same as the ambient noise, the *frequency*, measured in Hertz (Hz), may be different. Dogs, traffic, radio, and kids could be at the same decibel level, yet at a different sound frequency. Therefore, wind turbine sounds may be distinguishable from ambient noise even though they are not louder. Does this necessarily mean they are annoying? That's a subjective question, and something everyone has to experience.

Now, not many people live at the base of the tower, so distance plays an important role as well. As you double your distance from the origin point of sound, you reduce the intensity by a factor of four—the inverted square ratio (or I/d^2). Therefore, any audible sound emitted from a wind turbine will quickly blend into the background noise with increasing distance from the tower.

Sound from a wind generator normally originates from two sources: the drivetrain and the blades. In machines that have a drivetrain—mostly utility-scale turbines—there is a gearbox transmission, high-speed generator, and mechanical brakes, all of which can make noise. However, most newer small wind generators are *direct drive*, meaning the rotor is directly connected to the alternator/generator, without a transmission, similar to computer disc drives. Because they are direct drive, the generators used in these systems are slower speed, and they do not usually produce mechanical or tonal noise.

Most of the sound that comes from a wind turbine is aerodynamic noise caused by the air resistance, or lift drag, as the fiberglass or carbon fiber blades pass through the air. As wind speed increases, so does the speed of the blades, and they get louder.

Aerodynamic noise is also a function of the speed of the blades as measured from the blade tips. If you are selecting your turbine, the data sheet should mention its *tip speed ratio* (TSR). This is a term that refers to the speed of the tip of a wind generator blade in relation to wind speed. For example, a wind system that operates with a TSR of 10 means that when the wind speed is 12 miles per hour (mph), the tips of the blades are moving at 10 × 12, or 120 mph. Increasing tip speed results in more noise. A well-designed, three-bladed rotor with a TSR of about five to seven (125 to 175 mph) will emit sound that is barely discernible from ambient noise with a decibel meter.

Large commercial wind turbines can have an additional aerodynamic noise that is often referred to as "blade thump." Some observers say this sound is usually only audible within 300 feet of the turbine, but blade thump has been reported at distances of up to 1.5 km away. This sound is caused by the large blades passing the tower of the turbine. Although measurements indicate the noise is not particularly loud, recent surveys from people who live near wind farms claim it can be irritating over time because of its heartbeat-like, rhythmic nature.[2]

For small turbines, depending on the design type, there are three scenarios in which sound can increase above normal operating conditions:

- Loss of grid connection
- When the batteries are full
- Exceedingly high wind speeds

When a utility blackout occurs with a grid-tied wind system (without backup), the wind system loses its load, the grid. Some turbine systems are still designed to "freewheel" during such an event; that is, the rotor is temporarily disengaged from the generator, permitting the rotor to spin freely in the wind. At this point, it can become noticeably louder. The system resets to normal as soon as the grid is reconnected. If you are home, the solution is easy: if you don't want the neighbors to complain about a whining machine, like bringing your barking dog in the house, simply shut the wind generator off with a manual braking mechanism that is included with these types of purchased models until the grid is back on.

Note that freewheeling will damage or even destroy many turbines, and wind experts like Ian Woofenden often train people to design their systems to avoid it, so in all likelihood this potential source of noise won't apply to you. Some models, like Southwest Windpower's Skystream 3.7, have protection systems that engage a dynamic magnetic brake if power is interrupted. Strong winds could not budge those blades.

The second scenario involves off-grid wind systems when the batteries are full and cannot accept more energy, resulting in an overvoltage in the line. This most often occurs during prolonged periods of sustained high winds. Some systems have a safety feature designed to freewheel when this event occurs. Or, as is more commonly recommended, the excess energy can be diverted, or "shunted," to a dump load, converting the electrical energy to heat. This effectively keeps sound within normal operating parameters.

The last scenario is achieved when the wind speed exceeds the turbine's cutoff speed. At this time, the turbine has already furled as far as it can go and the rotor will freewheel, creating aerodynamic noise not only from common drag, but also from the air turbulence as a result of the wind hitting the blades at an angle of more resistance.

According to Karen Sinclair, a senior project leader with the National Renewable Energy Laboratory (NREL)'s National Wind Technology Center in Colorado, consumers should shop for turbines that have been tested to international standards, which she says include noise standards. (American standards for small wind have recently been developed, but as of this writing few turbines have yet finished qualifications.) "But even with standards, if turbines are not properly sited they can be annoyingly loud," Sinclair told us via phone. Similarly, some home-brew turbines can be quite noisy.

Small-scale wind turbine manufacturers are improving their technology and making it quieter. For example, sound insulators are now commonly installed on the tower stub to minimize sound conductance from the turbine. Dynamic braking, which controls the rotor in high speeds, effectively increases resistance via electromagnetic force applied onto the spinning rotor, slowing it down. Consequently, it also increases its efficiency.

Boosters of vertical axis wind turbines (VAWTs) often argue that their products are quieter than horizontal axis wind turbines (HAWTs). In fact, it's one of the reasons why VAWT manufacturers sometimes promote them for urban areas, although that's also controversial.

Colin Malaker, a dentist in Columbia, Missouri, recently installed a small WePOWER VAWT in front of his practice, which he says delights his patients (Figure 2-5). "It's very quiet, I never hear it," Malaker told us. "When it's spinning really fast it maybe sounds like a B-52 at 30,000 feet. But a microwave oven is twice as loud," he added. In addition to dentistry, Malaker runs a small renewable energy business on the side, and he plans to sell WePOWER products.

Figure 2-5 Colin Malaker, a dentist in Columbia, Missouri, recently installed a small WePOWER VAWT in front of his practice. He says the turbine is very quiet. *Colin Malaker.*

We'll get into the differences between horizontal and vertical wind turbines later, but it's worth noting that noise is often brought up in the discussion.

 Power Up! *We recommend that you try before you buy, especially since perception of sound can be fairly subjective. It's a good idea to visit several installed examples of the turbine you are considering. Go during different wind and weather patterns, and walk around it. Measure out the distances on your site from your residence to the tower and to any other important features. Make sure you have detailed conversations with your family and neighbors before you pour the concrete and start moving electrons. Otherwise, you could end up with an angry family and angry neighbors.*

Infrasound and Low-Frequency Sound

Infrasound is a sound that is lower in frequency (<20 Hertz) than human hearing can perceive. Think the opposite of a high-pitched dog whistle.

Infrasound is generated internally in the body (by respiration, heartbeat, coughing, etc.) and by external sources, such as wind and breaking waves, work environments, industrial processes, birds, vehicles, and air conditioners. In short, it is ubiquitous. Turbines are also capable of generating infrasound, but usually at levels that are similar to what's already prevalent in the environment.

Studies suggest that the levels of infrasound near modern commercial-scale wind turbines are in general not perceptible to humans, either through our ears or any part of our body. Putting it in perspective, the levels as measured from dwellings in proximity are no different from what's measured from natural gas compressor stations, industrial sewage pumping stations, and other power plants. In addition, there is no evidence of adverse health effects due to infrasound from wind turbines.

No obvious health effects are found among those who are exposed to infrasound below 100 dB(A). However, those who are chronically exposed to infrasound above 110 to 120 dB(A) "present notable subjective sensation of autonomic neurobehavioral dysfunction (somatization, depression, hostility, phobic anxiety, and psychotism)".[3] Thankfully, all modern wind turbines are typically designed to well below half of this level as measured to any occupied building.

Dr. Nina Pierpont, a population biologist, recently completed a self-published report speculating that a constellation of physiological symptoms, which she christened "wind turbine syndrome," are caused by low-frequency noise and vibration, as well as shadow flicker, from large wind turbines, affecting the body's various balance organs of the inner ear. Low-frequency sounds are caused by eddies of turbulent air that form off the blades, and they travel much farther than infrasound and high-frequency sounds.[4] Pierpont's small case study obtained reviews and notices from several practitioners in related fields, but there are as yet no reports in the peer-reviewed clinical literature linking wind turbines to these or any other symptoms. British sleep expert Christopher Hanning, however, has noted that sleep deprivation can cause most of the reported symptoms.[5]

All that being said, it is important to remember that these effects are greatly diminished when it comes to small-scale wind. In conclusion, there is scant evidence of harm from infrasound produced by wind turbines, although it is worthwhile to keep abreast of current research and common perceptions. In order to give the benefit of the doubt, turbines can be sited a sufficient distance from human habitation. It may also be possible to further modify the blades to reduce turbulence.

Perception and Annoyance

Wind turbines are among the more recent technologies encroaching onto our landscapes. The public has already adjusted to power lines, water towers, and cell towers. Those things are so common that we usually don't take notice. But even some of the best advocates of wind power and renewable energy are not always enthusiastic about seeing wind turbines close to home. (Just ask the local environmentalists who have spent years opposing plans for a wind farm out in Nantucket Sound.)

Unfortunately, if someone already didn't like turbines, having to see one or more from their residence tends to make the person like the technology even less.[6] In addition, in several surveys of homes near large wind farms, turbine noise was said to be more annoying than transportation or industrial noise at comparable levels, possibly due to the intermittent nature of the sound. The exception is people who benefit economically from wind turbines, who tend to have a higher threshold for annoyance, despite exposure to similar sound levels.[7]

Perception and annoyance are also linked to terrain and urbanization:

1. There is some evidence that people are more annoyed with turbines in rural areas versus suburban areas.
2. In a rural setting, hilly terrain may increase annoyance versus flat terrain.[8]

Therefore, it may pay to take the unique local environment into account when planning a new wind turbine.[9] However, it's also important to remember that any annoyance issues are greatly reduced when it comes to small wind turbines, and that most studies on the issue have only looked at large systems.

Flicker Effect

Shadow flicker effect can occur when a wind turbine's rotor casts a moving shadow on an observer. The phenomenon only occurs when, to the viewer, the sun appears to set or rise behind the spinning blades of a turbine. The effect creates a pulsating light that can last for 15 to 30 minutes.

FIGURE 2-6 Most of the issues people bring up with regard to wind turbines relate only to utility-scale installations, such as flicker effect and low-frequency noise. *Brian Clark Howard.*

Many factors must be present for a shadow flicker effect. It occurs only in specific directions, depending on the latitude (west-northwest and east-northeast in the Northern Hemisphere, east-southwest and west-southwest in the Southern Hemisphere), and needs penetrating sun at certain specific times of the day. In addition, the size of the turbine blades has to be substantial (Figure 2-6), meaning that micro- and residential-size systems are not large enough to cause the effect.

Shadow flicker from commercial-scale turbines has occasionally been blamed for health effects, ranging from distraction while operating equipment to vertigo, headache, and nausea. Scientists are currently researching whether any shadow flicker from wind turbines might cause a response, such as seizures, in the approximately 0.025 percent of the population that suffers from *photosensitive epilepsy*.[10]

A few studies found there is a negligible risk that three-blade, commercial-size turbines in specific conditions will have the blade interrupt sunlight to your eyes more than three times per second (60 revolutions per minute). This factor is known as the *flash frequency*. For this effect to occur, the viewer would need to be located in alignment with persistent sunlight and the turbine rotor, with the sun appearing at a position relatively low in the sky.

However, until such results are irrefutable, researchers at the Neurosciences Institute in Birmingham, United Kingdom recommend that, regardless of the viewing distance, shadows cast by turbines "should not be viewable by public"[11] if the flash rate exceeds three per second (60 revolutions per minute) and when the sun appears low in the sky (usually sunrise and sunset).

Fortunately, morning events rarely coincide with waking hours and usually pass unnoticed, but evening events can cause nuisance, similar to the sun's glare in the car window. Large wind farm planners can take steps to reduce the flicker effect by proper siting in relation to occupied or traveled areas or with tree plantings in the area of the shadow.

The Danish Wind Industry Association has graciously offered a shadow calculator to use for your assessment guidedtour.windpower.org/en/tour/env/shadow/shadowc.htm.

Resource

Online Shadow Calculator   bit.ly/shadowcalculator

Visual Influence

Perhaps the most commonly voiced opposition to wind turbines is along *viewshed* lines.

Technically, a viewshed is an area of land, water, or other environmental element that is visible to the human eye from a fixed vantage point.[12] In other words, a viewshed is what people see. Of course, what people like to look at is highly subjective and individualized, although many people can agree that natural is good, especially as the world becomes more urbanized.

Many people report that they find wind turbines aesthetically appealing, even awe inspiring or beautiful. To these folks, wind turbines fit right into the landscape. Others, however, just plain don't want to look at them sticking up in the air—spinning continuously and "making noise." Other factors also affect the visual appearance of a wind generator. Larger turbines rotate more slowly than smaller ones, and a wind farm of fewer larger turbines is often preferable to an installation of many smaller ones. In some cases, arranging the machines in a straight line makes them more favorable to the public.[13]

Occasionally, neighbors ask if a turbine could be painted a certain color, say blue, to "blend into the sky," or green, to "blend into the hills" (Figure 2-7). Whether you seriously consider such a request is up to you, though note that there's little

FIGURE 2-7   After Rosalie Bay Resort installed a wind turbine in Dominica in 2008, one neighbor asked if it could be painted blue or green. However, there's little evidence this will make a difference. *Rosalie Bay Resort.*

evidence to suggest that any particular color ultimately helps a turbine blend in. Some people have made up turbines to look like spinning flowers, and Kevin had the idea to paint a nacelle to appear like a propeller airplane. Obviously, blade size ratio to the plane would be exaggerated, but what a creative solution for alternative energy at an airport or aviation museum!

Opinions definitely change, although it can take diligence, time, and patience. Commonly, public support of wind turbines tends to rise once the machines are installed and operating, according to several surveys carried out in the United Kingdom and Spain.[14] Like other improvements—high tension wires, water towers, cell towers, oil refineries—all utility landmarks eventually become ubiquitous within the community. If someone gave you local directions in your own town and used the city's water tower as a landmark, you might say, "Where is that?" The opposite can also be true for wind. It could be a landmark that represents the community's stand on some ideal lifestyle. In fact, in 2008, the city of Rock Port, Missouri, proudly announced that it was America's first town to be wholly powered by wind energy.

Sometimes, attempts to oppose wind turbines may have their roots in a feeling that the community has had no control over previous developments. Being that wind turbines don't yet have the same acceptance rate as other tall structures like electric, cell, and water towers, some people feel they have more opportunity to oppose them at zoning hearings. Did you know that when windmills were first introduced in Holland, many people considered them undesirable?[15] People have long been resistant to change. But then things change.

Property Value

As the pace of wind project development increases, opponents raise claims in the media and at siting hearings that property values will be hurt. As we have pointed out, wind turbines do generate some noise, can produce flicker effect, and are visible with moving parts (Figure 2-8). But again, such issues are greatly reduced, or even eliminated, when it comes to small turbines. So do wind turbines affect property values, positively or negatively? This is a serious question that deserves to be empirically examined.

The Department of Energy sponsored a study in 2009 that involved collecting data on almost 7,500 sales of single-family homes situated within 10 miles of 24 existing wind facilities in nine states. The conclusions of the study are drawn from

FIGURE 2-8   Are wind turbines beautiful or eyesores? It's a highly subjective question, but one worth asking in your community. *Brian Clark Howard*.

eight different pricing models, as well as both repeat sales and sales volume models. The research concluded that

> "Neither the view of the wind facilities nor the distance of the home to those facilities is found to have any consistent, measurable, and statistically significant effect on home sales prices."

This conclusion does not dismiss the possibility that individual homes or small numbers of homes have been or could be negatively impacted. The results suggest that if these impacts do exist, they are either too small and/or too infrequent to result in any widespread, statistically observable impact.[16]

Another study done by the Renewable Energy Policy Project (REPP) found no evidence that property values decreased as a result of large wind farms. Quite the contrary:

> "For the great majority of projects the property values actually rose more quickly in the viewshed than they did in the comparable community. Moreover, values increased faster in the viewshed after the projects came online than they did before."[17]

In Europe, several countries have conducted similar studies and came to similar conclusions.[18,19]

However, there still can be a difference between actual property values and the perception of what will happen to property values. According to an economic analysis of a proposed (and controversial) wind farm in Nantucket Sound in Massachusetts, a survey of tourists, residents, and real estate agents suggests

> *"Homeowners expect the project to decrease their home values by an average of 4 percent."*[20]

That proposed wind farm hasn't been built yet, as of this writing, so the actual effect cannot be reported, although historical trends suggest that the homeowners will be pleasantly surprised that their fears were unjustified.

Noise has also been the bigger factor in determining property value. In the United States, lawsuits and complaints have been filed in several states, citing lost property values in homes and businesses located close to industrial wind turbines due to noise and vibrations. As of this writing, these cases are unresolved.[21] In one case in DeKalb County, Illinois, at least 38 families have sued to have 100 turbines removed from a wind farm there. A judge rejected a motion to dismiss the case in June 2010.

Opponents and proponents have surmised that negative environmental effects may be abated by better placement, new sound reduction technology, corporate buy-outs, and better municipal policies.

Interestingly, Denmark adopted a policy in 2009 that requires developers to pay compensation for alleged loss of property value following erection of a wind turbine more than 25 meters (82 feet) high. The value must be determined by an appraisal authority.[22]

Still, there is some evidence that wind turbines placed in strategic locations may actually improve property values of the community. It can represent to a new homebuyer savings in energy costs, and holds the image of reducing their carbon footprint. In fact, a number of residential developments have been announced across the country that will feature small wind turbines, with that feature prominently promoted in advertising materials. As Rick Rosa of the Solarium apartment building in Queens told us, a wind turbine can help a property get noticed.

Indeed, many wind turbines are themselves tourist attractions. In a popular vacation area of Scotland, a poll found that 80 percent of tourists said they would be interested in visiting a wind farm if it were open to the public. An impressive 91 percent of respondents said they would not be put off from visiting an area because of the presence of wind farms. In Denmark, many tour agencies run boat trips to take visitors to see the offshore wind farm at Middelgrunden, near Copenhagen. In the United States, a National Renewable Energy Tour visits homes that showcase clean technology. In Swaffham, Norfolk, over 50,000 tourists have climbed the wind turbine tower to see the spectacular sights from the viewing platform, 65 meters above the ground.[23] If you are seeking more than increasing your property value but are into some passive income, wind turbines evidently have a particular market appeal, and certainly are more visible than the towering large panel signs.

In summary, it's undeniable that a small wind turbine *could* affect your property value or that of the neighbors around you. This is highly dependent on many factors, although the evidence we do have, particularly from studies of large wind farms, suggests that fears are overblown. With proper design, placement, and diligence in educating others within the community, you are promoting a positive image that few can refute, and buyers might find of value. If recent studies and historical trends are any indication, chances are good that your turbine will not harm property values, and may actually contribute to them.

INTERCONNECT

Kevin's Unique Dome Home

When we talk about property value and visual appeal, I must insert my own unique structure. In 2004, I directed the construction of my "green" dome home (Figure 2-9). It didn't receive positive feedback from the neighbors. I got yelling and suggestions of where I could place my dome.

When I planned to build my wind turbine, I was well aware of what I would expect. But I got the opposite reaction. I have found eager guests in my driveway, asking about the

Figure 2-9   When Kevin started building his dome home in 2004, some of his neighbors were initially upset. But he found that they were more accepting of his small wind turbine. *Kevin Shea.*

wind turbine and how they can get one. When passersby see the 70-foot-wide dome, they might think it is a salt or grain storage facility (I do live next door to a farm), but put up a wind turbine and they were attracted like bees to a flower.

I would like to absolve myself of any responsibility for any vehicle accidents that were caused by drivers who inappropriately stopped in the middle of the busy road to view the turbine. You know who you are. In addition, I don't know how these accidents affect the property value.

Electromagnetic Transmissions

Some people ask if wind turbines might interfere with radio, television, or cell phone signals. Your neighbor might want to know if your proposed turbine is going to cut off that critical Superbowl play or the finale of *Lost*, or that emergency phone call or Pandora Internet radio streaming from their Wi-Fi.

Some studies have looked at such issues as:

- *Electromagnetic interference* (EMI), which happens when a wind turbine creates and radiates a frequency in the radio band that is being used
- *Near-field effects*, which is the change of the characteristics of an antenna when in close proximity to another object
- *Diffraction*, which is a signal loss when a radio wave is partially blocked

- *Scattering* (reflection), which is when the radio waves are reflected from a metallic surface[24]

All of these issues sound serious. Fortunately, small wind turbines have never been found to create interference with signals.[25] Importantly, the materials used to make the blades are typically nonmetallic (composites, plastic, wood) that don't cause reception problems. Small turbines are too small to "chop up" any signal. In fact, small wind systems are commonly used today to power remote telecommunication stations for both military and commercial uses. Many years ago, a few wind turbines equipped with long, metallic blades did cause some localized problems, but they are no longer commonly used.

Today, it is only large wind turbines, which may use metal in their blades, that can interfere with radio or TV signals, but only if a turbine is in the "line of sight" between a receiver and the signal source, and that's something installers can plan against.[26]

Radar

Radar is basically designed to filter out stationary objects and display moving ones, so moving wind turbine blades can create radar echoes. Only the newer large wind turbines have been noted to affect the 50-year-old defense radar technology, due to their enormous coverage. They tend to create a "mirage" on radars. Peter Drake, technical director at Raytheon, a major provider of radar systems, has said, "A wind turbine can look like a 747 on final approach."[27]

After elements in the British and U.S. military expressed concern, some wind projects were halted until a study was conducted. Research by the British government concluded that some radar installations may require no change, while others can be modified to ensure that air safety is maintained in the presence of wind farms.[28] It's a purely technological issue—potential fixes include new coatings for wind turbine blades, software patches for existing radar systems, and overhauling obsolescent radar systems.[29]

If your wind project is proposed near an airport or military airfield, this issue might attract some wind turbine opponents. However, rest assured that interference is generally limited to small, low-flying airplanes that are physically shadowed by your turbine. So, even if the radar technology is outdated, there is low risk that a small-scale turbine would have any effect on radar.

Ice Throwing

Some people have a vision of wind turbines in winter as spinning pinwheels of death, flinging dagger-sized icicles off their blades at 100 miles per hour (Figure 2-10). However, in reality, wind turbines shut down during icing events, due to either imbalance detection or the control anemometer icing (the control anemom-

FIGURE 2-10   Although many people are afraid wind turbines will fling shards of ice, we aren't aware of any damage. Pictured is the 1.5 MW wind turbine at Jiminy Peak in Massachusetts, which provides about a third of the ski resort's power needs. *Brian Clark Howard.*

eter tells the turbine how fast the wind is blowing). Any ice buildup tends to shed in thin pieces while the turbine is at rest or while it is starting up. Not too much different from the eaves of a roof.

Current turbine setbacks from roads and residences should be sufficient to protect the public from ice shed. To date, we are not aware of any insurance claim for injury due to ice shed from turbines.[30]

Ecological Impacts of Small Wind Turbines

Small wind advocates sometimes brush off the potential environmental impacts of their machines, arguing that their devices are too small to make much of an impact on nature. However, we feel that it's worthwhile to give a thought to the entire lifecycle of the project, from the energy it takes to produce and transport the equipment to preparation of the site and operation and maintenance. Not only will you be participating in a process that's as "green" as possible, but you'll also do credit to the industry, which has its fair share of critics. And we can follow that with a more socially and politically popular topic—we shall teach you about the birds and bats!

Embodied Energy

As with anything manmade, there is going to be some environmental impact involved with the hardware itself. According to Ian Woofenden, freestanding manufactured towers require more energy to produce than other tower types, largely because they require more steel and concrete, the latter of which is particularly energy-intensive.

"You could also make a case for the cheapest tower for the job being the most eco-friendly, because it will probably require the [fewest] resources," Woofenden told us. "That would probably be tilt-up for the tower, since they need less concrete," he said.

Wind turbines will take some time to "pay off" the amount of emissions that went into setting them up, so it's good to get something in place that has a shot of actually producing energy. The more clean power you make, the greener your machine. Similarly, aim for longer-lasting equipment, as that means fewer resources are needed.

The better tuned you keep your machine, the fewer problems you'll have down the line. Where possible, choose locally produced equipment, as that decreases transportation costs and impacts. Hire help locally, too, as that cuts down on mileage and builds local expertise. When a part is no longer useful, recycle it at your nearest scrap yard.

Site Preparation

Setting up a wind turbine can cause disturbance to soil and vegetation, particularly if large earth-moving equipment or cranes are needed (Figure 2-11). Still, since the scale is relatively small, the impact should be pretty minor and relatively temporary.

Even so, you can do your part to try to minimize impact by making sure you don't cause problems with erosion or soil compaction. Occasionally, when turbines have been set up in very sensitive, or very remote, areas, they are installed via helicopter. That reduces impact of overland trucking, but it can be expensive.

Potential Impacts on Birds

Kevin would like to preface this section by mentioning that he is building a wildlife reserve, which will hopefully support an abundant bird and bat population. In our opinion, mitigating bird mortality is a serious matter.

In fact, for every person who asks if wind turbines "throw" ice, we'll introduce you to four who insist the spinning blades chop up birds like the Iron Chef whack-

Figure 2-11   Preparing the site for a turbine tower can be fairly invasive, and can require large earth-moving equipment. Take care to minimize erosion and long-term impact to the ecosystem. *Paul Anderson/Wikimedia Commons.*

Figure 2-12   A great horned owl perches on a 65-kilowatt Micon turbine during a low wind period at Altamont Pass in California. Scientists have learned to reduce the impacts of the technology on birds. *Shawn Smallwood/DOE/NREL.*

ing at an onion. Part of the reason for that perpetuated perception is that, yes, some early wind farms did have a noticeable impact on birds, namely the famous installation in Altamont Pass. However, a number of specific factors were at play, most of all turbine placement in proximity to existing bird migration, mating, and nesting areas. We shall say this for energy production as well as avian life preservation: location, location, location. The utility-scale industry has learned a lot about mitigating effects on birds by conducting extensive analyses, raising the heights of towers, and slowing down the blade revolutions (Figure 2-12).

Do small wind turbines kill birds? Occasionally. Sure, that's unfortunate, and not just because people tend to like the pretty winged things. Birds play critical roles in the ecosystem, and many species have seen stark declines over the past century or so. In working on this book, we did an informal survey of small wind owners and asked if any of them had ever encountered a dead bird at their tower. Most said no, although one person admitted to "seeing one," and another "knew someone who had."

What kills them? Collisions with turbine blades have been recorded, as have occasional collisions with towers. How much of a problem is this when it comes to small wind, and can it negatively impact the health of avian populations?

According to Karen Sinclair of the National Wind Technology Center, there is almost no reliable data that can answer those questions. Sinclair is one of NREL's top researchers on the environmental impacts of wind turbines, and she told us that small wind owners should give the issue some thought.

"Everything has an impact, so the question from a societal standpoint is, what's a tolerable impact?" Sinclair asked. "If you knowingly put something up that could impact a threatened species, then that's a problem," she added. In fact, Sinclair pointed to some small wind turbines that have been sited in important habitat for the endangered Indiana bat.

If you are considering a turbine, Sinclair says it's first important to understand the nature of your site. "There's a lot of literature out there on what species might be of concern in that area," she explained. This is important, because species can differ in their vulnerability to collisions, as well as in the health of their local populations. "It's not likely that the flight patterns of condors would put them in conflict with small wind, but that could be an issue with commercial wind farms," Sinclair explains as an example.

Sinclair suggests checking in with your local Audubon Society chapter and natural resources office to see if there are any known species at risk. "Maybe soon there will be maps available to small wind installers [that show species of concern]," said Sinclair. "We know that thousands of small turbines are installed and they don't go through the rigors of commercial wind farms. But there is concern on the part of the U.S. Fish and Wildlife Service."

Sinclair points out that when her U.S. team installs a small wind turbine with federal money, the project has to go through a National Environmental Policy Act (NEPA) review. This involves a pretty comprehensive checklist, covering any potential impacts on wetlands, cultural history, water, and a host of other facts. "We're not saying people have to do a full NEPA review, but they should think about what the impacts could be," said Sinclair. "Remember, all impacts are site specific. These turbines are not a benign technology, and the industry needs to be thoughtful and proactive."

If we look at birds, its helpful to note that total bird mortality in the United States is estimated at 1 billion animals per year.[31] Recent data show wind turbines account for less than 0.003 percent of all annual U.S. bird fatalities—and most of that tiny percentage is attributed to utility wind farms. So clearly, bird deaths caused by wind turbines are a minute fraction of the total anthropogenic bird deaths.

Having said this, in Table 2-1 we provide U.S. estimates summarized by Erickson et al. (2005) and the U.S. Fish and Wildlife Service. Those sources emphasize the uncertainty in the estimates, but the numbers are so large that they are not overly obscured by the uncertainty.

TABLE 2-1   In the scope of human influence, wind turbines represent a vanishingly small threat to birds. That doesn't mean the industry isn't taking the issue seriously, however, and taking steps to reduce impact further.

Type of Bird Death	Estimated Death Count Annually in the United States
Collisions	
Buildings	97 to 976 million birds
High-tension lines	130 million to 1 billion
Cars	80 million
Communications towers	5 to 50 million
Wind turbines	20,000 to 37,000
Toxic chemicals (pesticides)	72 million
Domestic cats	Up to hundreds of millions
Fishing "bycatch"	10,000s to 100,000s
Oil spills	100,000s
Oil and wastewater pits	2 million

There is currently no evidence that fatalities caused by wind turbines result in measurable demographic changes to bird populations, with the possible exception of raptor fatalities in the first wind farm in Altamont Pass, California. There an estimated 1,300 raptors are killed annually, among them 70 golden eagles, which are federally protected.[32]

The probability of bird fatality is affected by both abundance and behavior of the species. Small songbirds that migrate at night are the most common fatalities at wind-energy facilities, probably due to their abundance and the fact that they can't see obstacles as well after dark. However, fatalities probably have greater detrimental effects on raptor populations because of their characteristically long life spans, low reproductive rates, and low abundances. Interestingly, long-lived geese and ducks have seemed to do a good job detecting and avoiding wind farms.

Experts agree that more research needs to be done on the potential impacts of turbines on birds. Remote technology could be used to monitor collisions, which could lead to new mitigation measures.[33]

Some progress has already been made. Newer, larger turbines appear to cause fewer raptor fatalities, likely due to higher towers and larger blades, which move slower.

Going further, some research has been done on making blades more visible to raptors. "There was some indication that one black blade and two white blades made them more visible," Sinclair explained. However, she cautioned that the concept may not apply to small turbines, which spin relatively faster.

It has long been known that wind turbines attract insects, which in turn bring birds within range of the spinning blades. Scientists have discovered that by tinting the blades the color purple, it seems to reduce the number of insects that are attracted during the day and at night.

More work is also being done to study the flight pattern of coastal birds, including a project at the Rødsand II wind farm in Denmark. Researchers have generated what is called a theoretical maximum distance (TMD) from the coast for raptor species. If the nearest turbine of a marine wind farm is placed farther away from the coast than the estimated TMD, the collision risk would be reduced.[34] Again, this doesn't necessarily apply directly to your backyard wind turbine, but it's good to be aware of.

It's worth revisiting the infamous example of Altamont (Figure 2-13). It is commonly repeated in the popular press that one of the reasons for the wind farm's

Figure 2-13   A golden eagle soars above a 100-kilowatt wind turbine at Altamont Pass in northern California. One of the oldest wind farms, Altamont has a reputation for killing raptors, although engineers have been redesigning the site. *Shawn Smallwood/DOE/NREL.*

high reported bird fatalities was the lattice structure of the towers, which allegedly attracted birds to nest and perch on them. However, according to Sinclair, more recent analysis of the data suggests that was never the problem. (Incidentally, we also asked small wind turbine owners if they had ever seen birds nesting in their towers, and they all said no.)

According to Sinclair, the reason abundant turkey vultures were not being killed at Altamont but raptors were boils down to behavior. "Red-tailed hawks were at greater risk at Altamont because there were rock piles that had been moved to make way for the turbines, and they provided habitat for California ground squirrels, which the hawks hunt," explained Sinclair. "During foraging, they collided with the turbines."

In sum, wind turbines seem to have a relatively insignificant effect on bird populations, predictably more so with small turbines. If you are comfortable keeping your cat, living in a building with windows, using electricity supplied by high-tension wires, driving in your car, and talking on your cell phone, you can rest assured that your wind turbine is vastly less harmful and provides more ecological benefit.

Still, one more thing to keep in mind is that if you are considering an installation, it's a good idea to factor in how construction and maintenance may alter the ecosystem through vegetation clearing, soil disruption, and potential for erosion. This could result in fragmentation and loss of habitat for some bird species or their prey. Try to avoid creating rock piles next to the turbine that might attract ground squirrels.

If bird fatalities still remain a sensitive issue for you or your community, you could conduct pre-siting and post-siting studies. It is relatively expensive and time-consuming, but it would provide helpful information. At a minimum, before you build, try to find out what species are in your area and if their behavior may put them at risk (Figure 2-14).

Impacts on Bats

Although bats might not be everyone's favorite flying creatures, it is important to remember that these bug vacuums reduce insect populations during birds' off-hours, making life more bearable for us (and saving you money on bug spray). Although there isn't a lot of conclusive data to adequately evaluate the effects of turbines on bats, there is some evidence that commercial projects may be having a negative effect, and some scientists are concerned (Figures 2-14 and 2-15).

Recent studies have found significant deaths of migratory, tree-roosting bats at utility wind power facilities in forested regions of eastern North America.[35,36,37] A study in Pennsylvania and West Virginia reported 466 dead bats in 42 days (11.1 bats/night or 0.26 bats/turbine/night) as a result of collisions with large wind turbines.[38] This does not suggest that all areas and all turbines may see similar results.

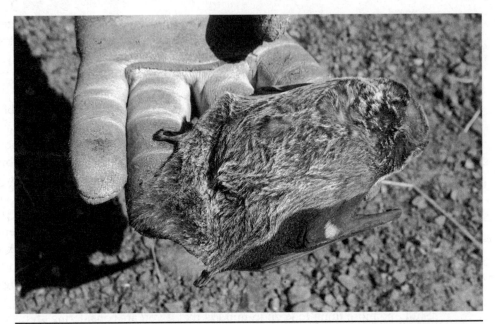

FIGURE 2-14   Hoary bats (*Lasiurus cinereus*) seem to be the most common species killed by commercial wind turbines, and scientists think the culprit may be near-contact barotrauma—lung failure from abrupt pressure change. *Paul Cryan/U.S. Geological Survey.*

FIGURE 2-15   U.S. Geological Survey (USGS) biologist Paul Cryan is studying bat fatalities at large wind turbines. Little is known about potential impacts of small turbines, but so far there is little evidence of harm. *Paul Cryan/U.S. Geological Survey.*

However, these fatalities have a greater detrimental effect on bat populations because of their relatively long life spans and low reproductive rates.[39]

Data on small wind effects on bat populations are sparse. According to Sinclair, the U.S. Fish and Wildlife Service is concerned about potential impacts of small wind turbines on the endangered Indiana bat. "They're taking a very conservative perspective and saying, 'Don't put small turbines in these areas,' even though there's no data," Sinclair explained. She added that there is some concern that bats may be more at risk than birds because of the altitude at which they commonly fly. (They also tend to fly at night.)

Given that bats use echolocation and generally detect moving objects better than stationary ones, it may seem surprising that they would run into turbines. A number of theories have been proposed to explain it, including that bats may be attracted to ultrasonic sounds produced by turbines, to bugs that congregate around turbine lights, or to the cleared land around towers. Some scientists have suggested that many bats may be killed not by striking turbines directly, but by the sudden differences in air pressure around them, in a process called near-contact barotrauma. The change in pressure is thought to result in internal bleeding, and bats are thought to be more susceptible than birds. Further, it is thought that bats are unable to detect the sudden pressure changes.[40]

In any case, migratory tree-roosting (they-sit-in-trees) and insectivorous (they-eat-insects) species of bats—such as hoary bats—appear to be the most susceptible to trouble around wind turbines (Figure 2-16). Another hypothesis is that bats view these tall structures as roost trees.[41,42,43]

Clearly, more research is needed on the impacts of wind turbines on these bug vacuums. In the meantime, there are some things you can do to minimize risk. Contact your local bat conservatory association for migrating patterns in your region, and change the location of the turbine if need be. In addition, you might also consider looking into one of these emerging strategies: (1) color, (2) electromagnetic frequency, (3) periodic shut-down, or feathering.

As we mentioned in the section on birds, painting or tinting the turbine system purple may help make it less attractive to insects, and therefore bats.[44] In other studies, applied electromagnetic fields seemed to keep bats away (but not pesky insects).[45] Bat deaths might also be reduced by feathering the turbine blades (changing their pitch angle) or slowing or stopping rotation during short periods of high bat activity.[46]

Another interesting idea comes from a biologist with Bat Conservation International, Ed Arnett, who suggests raising the cut-in speed, or the wind speed at which wind turbines switch on, to 11 mph. According to Arnett, bats usually don't fly when it's too windy. In his trial, bat fatalities were reduced up to 93 percent, while annual energy output loss was reportedly less than 1 percent.[47] Admittedly, you'd need to have a great wind resource to have the flexibility to do that.

FIGURE 2-16   Most of the concern for birds and bats revolves around large utility-scale wind farms like this one near Bickelton, Washington, but that doesn't mean small wind installers shouldn't give any thought to potential impacts. *Iberdrola Renewables/ DOE/NREL.*

Summary

Wind turbines have many benefits, but they also have some limitations and challenges. There is also a lot of misinformation about them. As author Paul Gipe told us, "They'll always be somebody who says it's going to electrocute them or make their hair fall out. Twenty-five years ago, people in the Mojave Desert were afraid wind turbines would make rattlesnakes come out. Others have accused wind turbines of shocking the teats of their cows [through automatic milkers]." In case you are worried, according to Gipe, the latter effect is the result of improperly grounded barn electricity, not renewable energy.

It's more than likely that you now know more about wind power than most of your neighbors, angry though they may be. They will need some form of education, and you can get them started, with this book as your primer. It's definitely a good idea to be respectful of your neighbors. "If you create a bad vibe, it just takes one person to create a major barrier down the road," Sinclair warns.

The next page includes this information in a brief table (Table 2-2), and as noted, most of these issues will not apply to your small wind turbine. But as an enthusiast of wind power, you have absorbed quite a bit of information in these pages, and found a few kernels that might have surprised you. What kind of images do your conscientious neighbors and community have about wind turbines? The probability is that they are more likely to generalize all wind turbines in a bunch. They will need some form of education, and you can help.

So, it has been a long trip on your second wind.

TABLE 2-2   Environmental Impacts of Small Wind Turbines

The following is a quick list of common challenges of wind turbine projects. Please use this information when issues arise about the installation of micro- to commercial-size wind turbines.

Energy Storage and Transmission	Grid-tied small wind turbines serve no problem to the power grid because small generators are distributed, meaning they produce power right where it is being used. The grid effectively acts as a small turbine's storage bank.
Noise	Sounds emitted by small wind turbines are not unique from other machine noise or particularly loud. Certified turbines can comply with local noise ordinances.
	Although turbines can occasionally be perceived as annoying, there is no conclusive evidence that the audible or subaudible sounds emitted by wind turbines, large or small, have any direct adverse physiological effects on humans.
Shadow Flicker Effect	A shadow flicker can occur with large wind turbines in an extremely limited time frame if conditions are just so. But proper siting can avoid this, and small turbines aren't affected.
Visual Influence	One person's eyesore is a highly subjective and individualized issue, with no clear resolution. However, there is no evidence of negative impact on property values from wind turbines, large or small. Some studies have shown a small increase in property values, possibly due to a perception of energy savings and eco-friendliness.
Radar	Electromagnetic interference from modern turbines is highly improbable, especially when it comes to small turbines. Interference is generally limited to small, low-flying airplanes that are physically shadowed by the turbine, but this should never be an issue with small turbines.
Ice Throwing	Many small wind turbines shut down during icing events due to blade imbalance. Most turbines shed ice gently, in the manner of roofs. Setback distance should be more than enough to compensate.
Site Preparation	May cause disturbance to soil, vegetation, and fauna, particularly if large earth-moving equipment or cranes are involved.
	No new expensive transmission lines, high-voltage lines, or any major undertaking that disrupts the environment is required for small wind turbines.
Potential Impacts on Birds	Relative bird mortality from large wind turbines is extremely low in comparison to other causes (e.g., cats and electrical lines). Mortality from small wind turbines is vanishingly small, but impact can be mitigated by (1) proper siting and planning for any species at risk in your area, (2) painting blades purple to deter insects, (3) periodic shutdown or feathering during known periods of high bird activity.
Potential Impacts on Bats	Some studies have suggested some impact with large turbines, but there is no evidence small turbines pose a threat to bats.
	Impact can be mitigated by (1) proper siting and planning for any species at risk in your area, (2) painting blades purple to deter insects, (3) periodic shutdown or feathering during known periods of high bat activity.

CHAPTER 3

Electricity, Energy, and Wind Science

The answer, my friend, is blowin' in the wind. The answer is blowin' in the wind.

–BOB DYLAN

Overview

- What do I need to know about electricity for a wind system?
- What are the most important parameters of wind energy production?
- How do I determine how much energy my site can produce?
- How can I maximize that energy?

In this chapter, we will provide a brief overview of electricity and the core principles of wind energy (Figure 3-1).

We will brush over the basics, but if you need more explanations, follow the footnotes.[1] We don't provide ludicrous imagery like the writers from the *South Park* television program, but we do use speeding cars, sponges, and aquariums to help illustrate a topic that is usually hard to explain.

Basics of Electricity

In a nutshell, William Beaty writes on amasci.com:

"Conductive objects [such as wire] are always full of movable electric charges, and the overall motion, or flow, of these charges is called 'electric currents.' Voltage can cause electric currents because voltage differ-

FIGURE 3-1   In order to understand wind energy, we first need to know a little about electricity. *Kevin Shea.*

ences act like pressure differences, which push the conductors' own charges along. A conductor offers a certain amount of electrical resistance, or 'friction,' and the friction against the flowing charges heats up the resistive object. The flow-rate of the moving charges is measured in amperes. The transfer of electrical energy (as well as the rate of heat output) is measured in watts. The electrical resistance is measured in ohms."[2]

So let's take a closer look.

Electrical Charge (Voltage)

You want to know what is inside the wires doing the work? Well, inside every wire is a sea of electrons—the *charge*. Beaty explains that charge in a wire can be thought of as similar to water in an aquarium, since the outer electrons of metal atoms are free to move through the entire mass of material, similar to water molecules in a tank. A charge can move, and usually needs little more than a push, like rubbing your foot on a rug in the case of static electricity. Without that "push," the charge inside a metal remains neutralized because each movable, negatively charged electron is near enough to a corresponding positively charged proton within a nearby atom, keeping the charge paired among its atom friends. Beaty suggests that's like a pipe stuffed with positively charged sponges, which are in turn filled with a negative liquid.

What gives the electrons a push? It takes a separation of positive and negative charges within the material, which leads to something called *electrical potential.* The actual driver of charge separation is called *electromotive force* (emf), so technically emf creates potential, which is measured as *voltage*.

Remember that sea of electrons and Beaty's sponge analogy? Well, if you put pressure on that sponge, it would send the negative liquid flowing through the pipe of sponges. This would leave the positive sponge left behind and an oppositely imbalanced charge. The haves with the have-nots. The voltage is, therefore, an electric field that extends across the space between the empty sponge and the full sponge. Electrostatic fields are measured in terms of volts per distance, and if you have an electric field, you always have a voltage. It's part of their "pumping" action.

A battery acts as a charge pump. If we have a ring of copper wire connected to the positive and negative ends of the battery, it will pull charge out of one side and push it into the other side. This causes a voltage difference to appear between the two sides of the ring. It also causes an electrostatic field to form around the ring. The voltage can be thought of as pressure.

So, in practical terms, the voltage between two points is a measure of the electrical force that would drive an electric current between those points. The higher the voltage, the more "push" that is available. Voltage is represented by energy per unit charge, and is measured in units called *volts* (V). Technically, one volt is equal to one joule per coulomb. We define joules in the section under power, and a *coulomb* represents charges approximately equal to 6.24151×10^{18} protons or -6.24151×10^{18} electrons.

One final note before we continue. Charge isn't "used up" in wires. It isn't like a limited number of gumballs in a dispenser that have to be refilled. The electrons return to where they were before, or they settle into the nearest convenient spot, but there is always enough to fill the structure of the wire. Perhaps it is more like the Willy Wonka everlasting gumball dispenser.

Current (Amperes)

In the spec sheets of turbines, you might see listed a rate for a particular number of amp-hours. For the sake of deciding what turbine you want to get, it is good to have an understanding about how to measure the energy flow that is coming from the generator (Figure 3-2). Whenever the charge "stuff" within metal wire is forced to flow, we say that *electric currents* are created.

We normally measure the flowing charges in terms of *amperes* or *amps (A)*, though the concept of current is often represented as the letter I. As stated earlier, charge is actually measured in units called coulombs, and the word ampere is really shorthand for the rate of one coulomb of charge flowing per second. If we were talking about water, then coulombs would be like gallons, and amperage would be like the rate of gallons per second, according to Beaty's helpful analogy.

Note that people often expect electric current to be at light speed from the source to the receiving point. After all, when you flick on a light switch, it gets brighter pretty darn quick. But actually, charge-stuff flows slowly through wires,

FIGURE 3-2 This Bergey 10-kW Excel wind turbine provides supplemental power for a farm in Scott City, Kansas. Excess power is fed back into the utility grid. *Warren Getz/DOE/NREL.*

slower than centimeters per minute. So, do not confuse this with the electrical energy that powers your light; we will discuss that later with watts.

The amperage or rate can be affected by two things: (1) speed of the charges, and (2) size of the wire. Here are some simple correlations:

- Double the speed of charges in a wire and you double the current (amps).
- Increase the size of the wire, and you increase the amperes.

Ampere-Hours (Amp-Hours)

While perusing for wind turbines or batteries online, you will likely see references to *amp-hours*. One amp of charge flowing through a wire for an hour is one amp-hour. As an example, a typical flooded lead-acid battery might store 420 amp-hours of electricity.

Resistance (Ohms)

Wherever it travels, electricity encounters *resistance* to its flow. It can be helpful to think of resistance as friction, since the result is similar: impeding action of a force. Resistance is usually represented as R and measured in *ohms*. The more ohms something has, the more resistant it is to transfer of current.

If we may borrow another analogy from Beaty, let's imagine a pressurized water tank. Connect a narrow hose to it and you'll get a specific flow of water, measurable in gallons per minute. It stands to reason that if you increase the thickness (cross-sectional area) of the hose, then you'll increase the flow rate. The same relationship holds to resistance and wires: thicker wires have less resistance, because there are more electrons available for moving charges. Now, double the length of the hose, and the flow of water decreases by two times. This is because we doubled the friction. Make the hose shorter and the reduced friction lets water flow faster.

Similarly, if we double the length of a wire, we double the friction and cut the charge flow (amperes) in half, assuming the voltage (pressure) is constant.

All Together Now: Ohm's Law

We can also change the flow by changing the pressure. Add a second battery to the circuit in series and this gives twice the pressure difference, which doubles the charge flow. *Ohm's Law* says if the pressure (voltage) goes up, the flow (amps) goes up in proportion. It also says that the resistance affects the charge flow in a predictable way. If the resistance goes up while the pressure difference stays the same, the flow gets less in "inverse" proportion. Simply said:

The harder you push (voltage), the faster it flows (amps).

The bigger the resistance (ohms), the smaller the flow (amps).

Let's take what you now know and write it shorthand:

$$\text{Volts}/\text{Ohms} = \text{Amperes}$$

Voltage divided by resistance equals current. This is the mathematical expression of Ohm's Law. So, make the voltage twice as large, then the charges flow faster, and you get twice as much current. Make the voltage less, and the current becomes less.

If you keep the voltage the same and you double the resistance, say, by adding another light bulb, you get another law:

The more resistance you have, the slower the flow (current).

The more light bulbs, the more resistance, which means that current is less and each bulb glows more dimly.

So that explains the relationships between volts, ohms, and amps. However, wind turbines usually list *rated wattage*. So what about watts?

Power (Watts)

A *watt* (W) is a concept that people often have trouble with at first. A watt is a rate, and it measures the rate of transfer of electrical energy at a given moment. This is a similar concept to your car's speedometer or a police radar gun, which measures the rate of speed of a vehicle at a given moment. Except electrical energy is measured in units called *joules* (J) (technically, one joule results from passing a current of one ampere through a resistance of one ohm for one second). And when you transport one joule of energy through a channel (a wire, say) every second, the flow rate of energy is one joule per second. That amount is defined as one watt. In other words, 1 watt = 1 joule per second, just as 1 mph = 1 mile per hour when it comes to your car.

So, if a cop were to pull you over for using 400 joules per second, that would be like saying you were caught pushing 400 watts of electrical energy at that moment. Remember, this is a rate, so it does not necessarily tell you the total result of your actions, since you would also need to know length of time. There's a big difference in terms of distance traveled between going 60 mph for one second and going 60 mph for four hours. However, knowing the rate lets you shop smarter for appliances, lighting, and other devices, because the higher the wattage of a device, the more energy it uses at any given moment. Similarly, the higher the wattage of a wind turbine, the faster it can produce energy.

In classical physics, wattage refers to *power*, which can be another confusing concept. That's largely because, in common usage, people tend to use the words *energy* and *power* interchangeably. (In fact, you will even see that occasionally in this book, because there just aren't enough synonyms to go around to make things readable. However, if we feel the difference between power and energy is important to understanding, we will use the terms precisely.)

So, what is the difference between energy and power? When it comes to the energy sector, you will see power represented in watts and energy represented in something called watt-hours, which we explain next.

 Tech Stuff   *Technically, power is the rate at which* work *is performed. In the case of electrical work, we express that rate in joules per second. As you may know, "work" also applies to mechanical processes, in which case one joule is equal to the work done in applying a force of one newton through a distance of one meter (a newton is a measurement of force; technically, the amount needed to accelerate a one-kilogram mass at a rate of one meter per second). The amount of work done by a joule is sometimes approximated by the energy needed to lift a small apple one meter straight up, the amount of kinetic energy of a tennis ball moving at 14 mph (23 kph), or the energy released as heat by a person at rest every hundredth of a second.*

Energy (Watt-Hours)

We save the most important measurement of all for last: *energy*. For the sake of making an important distinction, let's continue the police car analogy. If we switch

FIGURE 3-3 Weather station data is a valuable resource that aids smarter wind energy prospecting. Pictured are two aerovanes (right) and sonic anemometers (see Chapter 5). *Warren Gretz/DOE/NREL.*

the rate of joules per second for miles per hour, we can observe a cop in an unmarked car following you as you speed at 400 joules per second on the highway at a constant speed. At the end of one hour, the police officer pulls over your car. Not only does he say that you were moving 400 joules per second (the rate), or 400 watts, you would have successfully moved this electrical energy for an hour, so we would say you used, or consumed, 400 *watt-hours*. Note that even though it has the word "hour" in it, a watt-hour is no longer a rate; it is an actual amount of something— and that something is what we call energy (Figure 3-3).

The reason energy (watt-hours) is so much more important than power (watts) when it comes to wind systems is because energy is what actually provides what we want: light and the operation of electronics and machinery. Energy is what is bought and sold, not power. So when you look at your electric bill, the utility doesn't really charge you by the rate; they charge you by how much energy you actually used, typically measured in kilowatt-hours. Kilo = 1,000, so that refers to how many hours you consumed at a thousand watts.

Here's the basic formula:

Watts (the rate of energy generation, movement, or use)
× hours (time) = watt-hours (a quantity of energy)

Remember, light bulbs, microwaves, toasters, and TV sets are rated by watts, the *rate* at which the device consumes electricity. Even turbines are listed with rated power of performance at a certain number of kilowatts. Although we will soon explain that this alone is not a good way to evaluate the energy production of a turbine, it can be calculated if you know the volts and the rated amps. The formula for this is:

Volts × Amperes = Watts

For example, if you come across a turbine that is said to output 120 volts at 30 amps, the rated wattage would be 3,600 watts, or 3.6 kilowatts. However, if you don't have a good wind resource, or your turbine isn't installed and operating correctly, it doesn't matter what the power rating is; it won't produce any meaningful amount of energy, and it won't "pay back" your investment. You earn credits from your utility, or drive operation of local loads, only by generating usable energy. And as you'll soon see, peak or rated power tends to be a very poor predictor of actual energy production rates in real-world applications.

Here are a couple of discovery tools for you to further explore electrical energy and current. Please follow the instructions, and feel free to experiment with various variables.

Resources

Discover the Charge
bit.ly/efield

Explore 3-D Electrostatic Activity
bit.ly/3dfield

How Are Watts, Ohms, Amps, and Volts Related?
amasci.com/elect/vwatt1.html

Hazard   *It's worth noting that amps are the part of electricity that is most dangerous to your health. You can have 20,000 volts go through your body and leave you virtually unaffected (although your hair might stand on end). But it only takes 0.5 amp to do some serious damage to your body in a very short time.*

Of course, it's always important to stay safe when working with electricity, and this book alone is not enough to prepare you for doing all necessary wiring on a complete turbine system. Even certified electricians are sometimes hesitant to work on wind turbines, unless they have received hands-on training with the equipment. There is no substitute for real-world experience.

INTERCONNECT

From Failed Crop Farmer to Idaho's First Successful Wind Farmer

For four decades, Bob Lewandowski and his family tried to scrape out a living as farmers on 20 acres of windswept plain between Boise and Mountain Home, Idaho. It was hard going. "I used to curse the wind," Lewandowski told *The Idaho Statesman* in 2004. "Every time we put seed in the ground the wind would come up and blow it away."

Lewandowski decided to try to put that steady wind to use. Literally betting the farm, he spent $120,000 to buy, ship, and refurbish three old 100-kilowatt turbines from a wind farm in California (Figure 3-4). Although Lewandowski had no formal training in the field, he did most of the work himself, raising the first 150-foot tower and even inventing some of his own performance-boosting repairs.

After four years and 19 permits, Lewandowski's first turbine started spinning, and he began to sell power to an Idaho utility, becoming the first outfit to do so in a state that is ranked 13th in potential wind resource.

Lewandowski eventually got three turbines working, each providing enough power for 15 to 20 homes, and his wind farm was a success. He passed away in 2005, and a local company proved unable to keep the turbines working profitably, largely because the makeshift work Lewandowski did over the years required too much maintenance to keep going. However, the turbines were recently moved to the College of Southern Idaho and Idaho State University, where they now serve as teaching aides for the next generation of wind installers.

FIGURE 3-4   Bob Lewandowski installed and maintained these used 100-kilowatt wind turbines on his property outside Boise, Idaho. Today, the turbines serve as teaching aids for local colleges. *Brian Clark Howard*.

FIGURE 3-5   Lewandowski's turbines provided enough power for up to 60 homes in total, making his small wind farm a success. He helped inspire the state of Idaho to launch several big wind projects. *Brian Clark Howard.*

Lewandowski's turbines were visible from a major highway across the West (Figure 3-5), and they inspired many people to consider the technology. Today, Idaho has several big wind farms, with more in development. Lewandowski showed how one man truly can build his own power plant, with a little ingenuity, elbow grease, and a relatively modest financial investment.

Wind Energy Principles

Soon, we will help you begin an assessment of your site and your situation. But before we can begin, we need to go over some wind energy basics. In Chapter 1, we discussed where wind comes from. But what power is available in the wind if we were to cut a slice out of it and examine its properties? What variables affect how much energy a given turbine can actually harvest out of the wind?

Let's call in an *Energy Special Unit: Windy Acres,* and use physical evidence to solve the invisible mysteries of wind energy.

Understanding Wind Speed Terminology

It turns out that most people are quite bad at estimating wind speed, especially when it comes to their own potential sites, where they tend to be overly optimistic about annual averages. In truth, wind speed varies a lot, depending on where you are and even what you are using it for. If you are on a boat or listening to weather reports, it is common to hear the term *knots.* "Aye aye, Captain, she is blowing at 3 knots." But for the rest of us, at least in many countries, the standard measurement is metric, so wind is measured in meters per second (m/s). Within the U.S. market of small wind turbines, you will often see this standard interchanged with miles per hour. The following conversion shows the change from miles per hour to meters per second:

$$(1 \text{ mph} / 3{,}600 \text{ sec}) * (1.61 \text{ km} / 1 \text{ mile}) = 1 / 3600 * 1609.344 = 0.45 \text{ m/s}$$

Therefore, there are 0.45 m/s for every mile per hour. We can approximate that there are about 2 miles per hour for every 1 meter per second, so any value on an analysis chart in meters per second can be doubled to get an approximate wind speed in miles per hour. For example, 6 m/s is approximately 12 mph. From now on, we will display speed as mph (m/s), so we can accommodate everyone.

There are three basic types of wind speed: (1) instantaneous, (2) average, (3) and peak. Luckily, these are relatively self-explanatory. Note that the least useful of the three is the instantaneous speed, or what the wind speed is at that moment. Wind does not come across your property like an extra-long power train. It is not a steady flow. It varies. You can sit all day and watch or measure the wind, and it won't give you any idea of your wind resource. In fact, the only thing useful about measuring instantaneous wind might be to find out the *peak wind speed*.

Peak wind speed refers to the highest gust expected on a site. This is important because it gives us an idea of how strong a turbine system will need to be in order to withstand peak forces. Those who are new to wind energy might reasonably assume that the strategy should be to design a system that can withstand the strongest winds ever recorded. However, that is a little like asking for a car to be built to last for 100 years, or for a laptop battery that will power your computer for a month. Such overengineering is going to make the system too expensive, too heavy, and probably inefficient at normal operation.

Put bluntly, small wind turbines are not going to be designed to generate power from, or even to withstand, tornadoes and hurricanes—at least for the foreseeable future. Currently, most small turbines and towers are advertised to withstand winds of 120 mph (54 m/s). However, there are a number of variables that can be involved, including quality of installation and the nature of the site. Unfortunately, most manufacturers can't afford to do a large sample of real-world failure tests, so there tends to be some extrapolation. We think that's another good reason to go with a time-tested turbine design from a reputable company with a track record of thousands of successful installations.

The most crucial measure for wind systems, as you may have guessed by now, is the site's *average wind speed*. This is because you need to know the average energy available over a specific period. What we need to start to evaluate is the percentage of time that wind flows at various speeds, or the *wind distribution* for the site.

FIGURE 3-6   This research station in Antarctica is powered by a wind turbine, solar panels, and a diesel generator. *Northern Power Systems/DOE/NREL.*

Power Up!   *Unless you are running a remote Antarctic or mountain research base (Figure 3-6), places with livable human conditions tend to see average wind speeds within the range from 0 to 15 mph (0 to 6.75 m/s). Residential sites and farms might see a range of 7 to 12 mph (2.8 to 5.4 m/s), and urban regions usually see less than that.*

The Basic Wind Energy Formula

Now, we give you the core of the formula that is really the holy grail of wind energy, and which is arguably the most important equation in this book (Figure 3-7). This brief formula can bring you closer in your journey of producing your own energy, or it can be used as a shield of defense against the marketing wizards that will dazzle you with numbers that, with a little education, do not add up.

$$\text{Power} = k\, C_p\, 1/2\, \rho\, A\, V^3$$

where

P   =   Power output, kilowatts

C_p =   Maximum power coefficient, ranging from 0.25 to 0.45, dimensionless (theoretical maximum = 0.59)

ρ =   Air density, lb/ft^3

A   =   Rotor swept area, ft^2 or $\pi D^2/4$ (D is the rotor diameter in feet, π = 3.141)

V   =   Wind speed, mph

k   =   0.000133; a constant to yield power in kilowatts. (Multiplying the above kilowatt answer by 1.340 converts it to horsepower [i.e., 1 kW = 1.340 horsepower].)

FIGURE 3-7   Arguably the most important equation in wind energy. *Kevin Shea.*

Power

So, what does the wind formula mean and how does it work? The P stands for power. You want to know the instantaneous power that your turbine is going to collect. As we stated previously, unlike energy, which is measured in watt-hours, power is measured in watts, the rate of energy flow. So, power equals watts. And what determines the power available to your turbine as the wind hits your blades is air density (D) and the swept area (A).

Air Density (D)

All air has density, just as solids and liquids do, although it's not a property we often associate with air, unless we go mountain climbing. Density is simply mass (measured in kilograms, kg) per unit volume (measured in m³, meters cubed). At sea level with a temperature of 60 degrees Fahrenheit (15 degrees Celsius), the Earth's air has a density of 1.225 kg/m³. Why did we provide all those conditions? Because, as you will see, air density matters in how well your turbine will perform.

When air molecules are more tightly packed, meaning they have greater density, they inflict more force when they move past your turbine's blades. Let us illustrate by asking which would place more force on your face, a "pancake" of whipped cream or the pie that it was on? If you're talking about a turbine, you'd prefer the pie. With higher density comes more energy.

Can we increase the air density? With all other factors remaining the same, lower elevation will have greater air density, because there is more "weight" of air above. That means lower elevations will have more potential wind energy available. For every 1,000 feet in elevation you climb, you can harvest 3 percent less energy with a wind turbine. So at 10,000 feet, you are left with 30 percent less potential energy. (At 10,000 feet, you will also probably notice that the air is a bit "harder" to breathe, because it is less dense.)

In addition, temperature affects air density. Specifically, air is denser when it is cold. That's because the molecules are packed closer together. That means there is more energy available out of cold air. In most temperate areas, this is an added benefit in the winter. Winter brings you more potential energy when you are typically inside more, using more of it. Then summer comes and the air is thinner, so there is less energy.

It's instructive to understand the relationships of elevation and temperature to wind energy, but for most purposes these variables are not really very important, and will rarely make or break a project. So if you live in a warm climate and up on a high mountain, don't move just yet. All is not lost. In fact, many productive wind farms are being installed on mountain ridges around the world, often even near the equator.

Swept Area (A)

The single most important factor of performance of a wind turbine that *we have total control over* is the *swept area*, the area that the blades sweep as they rotate (in a circled plain). You could periscope a tower thousands of feet into the air to get into the ceaseless trade winds above, but unless you have an adequate collector size, you are not maximizing the potential energy available. And the math is simple. If you double the swept area, you double the size of the pie that will hit the turbine, subsequently increasing the power. Yummy. But don't stop reading, there's more.

To continue we need to mix squares and circles for a moment. Swept area is measured in square feet or square meters. But it is a circle. To calculate the swept area, you need to use the formula for an area of a circle, which you may remember from high school geometry: the square of half the circle's diameter (the radius) multiplied by π (roughly 3.14).

$$\text{Area} = \pi(D/2)^2$$

What you will find is that as you double the length of the blade, you quadruple the swept area! Congratulations. With adding a little length to your blades, you increase the amount of energy you can harvest by a lot. At this point, you should have a good idea why commercial wind turbines keep getting bigger and bigger: the larger the swept area, the more energy that can be produced, and the relationship is exponential.

Now with vertical axis wind turbines, the swept area is not measured in quite the same way. If you were to take a picture of a VAWT, on the photo, you could draw an outline around the rotor, and you will see that its area is more of a square (Figure 3-8). Accordingly, the swept area would seem to be width times height

Figure 3-8   A Windspire 1 kW vertical axis wind turbine (left) and monitoring tower at the National Wind Technology Center (NWTC) in Golden, Colorado. *Amy Bowen/DOE/NREL.*

(W × H). Right? Wrong! If the wind is hitting the entire square swept area, it is not only hitting the blades that are turning away from the wind, but also the blades that are returning *against* the wind. Hence, your swept area would be half the width. And you can't say that the swept area is all around the vertical axis turbine, either. If the force of wind were coming from any other direction at the same time, it might push the rotor in the opposite direction, right? Is this physics defied? No, but then that is one of the reasons why vertical axis turbines are made to be tall, to compensate for all the energy that is being wasted.

Now that you have bigger blades, you don't have to move. Problem solved. You just have to add a few feet to your blades, and you will be in action. Now, just one more thing to look at: the wind speed.

Wind Speed Cubed

Here comes the final ingredient, which also has the most dramatic effect on the power available for your turbine: wind speed. And guess what, you can control the speed … sort of. We will get to that when we start talking about site evaluation and towers in Chapter 4. But we'll give you a hint: height matters.

The power available in the wind relates to the cube of the wind speed (V^3 or $V × V × V$). What exactly does this mean? Let us show you in Table 3-1.

TABLE 3-1   Available Power in Winds of Increasing Speed

Average Speed mph (m/s)	V3	Available Power (Watts)	Percentage (Times) Increase from Previous Speed
1 (.45)	1 × 1 × 1	1	
2 (.90)	2 × 2 × 2	8	800% (8×)
4 (1.8)	4 × 4 × 4	64	800% (8×)
8 (3.6)	8 × 8 × 8	512	800% (8×)
12 (5.4)	12 × 12 × 12	1,728	333% (3×)
16 (7.15)	16 × 16 × 16	4,096	240% (2.4×)
20 (8.9)	20 × 20 × 20	8,000	200% (2×)

Notice a pattern: Every time you double the wind velocity, you don't just double the energy available; you increase the potential energy eight times. The reason we showed so many numbers is to illustrate the enormous change in energy. Do you realize that when your wind velocity increases from 0 to 8 mph, you have already increased the energy available over 512 times, or 2,400 percent? Wow!

Power Up! *This leads to an important point. If the power increases exponentially as the wind goes faster, that means we also are losing that enormous resource when the wind is slow. The rotor will spin, but it won't be producing much more than the heat on the bearing. So, when someone is trying to sell you a wind turbine that is designed to "capture wind energy at low wind speeds," realize that you will necessarily see only tiny amounts of energy at those speeds. Unless the wind turbine is somehow eight times more affordable, it may be wise to avoid them.*

"If someone tells me that their turbine can start producing energy in very low winds, I say so what," small-wind author and expert Paul Gipe put it bluntly. "There is so little energy available in low wind speeds that it is essentially worthless. Having something spin around doesn't change that," he told us. Author Ian Woofenden warned, "The fact that wind energy is a cubic relationship is something people repeatedly overlook."

But wait, isn't wind speed one of the factors that we can't change unless we change our location? Does that mean I am stuck with my average wind speed unless global climate change improves my condition? Yes and no. You *can* change the location of your turbine ... if you build a taller tower. Even though it is true that the air density gets less as we go higher, the higher wind speeds we see more than compensate, largely because of the cubic relationship. If you could afford to go up 10 feet you might gain 1 mph, and that would translate to 30 percent more available energy. Wouldn't you say that it is worth it? As Paul Gipe said, "If you want clean energy that helps save the planet, you need to make energy."

When is too much of a good thing bad? You noticed we didn't list very high wind speeds, like, say, 40 miles an hour and above. There's a good reason for that. Believe it or not, mad scientists design the system to protect the wind generator from such speeds, so normally the turbines are battening down the hatches when the wind is impeding on your walking. What, you ask? Wouldn't we want to collect *all* of the energy?

Whoa, there, thrill-seeker. Engineers are currently developing systems to tap into this region; however, there are two good reasons why there is not much feasibility yet: weight and frequency. First of all, what good is a dead turbine? All that energy in strong winds is going to affect the machine. If you were creating a turbine dragster that was to work only a few times, you might be able to get away with operating at high speeds.

Wisely, most manufacturers aim to make products that will harness wind energy for a very long time (this is especially important if you want to see any kind of payback on your investment). To design a machine that would work through harsh winds would take some very heavy duty stuff. This would tend to make the whole system bulkier and heavier, and consequently less responsive to moderate winds, because it would take more wind energy to push the blades. It would also make it more expensive.

Wind Distribution

The next reason for mitigating the effects of very high winds is their relative frequency.[3] As you know, the strength of wind varies, and an average value for a given location does not alone indicate the amount of energy a wind turbine could produce there. Frequency does. Again, if we lived in that place where the wind speed was constant at 40 mph, the speed at which walking in the street is impeded, think how different life would be. We're thinking not much outside playtime, and some serious eye goggles and Kevlar clothing to protect us from light projectiles every time we are outside. But alas, such high winds are not commonplace. In fact, as we can show, they generally occur less than 2 percent of the time.

To better assess the frequency of wind speeds at a particular location, especially when comprehensive data is unavailable, installers often run sampled data through a probability distribution function, such as that found in the Weibull distribution model and its companion, the Rayleigh distribution model (Figure 3-9). These models describe the range of possible values that a random variable, like our reliable friend the wind, can attain and the probability that the value of the variable is within any subset of that range. These models describe wind speed variability mathematically in an attempt to approximate the real world. Different locations may have different wind speed distributions, based on the average wind speed, elevation, latitude, and other factors, but they all mirror a bell-shaped curve.

In 2002, Sandia National Laboratories commissioned a wind study at Lee Ranch in New Mexico.[4] The results, summarized in Figure 3-9, give an idea of wind distribution in many terrestrial locations. The gray bars represent the wind energy available, measured in the number of megawatt-hours, at a particular wind speed over a year. The black bars represent the number of hours that a specific wind speed was

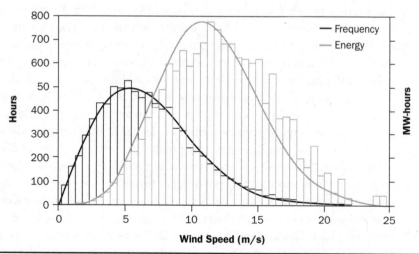

FIGURE 3-9   Distribution of wind speed (black) and energy generated (gray) in 2002 at the Lee Ranch test facility in New Mexico. The histogram shows measured data, while the curve is the Raleigh model distribution for the same average wind speed. *Wikimedia Commons.*

recorded. Picture these like stacks of coins. For every second, data received is a measurement of the amount of energy produced and the wind speed that produced it. A coin could be placed in the corresponding columns for every measurement.

For example, you can see that not many black coins (frequency) were placed in the black column for 20 mph. Barely a few hours. Yet, it produced a stack of gray coins (megawatts available), equating to 100 megawatts from those infrequent wind bursts. Contrarily, you can see this location recorded over 500 hours of wind at 5 mph. That appears to be nearly 500 times more frequent than at 20 mph. Yet, the energy available was nearly equal. That is the reason that most reputable professionals and manufacturers don't attempt to market a wind turbine that has a "cut-in speed" of less than 7 mph. Frequency alone doesn't always equate energy.

On the other hand, it appears evident from the chart that half of the energy available arrived in just 15 percent of the operating time, at around 8 to 20 mph (3/6 to 9 m/s).[5] Bring back that question of building a turbine to handle the stronger winds; if you designed a machine to handle the wind speeds above 20 mph that arrive only 2 percent of the operating time, its weight would tend to make the turbine less responsive in lower speeds, subsequently returning lower potential energy.

So, hopefully you can see that the best solution is usually to design a wind turbine that harvests the sweet spot of wind that is relatively frequent and also provides a useful amount of energy. In the future, with more data available for particular areas and more advanced technology, it might be possible to more fully customize a turbine, sort of like how sports shoes have become highly specialized based on activities. But for now, most turbines work within the range of moderate winds.

Let's return to calculating the potential wind energy using what we now know about wind speed distribution. We have discussed air density, swept area, and wind speed. We understand that the potential energy is proportional to the wind velocity cubed (V^3). But if it is true that every location with the same wind speed can have a different wind distribution, wouldn't that result in different power available?

For example, let's say that a property in Long Island, New York, has the same average wind speed as a property in Ostional, Nicaragua. However, Long Island has more volatility in its wind speed distribution, as it is in the path of trade winds during the winter, which pack a powerful punch. In contrast, Ostional doesn't have those fluctuations, due to the constant prevailing winds from nearby Lake Nicaragua. These steady winds produce less power over time than the temperate region. How can the formula for power availability take into account those differences?

That requires us to average the cube of all different wind speeds over time in a distribution chart. Remember those coins in the black columns? If you cube the wind speed (V^3) for each time there is a coin in each black column, then divide it by the total number of coins, you will get an average that better estimates the projected annual energy availability.

Power Up! *Meteorologists have characterized the distribution of wind speeds for many of the world's wind regions. With their results, they were able to determine that if you live in a temperate zone like the United States and Europe, the current Rayleigh wind speed distributions offer a good approximation, so it won't require extra work. In fact, a study of small wind turbine performance (Figure 3-10) found the AEO (annual energy output) estimates based on the Rayleigh wind speed distribution to be only 5 percent less than what was actually produced in the field, according to Paul Gipe in his 2004 book,* Wind Power *(Chelsea Green).*

Figure 3-10 Jim Johnson, a senior engineer at NREL's National Wind Technology Center in Colorado, conducts research on large and small wind turbines to make them more effective. *Scott Bryant Photography/DOE/NREL.*

Applying the Wind Equation

Tech Stuff *Let's sum up the wind formula with an example. We explored effects on the amount of power available from air density, the swept area, and the average wind speed. If we would like to know the average annual power available to a particular site that is at sea level, with average temperature 60 degrees Fahrenheit, average wind speed of 9 mph (4 m/s), and a blade diameter of 12 feet (3.65 m), we would grab our handy equation and plug in the numbers.*

For less confusion, we will use the symbol ρ^a in place of D for the air density. Please remember to convert feet to meters and miles per hour to meters per second.

$$P = \rho^a \times \pi(D/2)^2 \times V^3 = \tfrac{1}{2}\,(1.225\text{kg}/\text{m}^3) \times \pi(3.65\text{m}/2\,)^2\,(4 \times 4 \times 4)$$
$$= .6125 \times 3.33 \times 64 = 130\ \text{watts}/\text{m}^2$$

Let's now make energy from power, or take instantaneous wattage, and let the clock spin forward so that we can take a peek at your future electric bill, or your credit. Now if we carry this average over 24 hours, 365 days a year (24 × 3,365 = 8,760 hours), you will get the grand total estimate of 1,138 kilowatt-hours/m^2 for the annual wind *energy* density.

What more can you want? More power? More energy? That is next.

Wind Speed, Power, and Height

Here comes a key factor that can strongly increase the amount of energy your system can generate, but which is most often sacrificed by small turbine owners during the decision process. Let us tell you the general rule so you can get your dessert first, before diving into another equation.

- Wind speed and wind power are both affected by height.
- Doubling the height increases the wind speed by as much as 10 percent.
- This increases the wind power available by as much as 25 percent.
- Increasing the tower height five times doubles the wind power available (200 percent).

Now, back to formula basics: If velocity is V and height is H, we need to determine the effects if both of these are expected to change. So if we are changing our original height (H_O) at our original average wind velocity (V_O) to a new height (H), we should expect an increased wind speed (V) at our new height.

If the formula were as simple as:

$$V/V_O = H/H_O$$

it would suggest that there exists a linear relationship, or a 1 to 1 ratio. For example, if we wondered how much the wind speed would increase, if we measured the original average wind speed to be 10 mph (4.5 m/s) at 33 feet (10 m), and we doubled the height to 66 feet (20 m), the equation would look like this:

$$\text{(in mph) } V/10 = 66/33, \text{ or (in m/s) } V/4.5 = 20/10$$

The new wind speed would be 20 mph (9 m/s), and that would suggest that doubling the height doubles the wind speed. But something is being overlooked—the environment around us.

The reason that wind speed increases as we move up into the atmosphere is because there are fewer obstacles and less friction to disrupt the wind. As air rushes past the Earth's surface, it encounters resistance due to friction, which is called *ground drag*. This effect slows down the wind, all the way to a height of about 1,650 feet above the surface. Perhaps not surprisingly, the greatest slowdown occurs closest to the ground. Over an essentially flat surface, wind is most slowed by friction for the first 60 feet of height. The rougher the surface, the higher the effects persist.

If you were standing on a grassy plain and measured a 20-mile-per-hour wind at 1,000 feet above the surface, it would likely be at only 5 miles per hour at 10 feet. This difference in wind speed with height is often called the *wind gradient*. Some writers also call it *wind shear*, although we find that designation a bit confusing because wind shear can also refer to a related concept.

To pilots and meteorologists, *wind shear* refers to a change in wind speed or direction over a relatively short distance. Wind shear is closely related to turbulence, and often causes it. Many factors produce wind shear, including smog, differences in atmospheric temperature, the wake of jetliners, or terrain features, including mountains and forests. In general, the rougher the surface, the higher the wind shear. That can be dangerous, even deadly, to pilots, and it usually means more turbulence and less recoverable energy for wind prospectors (Figure 3-11).

The combination of wind shear and ground drag means that wind speeds are fastest over water and ice, then flat ground, then bumpy terrain. The next chapter covers ways to evaluate your site.

 Tech Stuff *The rate at which wind speeds increase with height can be estimated for various types of terrain with a value called the* wind shear exponent, α. *The lower the number, the less rough the surface, and the less dramatic the effect of increasing height. We show you how to calculate the wind shear exponent in Chapter 5, but suffice it to say that it is derived from another variable called* surface roughness: *the sum of the effects of vegetation*

FIGURE 3-11   A Bergey small wind turbine outfitted for field study by NREL, which is measuring wind speeds upwind and in the tail wake, tower bending, yaw rate, torque, blade bending, rotor speed, and other variables. *Dave Corbus/DOE/NREL.*

and topography. Table 3-2 shows wind shear exponents for the three main types of terrain that you may consider putting a small wind turbine on. Note that the values range from 0.43 for woodlands to 0.19 to flat prairie. (See a longer list in Chapter 5.)

TABLE 3-2   Wind Shear Exponents (α) for Various Types of Terrain and the Associated Effect on Wind Velocity with Changing Height

Site Location	Terrain	α	Δ Height	Vo mph (m/s)	V	% Velocity Change
Hunting Cabin	Woodlands	0.43			13.47 (6.1)	34
Suburbs	Suburbs	0.31	2 × Height	10 (4.5)	12.39 (5.57)	24
Farm	Crops, tall-grass prairie	0.19			11.40 (5.13)	14

Including this factor, the formula will change to look more like this:

$$V/V_o = (H/H_o)^\alpha$$

And since we are solely interested in calculating the unknown wind speed (V), it would be finely tuned to look like this:

$$V = (H/H_o)^\alpha * V_o$$

Notice that the change in height is multiplied by a fractional exponential.

So, if you are still with us, consider the same increase at three different scenarios for common installations: (1) in woodlands for a cabin, (2) on property in the suburbs, and (3) on a farm. In all three cases, we are doubling the elevation.

Do you want to calculate in fractional exponents? If so, it is easy. Get your handy calculator and type in the change in 2, then select the "to the power" symbol that looks like this "^", then type in the α value earlier, and then press the equal button. If your calculator can't do that, here are some links.

Resources

Online Fractional Exponent Calculator
bit.ly/fractexpo

Android calculator
bit.ly/androidcalc

iPhone/iPad/iPod calculator
bit.ly/iphonecalcapp

These results, as summarized in Figure 3-12, show that any doubling of the height will significantly increase the wind speed. Moreover, the more surface roughness present at the original height—as represented by the wind shear exponent—the more dramatic the effect.

But, let's bring it back to what you really want to know: how much more power (P) is available? (Just when you thought we were done with this equation.) First, we showed the basic equation; then we added the important wind shear exponent (α). Now, we are going to convert all of this into power.

Air density and the swept area are assumed to remain constant in this equation. In order to determine the power increase, we equate the power available (P) to the cube of velocity, where P_0 is the power at the original height of 33 feet and P is the power at the new height of 66 feet. For this example, consider a large area with a cut lawn—wind shear coefficient of 0.14.

$$2 \wedge (3*.14) \times P_0 \qquad\qquad 3 \wedge (3*.14) \times P_0$$
$$= 2 \wedge .42 \qquad\qquad\qquad = 3 \wedge .42$$
$$= 1.34 \times P_0 \qquad\qquad\quad = 1.57 \times P_0$$
$$= 34\% \text{ more available} \qquad = 57\% \text{ more available}$$

Doubling the tower height increases the power availability nearly a third! Just for kicks, what if we were to increase it to a 100-foot tower? That is three times the height. Just change the 2 to 3. The result is $1.57\,P_0$. This suggests that by tripling the tower we can increase the power available over 50 percent. This is an enormous change. Figure 3-12 graphically represents how increasing height increases the power in the wind.

Now, a little caveat here to our fun to ground us all. The increase in power with height is a rule of thumb. Remember the discussion that we had about wind speed distribution, where the average wind speed varies? This is for the location, including elevation, where your turbine currently is set. Well, as you increase the tower height above rough terrain, the wind is less impeded, turbulence decreases, and wind speed increases. This means the average annual wind speed will increase. So,

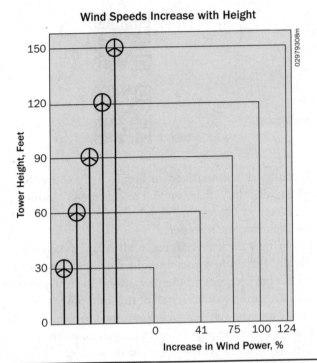

FIGURE 3-12   Wind speed increases dramatically with height. *U.S. Coast Guard.*

not only does the overall power density intensify, but it also increases the accumulative energy production.

Super Power vs. Super Energy: Annual Energy Output (AEO)

Now that we have spent some time covering power, it is time to look more closely at energy, which we have already pointed out should be much more important to you.

One of the first things you will see on the specifications of wind turbines will be *rated power*, accompanied with a *power curve*. In this section, we will familiarize you with these concepts, and introduce the *energy curve* as a better tool to predict turbine performance.

As we stated earlier in this chapter, the difference between power and energy is fundamental. Power (watts) is the instantaneous view of potential energy. It is that framed picture of energy we hope to see. Energy (watt-hours) is the accumulation of that power used or generated—it is the motion picture made from all of those frames. When you pay your utility bill, you are paying for the amount of energy consumed in time. You are particularly interested in 8,760 hours, the number of hours that are in a year. This is commonly referred to as the *annual energy output* (AEO).

Tech Stuff   *Consider a micro-turbine that has a swept area of 1 m² (10 ft²). Input a modest wind speed of 4.5 meters per second (10 mph). Remembering that there are 1,000 watts per kilowatt, then energy can be roughly determined from the power (P) in a specified time (t) like this:*

$$E = P(t)$$

If we want to determine the energy output for the year, it is provided here (in meters):

AEO = [½ Air Density × Swept Area (m²) × Average Wind Velocity(V)³ / 1,000 watts] × [hrs/yr]

AEO = (1.225 /2) × 1m² × (4.5 m × 4.5 m × 4.5 m)]/ 1,000 watts per kW × (8760 h)

AEO = 0.6125 × .09113 kW/m² × 8,760 h = 798.3 kWh/m²

If we want to calculate with feet, not meters, we can grab a formula that makes it even simpler:

AEO = 0.6125 × [Swept Area (ft²)] × [Average Wind Velocity (mph)]³/1,000 watts × (8,760 h)

AEO = 0.6125 × [10ft²] × [10(mph)]³ / 1,000 × (8,760 h)

AEO = 5,365.5 kWh/ft²/year

So, if all conditions remain the same, this micro-turbine can generate nearly 800 kilowatt-hours annually for every square meter, or 80 kilowatt-hours per square foot.

So, we can produce that much energy per year?! Uh, not quite. It is just about physics. Read on if you dare.

Powering Down: Losing 40 Percent of the Betz

That is not a misspelling of "bets." It's a person's name, and it explains the limits of wind power.

A turbine cannot extract all the energy from the wind passing through the rotor because of the simple fact that to absorb all the energy, it would have to be a solid disk (Figure 3-13). In this way, it would act no different than a brick wall. Brick

FIGURE 3-13 This 10-kW Abundant Renewable Energy 442 catches the wind at NREL's National Wind Technology Center in Colorado. *Lee Jay Fingersh/DOE/NREL.*

walls don't currently rotate or harness energy for our consumption. In fact, the rotor needs the spaces in between the blades to cause it to rotate. If there are too many blades closing up the open spaces, some portion of the wind would deflect around the swept area, or put enormous force on the rotor. This is largely why Charles F. Brush's massive 144-blade turbine produced so little energy in the 1880s (see Chapter 1). It is a delicate balance to maximize the use of the wind energy as it passes the blades.

In 1919, German physicist Albert Betz established a theoretical limit on wind turbine efficiency, claiming that no turbine can capture more than 59.3 percent of the potential energy available in the wind. This value is often called the *Betz coefficient* or *Betz limit*.

To add pain to punishment, this ideal limit can only be reached with a rotor without a hub and massless blades. Haven't seen anything like that around, have you? No research has emerged since, or designs have been tested, that contradict the Betz limit. In fact, some studies have resulted in less flattering percentages.

So, this loss in power subsequently reduces our potential energy output by at least 40 percent just at the rotor. Today's best commercial turbines are approaching the Betz limit thanks to sophisticated, computer-managed active controls and high-tech materials, though small turbines aren't quite so efficient. Plus, we still need to convert the mechanical energy into usable electrical energy before we can speak in terms of actual kilowatt-hours. The remaining energy is processed through a turbine generator, rectifier, inverter, controller components, sensors, and even any batteries, all of which add up to take up to an additional 40 percent from the net energy (59 percent) harnessed.

In fact, the most effective turbines today of all sizes have approached only a 35 percent total efficiency rating by the time the power is released. Calculating this from our previous example is simple enough.

$$798.3 \text{ kWh/m}^2 \times .35 = 279.4 \text{ kWh/m}^2$$

So, at 10 mph (4.5 m/s) we should be receiving roughly less than one-third kilowatt per squared meter (10 squared feet). With an average American house consuming 10,000 kilowatt-hours per year, that would mean about 3 percent of a typical home's usage would be provided by this micro-turbine. For all the spinning, it doesn't seem like it is cutting it.

Now you can blame the rotor, you can blame the generator, and all the other components that might drain some of the energy before it gets delivered. But before you do that, let's take a look at how other energy conversion devices are doing.

Converting one energy source into another always causes some loss. Table 3-3 illustrates the approximate conversion efficiency of several common energy sources into alternating current.

TABLE 3-3 Energy Conversion Efficiencies for Various Generating Technologies

Energy Conversion to A/C Electricity	Peak Achieved Efficiency
Water turbine	70%
Combined cycle gas turbine	60%[7]
Diesel turbine	51.7%[8]
Hydrogen fuel cell	50%[9]
Natural gas turbine	37%[10]
Wind turbine	35%
Coal turbine	33%[11]
Oil turbine	33%[12]
Solar-thermal turbine	31.25%[13]
Geothermal power plant	30%[14]
Solar PV	17%[15]

Adapted from table from alernative-energyguide.com[16]

For this table, we are referencing technology that is out of the lab and in production. And as the table specifies, this is conversion efficiency to alternating current. In the case of solar, it has a relatively high conversion efficiency to DC power, but common energy consumption is alternating current, so as with wind technology, it sees comparative losses during conversion.

On balance, wind turbines fare relatively well compared to other resources when it comes to efficiency. And in the data in the table, the energy required to make or transport the continual supply of fuel (natural gas, oil, coal, hydrogen) was not factored in.

But why stop there? Let's also take a look at the conversion efficiencies of common devices and natural phenomena (Table 3-4).

TABLE 3-4   Energy Conversion Efficiencies for Processes

Electric heaters	95%
Efficient household refrigerators	40 to 50%
Light-emitting diodes	Up to 35%[17]
Gasoline to mechanical energy (car)	20 to 25%
Muscle	14 to 27%
Compact fluorescent lamps	19.5%[18]
Incandescent light bulbs	5 to 10%
Photosynthesis	Up to 6%[19]

Adapted from table from alernative-energyguide.com.[20]

All processes have a loss when they convert one form of energy into another. The traditional incandescent light bulb has an efficiency as low as 5 percent to produce light, although you get a whopping 95 percent efficiency to produce heat.

Summary

We started this chapter with a review of electrical concepts that set the table for the serving of a few wind principles. Next, in our search for the power in the wind, not only did we find multiple factors determining what is available, but we also encountered limits and losses during the conversion of wind energy to household electricity.

While estimating what a wind turbine will produce, and noting that it might not be enough juice for your site, it may be helpful to know the three ways to increase output, listed in order of descending priority:

1. Increase wind speed.
2. Increase swept area.
3. Improve the wind turbine's conversion efficiency.

You may increase the wind speed by either selecting a better site with higher average wind speed or building a higher tower. In the next chapter, we will help you evaluate your specific site. With those resources, you will be able to determine the minimum height necessary to maximize whatever turbine you ultimately select. In a later chapter we will take a personal assessment, and you will have to reconsider if a high tower is practical for you.

Once you have the optimal site and height, you can consider the size of the swept area. We touched on the concept here, but in Chapter 8 we will cover the swept area of specific models. So you can begin to try on a few models and see if one fits your energy goals.

The third option is really a recommendation that you find a turbine that is the most efficient for the value. Some manufacturers have been accused of making outrageous claims of performance. You have been provided a tool to test their claims. Chapter 8 will better prepare you for this. But, know that it is important to understand what many overlook—energy output is most sensitive to wind speed and swept area. So, before you try to find a turbine that claims to beat the Betz limit, don't underestimate the value of reliability and cost-effectiveness.

CHAPTER 4

Personal Assessment: Is Wind Power Right for You?

All energy prices have gone up, and wind is a pretty hot commodity right now. There is more demand than there is supply right now.

—MICHAEL SLOAN

Overview

- Why do people install wind turbines?
- How does the relationship of your wind turbine to the power utility affect design and finances?
- How can small changes in a building's energy efficiency save hundreds of dollars?
- How can I reduce the up-front cost of a turbine?

The next few chapters will provide information on the tools, methods of observation, and mathematical formulas you'll need to start determining if a small wind system will work for you. Initially, we look at your personal assessment: a way to measure your needs, your home's energy load, and your personal limits. This is followed by assessing your site. Topics will include prevailing wind patterns, wind volume versus wind speed, wind power categories, where wind is likely the most promising within the continental United States and globally, and strategies to most effectively convert wind into usable energy. Determining the feasibility of any project includes consideration of personal lifestyle, costs, municipal regulations, grid compatibility, and social friction.

Before we bring out the tools to evaluate your site for wind energy, let's stop and evaluate your relationship with the wind and wind energy technology (Figure

FIGURE 4-1   There are lots of fun things that spin, but only you can decide if a wind turbine is right for you. *Brian Clark Howard.*

4-1). We're not talking about deep emotional stuff. We're talking about what you have brought to the table to choose wind over other options.

Before you put up your defenses, let us say that you can pass this chapter if you feel any of these apply to you:

1. You don't care how much it will cost you. You want a wind system and that is that!
2. Energy efficiency is not your game. In fact, you're hoping a wind system will allow you to use more energy.
3. You represent the municipality. You want it, and that is good enough reason.
4. You live in a very energy-efficient home already and are charged excessively high electric rates.
5. You live in a very high average wind region, and you want to lasso this sucker.
6. You live way off-grid or on a boat.

If you have one of these scenarios, be our guest and move on. You are probably ready to invest in a wind system. For the rest of us, read on.

So, are you thinking about why you leapt into wind energy? Think about it as we discuss a few things:

Namely, (1) your emotional needs, (2) your energy needs, (3) your relationship to the grid, and (4) your relationship with your community.

Nothing Personal: Your Motivations

We are not here to tell you that installing a wind turbine is a need, in the same way that food, water, and air are needs. A wind turbine is simply one way to get energy for all the things you need, or at least think you need.

We can't ignore a trend that exists in most developed nations: people buy things because they think they need them. It fills something missing, or it makes a statement. Some people buy wind turbines because they represent something, just like the house, the car, and the clothing. An equally valid reason to buy a wind turbine is to save or even to make money. This is a nonjudgment zone. It is not always important to know the real reason, but it can help in the design, because you can find the limits of what you are willing to do to satisfy this need.

First the wind. Sure, you see its power as it blows over your garbage cans, or tousles the branches of your trees, or sends waves through the crops on a farm. You hear it whistle around buildings. You've also seen the damage it can levy in storms. Wind moves around like it's still the Wild West. It is perpetually unleashed for its own pleasure (not really; it isn't a thinking thing, you know). But *you* want to harness this energy. You want it enough that you bought this book. Perhaps, this book was near other titles on solar, geothermal, and other renewable energy resources; or you may be reading this in the light from your wood pellet stove. Or perhaps, the energy from the wind is an option that stands out from the rest.

Then there is the technology part. Sure, it looks sleek standing there on your property. It is the most visible of the renewable energy options. Along with micro-hydro, it is a rare electric generation system that you can actually see working, without having to look at digital-read outs or your electric bill. A wind system commands attention, being that it towers above all other surrounding objects (unless it is on your boat or it is installed improperly). And if you have a plug-in hybrid or electric car, this sucker could cut your trips to the fuel station (and, by the way, it looks really sexy next to your ride).

Wind is the best sibling for a solar photovoltaic system. If you live in a temperate zone, sunshine is more plentiful during the summer. Meanwhile, the wind energy is not as strong and frequent. But in the winter, the wind is an unforgiving force, more than compensating for the shorter days and lower solar radiation. Wind technology almost seems natural, doesn't it?

What we are leading you to is this: what is this relationship to the wind turbine worth to you?

Let's play a simple game of priority to help you decide if wind is for you.

This is how the game works. First, we tell you what each term means, then you fill each box in Table 4-1 with a + or – symbol. The plus signifies that the item in the row is *more* important than the item in the column. Since we can't compare something against itself, those boxes are blacked out.

What is the goal? To determine the most important reasons in your life and cast them onto your choice for wind power. For example, if given a situation of receiving or saving money at the risk of losing a close friendship, which is more important to you? But the twist is that we relate it to wind electric systems and see if something stands out.

Here are the terms:

Money: Any financial benefit you would get for having a wind turbine. Be it a feed-in tariff, tax incentive, rebate from your power authority, or money saved.

Friendship: Any friends gained or lost due to this big decision. Remember that a wind turbine is a personal decision, but if you don't live alone or isolated from others, it could be an issue.

Popularity: Gained or lost attention from human beings (as opposed to birds and bats) as a result of this towering turbine. If you are a business, or you love to share, popularity should be a good thing.

Personal Achievement: Acknowledgment of meeting the challenge of putting up a wind turbine and keeping it maintained for years to come.

Independence: Freedom from your electric bills or the power authority, or if you moved to an isolated place, the rest of the world.

Acceptance: Tolerance of the wind turbine by your local community, particularly your neighbors (who hopefully won't be too angry). For many, there is the tendency to conform to others.

TABLE 4-1   This handy chart can help you figure out your motivation for wanting a wind turbine. Try it!

	Money	Friendship	Popularity	Personal Achievement	Independence	Acceptance
Money	▓					
Friendship		▓				
Popularity			▓			
Personal Achievement				▓		
Independence					▓	
Acceptance						▓

Now that you have completed the table, this is how we determine your priorities. Simply count the number of times you put a + for each term. (The maximum number for any item is 5.)

You and only you can interpret the results. Although the relationship between these values can be telling, remind yourself that this is only a sample activity, asking you to prioritize under extreme circumstances. We don't know of any divorces or church excommunications as a result of turbine installations. We do know of families who installed a turbine and then felt a sense of independence, yet didn't realize how much more invigorating it was to know that their community welcomed it, and they later became a local source of information on the technology.

If this kind of decision making doesn't appeal to you, we'd still encourage you to try to think about what you really want before you put money on the table for something that will be around for 20 or 30 years.

Determining Your Relationship to the Grid

What is going to be your relationship to the power grid? To some, Grid is God. In other words, they wouldn't think of anything but having their home or business hooked up to the grid. They grew up with electric poles on their street, so why would they want to do anything else? When they plug things in, they work. They may ask, "As long as I don't have to think about it, what is the problem?"

Then there is the polar opposite (Figure 4-2). Perhaps the ratepayers who say the big utility companies do not do them justice. These folks would rather go out

FIGURE 4-2   If you have a cabin in the wilderness or need to pump water for cows on a remote ranch, it may be more cost-effective to have a small wind turbine off the grid, like this 1 kW Whisper on a ranch in Wheeler, Texas. *Elliott Bayly/DOE/NREL.*

on their own, count their kilowatts, and wave to the naysayers who still pay high bills, as they sit comfortably behind their energy-efficient windows, watching their Superbowl or World Cup game from the satellite dish … during a blackout. There are also boaters who prefer to not have to charge up at the dock. Or consider that hunter's paradise in the Catskills mountain range, where stretching a copper electric cable from the nearest part of the grid could be more expensive than 50 years of round trips to the getaway bungalow.

Then there is the largest group of people, who are driving the most growth in the small wind industry: the best-of-both-worlds people. They want to hook up to the grid, but see no reason why they can't save money (net-metering) or even earn money (feed-in tariffs).

These clusters are in no way suggesting what type of person you are; they are merely a colorful way to illustrate what type of bond people commonly keep to their public utility. In sum, there are several options we will review:

1. *Keep the grid, and buy someone else's renewable energy.*   This option is available to many already, and if you call your power authority, they may be able to sign you up right away.
2. *Get me off the grid, and bring on the batteries or some other storage mechanism.*   This can be a rewarding achievement, but except for boats or separate special functions (pumping water), it certainly is not the easiest or the cheapest of the options.
3. *Share my surplus with the grid.*   This is a popular choice for those locations that offer you the option to offset your usage by banking your electrical surplus. This is one of the most popular options outside of feed-in tariffs, and it permits you to achieve savings as soon as you connect to the public utility.
4. *Be my public utility.*   This is an option that exists in many nations, especially in Europe. If you are fortunate to have such mechanisms as government-subsidized feed-in tariffs (FITs), you will basically be your own power plant, and will be paid handsomely for every kilowatt-hour that you produce. The hook-ups are as easy as net-metering, and if you keep your system maintained, you will earn income for 20 years or more from either your public utility or your government (see Chapter 6).

There are advantages and disadvantages to each option, and you must choose the relationship you feel most comfortable with (Figure 4-3). In many cases, you can change relationships if it doesn't work out, although that change may be expensive. Some equipment, like inverters, may have to be changed to comply with the public utility. With FITs, some contract agreements may have penalties if you opt to fold out or are not making enough power to meet your quota.

FIGURE 4-3 Like this California homeowner, you can choose to connect your small wind turbine to the grid or keep it separate. *Bergey Windpower/DOE/NREL.*

Now, let's move from your person to something you spend much of your time with: your home.

How Energy-Efficient Is Your Home or Business?

Welcome! If you are hoping to get your wind turbine to power your home or business, we can do that, and we can do that easy! The qualifications are simple: cut your energy demand by half, and then come back to us. By that time, you will have saved enough money to buy the turbine. Have a nice day. Don't let the foam-core door hit you on the way out.

If you really want to proceed without making your home or business efficient first, you are wasting a lot more money. We get into costs in detail in Chapter 6, but we think it's important to give you a glimpse now. We are making the assumption that, ideally, you would like to have a turbine that at least covers your energy costs.

Later, we go into more detail about the economic feasibility of wind turbines. But if we were to spend a few hours doing some paperwork and end up saving you half the price of installing a wind turbine that would power your home for 20 to 30 years, would you be interested? Of course you would.

Table 4-2 sizes wind turbines to the average home energy usage, exclusive of the actual average wind speeds of the regions. This is just the wind turbine, without the tower or the rest of the components and installation, such as the backhoe and crane, concrete and rebar, electrical components, shipping, and sales tax. KaChing!

All of us could probably reduce our energy demand significantly; however, as a society, we are not always conscious of the possibility. The household energy use

TABLE 4-2   Average Energy Use per Household by Region

Location	kWh/Year/ Household	Rotor Diameter	Manufacturer's AEO	Average Turbine Cost
Tennessee, U.S.[1]	15,600	20' (6m)	15,000	$22,900
United States	12,000	16' (5m)	12,000	$12,000–20,000
Ontario	10,000	15' (4.5m)	10,000	$12,000
California	5,600	12' (3.6m)	5,040	$10,000
Northern Europe	3,500	8' (2.5m)	3,500	$2,590

values noted here are affected by a number of factors, among them social, economic, political, climate, and availability of energy-efficient technology. Perhaps not surprisingly, in Europe and California, there are more robust incentives and pressures for conservation and efficiency.

Normally, we don't think in terms of counting kilowatt-hours. You may wonder what you are going to have to do—shut off the lights, not operate the pool, watch less TV—have less fun. Well, in this activity, you will get to count the beads, so to speak, and once you get to the "numbers," we don't necessarily presume that your lifestyle is going to change. If a few replacements or modest modifications can save up to 20 to 60 percent of your energy use, you will shave thousands of dollars off the purchase of your wind turbine. Why? Because you can size down the turbine to a model that meets your reduced load. KaChing coming in!

In this brief interlude, we will determine your electricity load analysis and current costs, and set some goals. The priority here is to find out your situation before you try to find solutions. No buying till we assess. Agreed?

It is only a three-step process:

1. Calculate your average annual consumption from your bill(s).
2. Identify how much energy each individual item uses.
3. Identify what you can do to reduce energy use cost-effectively.

Calculating Average Annual Consumption

First, you have to ask how much are you spending. To do this, start by looking at your electric bill. If you recently entered the building or did some renovations, we will cover that shortly.

On your bill, note the total for kilowatt-hours (kWh): that is how much energy you used. And it will fluctuate day to day, month to month, throughout the year. If you want to know your average energy demand for the year, go through your past bills or contact customer service for your power utility. Provide them your electric account number located on the bill, then tell them that you are considering installing a wind turbine and need: **12 months (a full year) of your electricity consumption in kilowatt-hours**.

Now you have taken the first step. Some people include more years just in case the most recent one was unusual. Perhaps there was construction, with high usage. of power tools, or perhaps you were vacationing in your summer home for three months.

This Thing Costs *How Much* to Use Every Year?

Where are all the costs coming from? Let's take a look at an average U.S. home (Figure 4-4). If you live in a developed nation in a temperate zone, it is likely similar.

And in 2008, a survey shows the U.S. electric consumption breakdown (Table 4-3).

TABLE 4-3   Average U.S. Home Electricity Consumption in 2008[2]

Space Cooling	16.5%
Lighting (indoor and outdoor)	15.4%
Water Heating	9.2%
Space Heating*	8.9%
Refrigeration	8.0%
Televisions and Set-Top Boxes	7.3%
Clothes Dryers	5.6%
Computers and Related Equipment	3.6%
Cooking	2.2%
Dishwashers†	2.0%
Freezers	1.7%
Clothes Washers†	0.7%
Other—Miscellaneous Uses	18.8%

* Includes fans and pumps.
† Excludes energy for water heating.

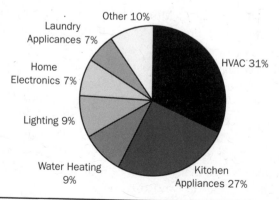

FIGURE 4-4   A summary view of average energy consumption in U.S. homes over the course of a year. *Energy Information Administration.*

This might give you a glimpse into how much your home or facility is using, but it doesn't escape the fact that if you are going to make a big purchase, it is best to have a more detailed list of your energy needs.

How would you know one item's usage from another? You see 100 watts, 1,000 watts. Made no difference to you in the past, so long as it worked. Then, the first year of operating the new pool, or a hot summer with your central air, you get your electric bill and … SLAP … you just found the hidden big cost. Hidden? It is right there for you to calculate, isn't it? I have to *calculate*? And so starts the dilemma.

Let's approach this with an analogy. Let's go to the supermarket. Take a cart, and select whatever you like for your family. Place it in the cart and then go home. Do this for a month. Don't even bother looking at the prices. Prices don't matter here. *What? It's free?* Oh, no. You get a cumulative bill at the end of the month. And it will tell you exactly how much you owe. But it won't tell you how much an item cost, the quantity, or even the name of the item. Just the amount you owe. That is exactly what you have going on with your electric bill.

If you're lucky, you may have received a *smart meter*, which is an advanced meter that records consumption in intervals of an hour or less and communicates that information to the utility for monitoring and billing purposes.[3] Even so, that meter will not tell you each appliance's individual usage (although that technology is currently being developed, and should become available in the near future). Today, even a smart meter can only provide a digital grocery bill of the sum kilo-watt-hours of your total house—just in a smaller time period.

To make it a little easier, many nations do require some form of energy-effi-ciency rating labels on a number of appliances. Figures 4-5 and 4-6 show a few examples, and there are many more listed in this footnote.[4] These labels show how many kilowatt-hours the device will use for a year of *average* usage and the cost based on the *average* electric rate. This is definitely helpful, but if you really want to get a handle on your energy use, you should still do the calculations. For example, Kevin owns an electric range oven but never cooks, so that can be calculated as zero watt-hours.

But perhaps Chef Chewy Munchalot, who lives in an area with the relatively high electric rate of 18 cents per kilowatt-hour, likes to make his culinary delights every day in his convection oven. According to the energy label, in small print, it states that the estimated annual cost is based on an average price of 11 cents per kWh. Chewy is wise to overlook the "average" calculation, estimate the number of hours he actually uses his oven, and plug the values into a handy-dandy calculator ($.18 × wattage of a range oven × 2hrs/day = his cost/day.)

Most electronics devices have a nameplate with the information you need to get started (Figure 4-7). If you find something with no wattage listed, you are not out of luck. You just need to make a simple calculation. As in Figure 4-7, many devices will list the rated voltage (V) and amperage (A). Simply multiply these values and you have the wattage. So, for example, if you have a phone charger that has an

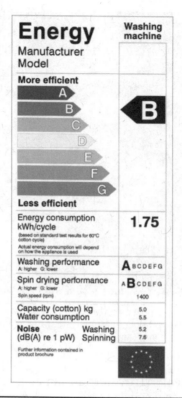

FIGURE 4-5   An example of an energy-efficiency label in the European Union. *Andrew Dunn/ Wikimedia Commons.*

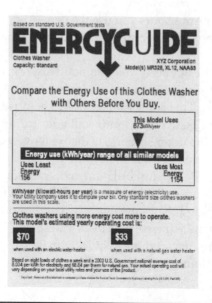

FIGURE 4-6   A sample energy-efficiency label for the United States' Energy Star program. *EPA.*

FIGURE 4-7   If you aren't sure how much energy a device takes to run, you can calculate it by looking at the nameplate. Multiply the rated voltage (V) and amperage (A) to get the wattage. *Brian Clark Howard.*

input voltage of 120 volts (120V) and an amperage of 5 amps (5A), 120 × 5 = 600, so the power would be 600 watts. If your charger says it is rated at 120V to 240V, and you take it to Europe, your input voltage would likely be 240V instead of 120V (handy that it still works, right?).

Next, determine the hours you use the item. And think year round. Winter days in temperate zones are colder and get less light, so that means more hours of lighting and heating.

Although it doesn't seem like glamorous work, without a good energy estimate, there is a high likelihood of underestimating your energy load and falling short of your ideal. This is a serious concern, because upgrading a wind system is not so easy. UPS or FedEx won't uninstall your turbine when you need to return it for a bigger one. The other scenario is that you overestimate and spend more than you need. Unless you have a feed-in tariff, the higher cost of the system could outweigh the benefits. Power authorities are not traditionally in the business of buying energy from small producers, and they usually pay you only a fraction of the retail price that they charge their customers. If you are off-grid, you now have to buy more storage to handle the surplus load. Yikes!

Wait! There are easy options to quickly make an energy load analysis. You can do this. There are four approaches, and one will be comfortable for you.

Auditing Your Energy Consumption

If you want to precisely determine how much energy you are using, you'll need an energy audit. Don't worry, it's relatively easy, and there are lots of folks who can help (Figure 4-8).

Option 1: Get a Professional Home Energy Audit

Getting professional help is good. Many localities offer free audits to identify energy-saving repairs. An energy audit is a comprehensive inspection, survey, and analysis of energy flows in a building, with the goal to reduce the amount of energy consumption. It is not unreasonable to pay up to $500 to have someone come into your home and give you a detailed report. But their suggestions will likely get you to save that back the first year. Many auditors offer to make the improvements themselves, although you may also want to get a second opinion.

If you want to know what auditors do and their tools of trade (blower doors, thermal cameras, etc.), there are many videos and webpages online that demonstrate their focus. Just search online for "home energy audit video."

Option 2: DIY Home Energy Audit

If you live in the United States, you can use the Department of Energy's Home Energy Saver calculator (at hes.lbl.gov) for a do-it-yourself (DIY) energy audit. Just enter your ZIP code and some personal information, including the size of your home, and you'll get ideas on how you can save money around the house.

Figure 4-8   The best way to get a handle on your energy use is to actually measure it with an energy audit. You can hire a pro, do it yourself with pencil and paper, or buy or rent some gadgets that can help improve your accuracy. *Kevin Shea.*

If you want to learn how to do a home energy audit by yourself, you can try this footnoted video link,[5] or check out the DIY home energy audit guide on The Daily Green (www.thedailygreen.com).

These methods attempt to be comprehensive, and rightly so. Heating and cooling systems generally require electric fans, blowers, and water circulators. The less you need to heat or cool your home, the lower your electric load.

For those who cannot find an online auto-auditor, the focus will be on electrical energy consumption. You will need to prepare a chart. The simplest way is to make a spreadsheet with the following columns (Table 4-4).

TABLE 4-4   Measuring Your Energy Use by Appliance

Appliances	Quantity	× Rated Wattage	× Hours Used per Day	= Watt-Hours per Day
Refrigerator*	1	535	8*	4,280
Next appliance				

*Fridges and freezers usually cycle on and off during the day

If you want to save time writing, use the free one we wrote just for you.[6]

Resource

My Energy (Microsoft Excel and Open Office Calc)
bit.ly/myEnergy

It is not much more than the pad and paper type, but it lists many appliances, and automatically calculates the kilowatts and the cost to you based on the numbers you enter. We even provide help to modify it to add items. Who knows—perhaps you have a 3,200-watt amplifier for those exquisite karaoke nights. Add that to the list!

The next thing you need to do is have an Easter egg hunt inside and outside the home or building. That is, you walk around and find the energy labels for each appliance. You might want to start in one room, and then move on to the next. In addition, appliances can include certain electric-hungry items that are part of a whole house system, such as the central air (or HVAC), wall vacuum, speaker, and security systems.

Resources

If you prefer to do the calculations on your smart phone, there is the Energy Costs Calculator for Android phones and the Wattulator for the iPhone.[7] They can calculate the approximate cost per year based on the watts and hours that you enter.

Wattulator   Energy Costs Calculator

When completed, you will have a fairly accurate calculation of energy usage for individual appliances. You will likely need to compare your total kilowatt-hours with your annual energy consumption from your energy bills. If there is a large discrepancy, either you didn't do a good job estimating the hours of usage or you overlooked a few Easter eggs.

Option: 3 (Recommended) Get the Tools and DIY

We saved this option for last, because we reasoned that it is probably the best option. Why? Because it is less work, less money, easy to do, and provides more accurate results. By the way, we do not make a commission off these tools (unless you count "thank yous" as a commission, which we will accept), but we do provide a discount to some devices if you mention the book. Scan the QR code, make the call, mention the book, and you can get up to a 10 percent discount off some devices. You're welcome.

Step 1: Buy an Energy Monitoring Device and Use It ($25–$120)   A watt-hour meter—such as the Kill A Watt from P3 International (Figure 4-9)—is a little device

FIGURE 4-9   The Kill A Watt is a handy device that can calculate the energy usage of a wide range of electronics or appliances. *Brian Clark Howard.*

that tells you how much electricity something uses, either at a given moment or over an extended period. Just plug the device into the meter, plug the meter into the wall, and read the display.[8]

One trick is to leave the meter and appliance connected for as long as you typically use the appliance: all day for a fridge, or an hour while a space heater warms a room, for example. If you provide your current electric rate per kilowatt, most of the devices predict how much you will spend today, in a month, and in a year. This is especially useful for finding the amount of kWh used in a month for devices that run intermittently everyday, like refrigerators. For seasonal equipment such as window unit air conditioners, just make sure you adjust the reading to compensate for the days it is not operational.

Measuring large appliances running on single-phase electricity (220 VAC), such as electric clothes dryers and most European appliances, requires a more versatile device, like the Watts up? from Electronic Educational Devices. This 240V meter ($100) comes with worldwide compatibility in the universal outlet version. Another option is the wireless e2 from Efergy Technologies.[9] This device also comes in handy with appliances that aren't plugged into an outlet (ceiling lights, water heater). There is a 10 percent discount from any of these manufacturers if you mention this book.

Resources

Kill A Watt
(120 VAC/60 Hz)
bit.ly/killAwatt

e2 (240 VAC/50 Hz)
efergy.com

Watts up?
(240 VAC/50 Hz)
bit.ly/h8UVxQ

Power Pro
(100-240 VAC/
50-60 Hz)
Prodigit.com

Step 2: Buy a Thermal Heat Loss Detector ($40)   This handheld device finds leaks, which can be the source of higher heating and cooling bills. Not only does it detect temperature variances with its infrared laser, but some devices have a flashlight mechanism that changes color to blue (cool) when it identifies problem areas around drafty windows and doors, and uncovers hidden leaks and insulation "soft spots" (missing some fiberglass there?). When Kevin was a firefighter, those devices cost thousands of dollars. But this baby does the job for the price of a week of *cafè macchiatos* and newspapers. That is cheap labor!

Resources

Black & Decker
bit.ly/bndthermal

MasterCool MSC52224A
bit.ly/mcthermaleak

Raytek MT6
bit.ly/raytekmt6

Kintrex IRT0421
amzn.to/kintrekirt0421

Step 3 (Optional): Buy and Install a Whole-House Meter ($120)   A whole-house meter tells you how much energy your dwelling is using at any given moment and how much you've used so far for the month—as well as how much it's costing you. These meters start at around $120 and can be installed by an electrician if you'd prefer (it's an easy, 15-minute job).

The Energy Detective and the e2 mentioned earlier both do whole house and individual monitoring. That would include monitoring your wind turbine energy generation as well. Another increasingly popular option is to monitor wind and solar electric generation from a personal computer or a smart phone.

Resources

The Energy Detective
theenergydetective.com

e2
bit.ly/efergye2

Wattson
diykyoto.com/uk

PowerCost Monitor
powercostmonitor.com

Such monitoring can, therefore, be done from anywhere in the world. No, they aren't spying on you! They are providing you access to information that you need. They can get this data from your smart meter, if you have one, or from devices like The Energy Detective. Check your availability at their website.

Special Note to New Building Owners Without Energy History

You are in luck, because there are tools that can help you estimate your energy use. For one thing, you can probably get some idea by looking at the average energy use of your current location. Start at www.EnergyStar.gov.

INTERCONNECT

Choosing to Stay Off the Grid in Vermont

A few years ago, Dale and Michelle Doucette decided to move their family into a new home in rural southern Vermont. The 22-acre property they settled on in Wilmington is wooded and gorgeous, and also not connected to the electrical grid. The Doucettes could have paid about $10,000 to run wires the quarter-mile to the grid, but instead they decided to become their own power plant.

According to *USA Today,* the Doucettes spent about $41,000 on a solar and wind hybrid system, with a shed full of 24 batteries and a backup propane generator. Their $500,000, 3,200-square foot house is built with efficient straw-bale insulation and lots of energy-saving features, like low-voltage lighting and high-efficiency appliances.

Dale is a wood carver who runs his saws on renewable energy, and Michelle is a chiropractor who works out of an office in their home. Their sons, 17 and 22 at the time of this writing, enjoy playing video games and watching TV, just like young people who live on the grid.

We gave Dale a call, and he told us he isn't sure exactly how much energy his Bergey 1K wind turbine produces annually (Figure 4-10). "We use it strictly as extra charging for the batteries," Dale told us. "It particularly comes in handy in the winter, when there's no

Figure 4-10   The Doucette family of Wilmington, Vermont, especially appreciates their 1 kW Bergey wind turbine in winter, when winds are strong and days of solar collection are short. Pictured is another Bergey (center) at New Life Evangelistic Center in New Bloomfield, Missouri. *Rick Anderson/DOE/NREL.*

leaves and we get stronger wind, and when we're using more energy, since days are shorter and lights are on longer."

Dale told us the turbine sits on an 80-foot, guyed tower. He said he did a lot of the installation work himself, though he hired some help as well. They didn't use a crane, getting along fine with a gin pole and winch. "I'm a pretty hands-on guy," he told us. "I have somewhat of an electrical background. And I like to know what's going on; I'm buying the system, it's my house, so I like to be informed."

Dale told us that incentives from the state of Vermont covered about half the cost of the wind system. "I wish I could have gotten a bigger one, but money and time were not going to allow that," he said.

Dale told us he is currently looking into adding a micro-hydro system on his property, since he has streams. "The object is to try and not have the generator running. We have two kids, and when they're both home a lot of energy gets used," Dale said.

We asked Dale what it has been like to live with a wind turbine, and he said it has worked great for five years, with only minimal maintenance. The only downside he could think of was that a "decorative nose cone has blown off a couple of times, so I don't even bother with that anymore. It's easy. They put it up, it spins, it makes power."

We asked Dale if he had given a thought to bats or birds when he installed his turbine, and he said he hadn't because he thought it was too small to make an impact. "The blades are only four feet long," he said, although he pointed out that the spinning turbine seems to keep deer out of his garden at night. When asked about the potential for angry neighbors, he said his closest ones are three acres away, and that they never seem bothered. "The only time we hear any noise from it is when there are really strong winds, like above 50 miles per hour, and it is going really fast. It has a shutdown feature, and at times it sounds like an airplane shutting off," said Dale.

When asked if he recommends small wind power to others, Dale said, "I recommend solar, hydro, wind. I recommend it all. We gotta make a stand here. Everybody takes power for granted." As far as what we need to increase adoption, Dale said, "Ultimately, price is what's the problem. Even my little generator was $5,000. In the society that we live in people aren't willing to make that sacrifice. They want everything instantaneous." For his part, Dale told *USA Today* he expected his system to pay for itself in about 20 years.

Dale's advice to those considering a wind turbine is to do lots of research, measure how much power you need, get familiar with your surroundings, and then go look at other installed systems. He cautioned that it's critical to make sure there is enough open space to support the tower (more on this later), and he stressed that it's a good idea to get a second opinion. He also warns people to do their due diligence on service providers. "There are a lot of so-called experts in the field, with multiple opinions. A lot of people are inventors, but can they install it or run a business? That's something you have to determine."

The Doucettes are among the roughly 180,000 American families who live off the grid, according to *Home Power* magazine. Interestingly, that figure has increased about 33 percent a year for a decade or so, although the number of people who are hooking up small renewable energy systems to the grid is growing even faster. The number of on-grid systems in the United States is expected to eclipse off-grid applications within the next few years.

Boosting Your Building's Energy Efficiency

Well, you got the results of your energy audit, and you have determined that some changes need to be made before you invest in a wind turbine.

Where to start: Cost evaluation. How much is it going to cost you? And what will be the annual savings and the payback period? Without further ado, Table 4-5.

TABLE 4-5   Common Improvements That Can Save Energy and Money

Energy Savings Improvements				
	Cost		Payback	Annual Savings (water, fuel, electricity)
	Low End	High End		
Energy Audit	FREE	$500.00	1 year	$1,100.00*
Caulking	$5.00	$250.00	1 year	$250.00
Ceiling Fan	$80.00	$80.00	8 years	$10.00
Compact Fluorescent Light Bulbs	$136.00	$136.00	2/3 years	$200.00
Refrigerator	$729.00	$729.00	8.1 years	$90.00
Duct Sealing	$1,350.00	$1,350.00	5.4 years	$250.00
Tankless Water Heater	$1,489.00	$1,489.00	8.5 years	$175.00
Insulation	$2,000.00	$5,000.00	5.7–14.2 years	$350.00
Storm Windows	$2,500.00	$4,750.00	4–7 years	$690.00
*Changes involve modifying settings on thermostats and lifestyle changes	Total Cost $8,588.00–$14,583.00		Payback 4–6 years	Total Annual Savings $2,015.00

Data source: National Trust for Historical Preservation[10]

The table shows projected savings generated by audit-recommended improvements to an average single-family home up to 3,000 square feet, with annual energy (electric, water, and fuel) bills over $2,200. The cost and total savings will vary from home to home, but this gives you an idea for your typical Jones' house. Finally, you can keep up with the Joneses by acquiring the stuff that pays you back in the long run.

First, let's look for low-hanging fruit that requires little elbow grease or purchasing.

Adjust That Thermostat

On most systems, you can cut heating and cooling costs by an additional 20 percent by lowering the thermostat during the winter by 5 degrees Fahrenheit at night and 10 degrees Fahrenheit during the day when no one is home, or by raising it during air-conditioning season. You can do that with daily adjustments of a simple thermostat, but few of us are so diligent.

Instead, you could install a programmable thermostat (as low as $25) so that you can set it and forget it. Some utilities even provide them for you. According to the EPA, the average family will save $180 a year with a programmable thermostat, and new models are getting easier to use.

Even when you are home, also consider turning down the thermostat in winter, or turning it up in summer. When it's cold, for every degree you lower the thermostat, you'll save between 1 percent and 3 percent of your heating bill. You can also make like Jimmy Carter famously did in the 1970s, and don a sweater. A light long-sleeved sweater is generally worth about two degrees in added warmth, while a heavy sweater adds about four degrees, according to The Daily Green.

Use Cold Water

Heating water is the second-largest energy hog in the home, accounting for 12 percent of your energy use and costing the average household around $250 a year. Short of replacing the water heater with a more efficient one, building owners can save 5 percent on their utility bill by dropping the water temperature to 120 degrees from 130 degrees F and insulating the water pipes. Also consider washing your clothing in cold water—you can save $60 a year.

 Power Up!   *Put your hand on your hot water heater. If it's cool, your unit is well insulated. If it's warm, cover it with a water heater jacket, available at hardware stores and home centers for a few dollars.*

Also consider taking shorter showers, because the more hot water you use, the more you have to pay to heat it up. According to the EPA, the average shower uses 25 gallons of water. Restrict yourself to only five minutes and you'll halve that.

Use Electricity When the Rate Is Lower

Some power authorities have varying price rates (called time-of-use, critical peak, or dynamic pricing), depending on the supply and demand for your region. Some rates may change by the hour, especially for large businesses. Check with your power provider for their policies. In many places, electric rates are cheaper at night, because demand is so much less (Figure 4-11).

Figure 4-11   The first step before installing any renewable energy systems is reducing your energy use through efficiency. *Brian Clark Howard.*

Taking advantage of lower rates could require only small changes in lifestyle, like placing your clothes in the dryer in the evening. But if the rates vary hour to hour, you might be left guessing. There are smart phone applications that do the thinking for you, such as *Power Stoplight Mobile* by Smart Power Devices.

With the program, you see a stoplight prompt on your screen that lets you know the best time to run major electrical loads, based on dynamic pricing. For those with renewable energy systems, a blue light indicates optimal times for net metering. The guessing is gone. As soon as you see the green light, wash those dishes. Currently, this app works in many U.S. states and Toronto, Canada. It is available on the Android, Blackberry, iPhone, iPod Touch, and iPad (cost: $0.99).[11] Again, we receive no commission, just thank yous.

Resource

Power Stoplight Mobile app
powerstoplight.com

In addition to such mobile apps, whole-house energy monitoring systems often have the same functionality built right in. Some utilities have also been testing pilot programs that distribute little plastic globes to elect customers. When the orb is green, the rates are lower, and the user might choose to run high electric loads, such as washing machines. If the globe glows red, that indicates higher pricing. Studies show that such active feedback encourages consumers to save energy and money, while making them feel more empowered and connected to energy use.

Now that you have taken these easy steps, you can afford to spend a little money for some significant cost savings.

Plug the Leaks

Get some caulking and fill in all gaps, cracks, and joints of windows, doors, skylights, and trim. Operating heating and cooling systems accounts for 46 percent of the average home's utility bill (more in cold climates), according to the U.S. Department of Energy. Yet drafts sap 5 to 30 percent of that energy.

For the best return on your energy-conservation effort, start by making sure all windows and doors have good weather stripping. If you have an attic, make sure there's sealant around pipes, chimneys, ductwork, or anything else that comes through the attic floor. Sealing those leaks is cheap and easy and can save 10 percent, or $190 per year, for the average home.[12]

Install Ceiling Fans

The annual savings in Figure 4-10 are based on one energy-efficient fan that is used only when the room is occupied. If you have more rooms and more fans, you can multiply the savings. Ceiling fans lower perceived temperature in summer, lessening reliance on air conditioning and saving energy. In winter, reversing the direction of fans draws warm air down from the ceiling, saving up to 10 percent on heating costs. In general, counterclockwise rotation produces cooling breezes while switching to clockwise makes it warmer. When you are fan shopping, be sure to look for Energy Star–certified models, and get one that has a reversible direction.

Resource

Energy Star Ceiling Fans
1.usa.gov/es_fans

Install Compact Fluorescent Light (CFL) Bulbs (or LEDs)

The annual savings in the energy savings chart are based on 40 bulbs illuminated for three hours each day, the average U.S. residential usage. Light in all buildings accounts for about 30 percent of the energy consumed in structures.[13] With proper

FIGURE 4-12   CFLs now come in many shapes and sizes, and they will help you save energy. If you don't like them, opt for halogens and dimmer switches or LEDs. *Brian Clark Howard.*

use, CFLs should last up to about 10 times longer than traditional bulbs, so you will be making that savings for at least five years. Each CFL that replaces an incandescent bulb should save $30 over its lifetime (Figure 4-12).

Go further and save even more energy with light-emitting diodes (LEDs), which are becoming much more widely available, for prices that are more competitive. LEDs can last up to hundreds of thousands of hours, and use up to 90 percent less energy than incandescents. They are also showing rapid advancements in color and dimmability, and are falling in price. A good LED replacement for a 60-watt incandescent can be picked up for $20 at Home Depot.

If you don't like CFLs and can't afford LEDs yet, get some halogens and dimmer switches. Halogens are up to 40 percent more efficient than standard incandescents and they last two to three times longer. Plus, the more you dim your lights, the more energy you save. Get occupancy and motion sensors to make saving even easier.

Learn a lot more about efficient illumination in the book *Green Lighting* (McGraw-Hill, 2010), part of this same series.

Install a Tankless Water Heater

Traditional water heaters end up wasting quite a bit of energy, because they pre-heat a large volume of water in the tank, even if you don't use it for a long time. This means heat is slowly leaking out. Plus, if you have guests over who all want to take showers, you may "run out" of hot water from the tank.

A tankless water heater, in contrast, solves both these problems by heating water only as it is needed, "on demand." Tankless water heaters can use electrical elements or natural gas burners to heat a stream of water, with both types having their advantages and disadvantages.

A seven gallons-per-minute (gpm) tankless heater can supply two or three major applications (shower, dishwasher, kitchen sink) at the same time. Tankless water heaters are also space saving and they tend to last longer than conventional heaters. They cost more up front, but they will save money over time.

Boost Insulation

Estimates on the cost and savings of boosting insulation vary widely depending upon locality and product used, but for homes with barely any, adding insulation makes a significant performance difference (Figure 4-13). This is especially true with central air systems because lots of air can be lost to the outside or unused basement, particularly at corners and joints. Insulate ducts wherever possible. If you have an infrared thermal device, you will know if you have done the job well.

FIGURE 4-13 Spraying insulation into your attic and behind walls can significantly reduce your heating and cooling bills. *Community Services Consortium/DOE/NREL.*

Install Storm Windows

The annual savings for adding storm windows to an average building, with 25 windows, is close to $700. Storm windows cost an average of $100 to $200 each and have a typical payback period of four to seven years. There is an easy way to select your windows based on your location, thanks to EnergyStar.gov specifications, which are included in the footnotes.[14]

Get an Energy-Saving Refrigerator

The Energy Star Program website can guide you to hundreds of selections of energy-efficient refrigerators. One of us recently bought a dented unit and didn't have to pay full retail (score!). We received a rebate, and started getting annual savings.

In fact, a good rule of thumb is to make sure any new appliance meets Energy Star standards or EU standards. A home that uses Energy Star products can save a significant amount of electricity and money each year compared with a home that doesn't.

Additional Energy Savers

There's no shortage of ways to save money on home energy costs. For more suggestions, go to your local or national energy efficiency office. We have provided a few.

Resources

Energy Star
energystar.gov

European Union Energy
energy.eu

Office of Energy Efficiency
bit.ly/ca_oee

Bureau of Energy
Efficiency
bee-india.nic.in

You can also opt for consumer websites like The Daily Green (www.thedaily green.com) and Earth911.com.

Here are a few additional ideas.

Give Your HVAC Equipment an Annual Tune-Up

Just as cars need regular maintenance, so does heating and cooling equipment. By keeping your system clean and in good working order, you can save an average of

5 percent on your bills and extend the life of your investment. Many utilities offer free or low-cost annual inspections if you call during the off season.

Change Filters Regularly

Dirty filters can restrict airflow and make your furnace work harder, so it's a good idea to change filters once a month during the heating season. If you get tired of that, consider investing in a permanent electrostatic filter. Not only does this reduce maintenance, but you'll also zap up to 88 percent of debris, compared with 10 to 40 percent for disposable filters. Plus, electrostatic filters are much better at controlling bacteria, mold, viruses, and pollen.

Put Up Window Plastic

If you can't afford storm doors and windows, or if they are old and leaky, consider putting up some thermal window plastic during the cold months. Don't worry—when installed properly it is virtually invisible, but it provides an extra layer of insulation by trapping still air.

Do it Now, While Incentives Last

Government incentives change all the time, but chances are good that you have something available to help you make efficiency improvements to your home or business. Energy efficiency is good for homeowners, local businesses that provide improvements, and the environment, so incentive programs are often politically popular.

Only a few years ago, we had to send in receipts to get power authority rebates that were sponsored by the government. But since energy infrastructure has come into the limelight, it seems many governments provide an increasing array of incentives. Many programs also target low-income homeowners with additional incentives.

The American Recovery and Reinvestment Act of 2009 provided a credit worth 30 percent of the cost of materials for a broad range of improvements, such as energy-efficient windows and insulation, up to $1,500. Unfortunately, as of this writing, that credit expired at the end of 2010. However, some renewable-energy systems, including solar panels, wind turbines, and geothermal systems, still qualify for a 30 percent tax credit on parts and labor through 2016. So if you are inclined to make an upgrade to your home, now is the time to do it, or to set money aside to do it before the credits expire.

Summary

You may have started this book thinking that all you needed for a wind system was a turbine, some wire, and a stiff breeze (Figure 4-14). But we hope you have an idea now that it's worthwhile to take a step back and look at the complete picture of how you use energy, and how you can use it smarter. If you spend a little bit of time and money improving your home or business, you can save 20 to 60 percent off your electric bill.

That means you can install a cheaper renewable energy system, if you choose to go that route. Richard Perez, the founder of *Home Power* magazine, wisely said, "Every $1 spent on energy efficiency saves at least $3 in renewable energy system costs." But whether you end up investing in a wind generator or not, these first steps toward energy efficiency are likely to yield the best return on investment.

FIGURE 4-14   A wind turbine is an iconic image that has captured the spirit of hope around the world. But before you take the plunge and buy a system, make sure you are saving as much energy as possible. *Brian Clark Howard.*

CHAPTER 5

Site Assessment: Will Wind Power Work at Your Location?

Everyone knows that you do NOT place a solar panel in the shade, and a similar rule should be obvious for wind turbines. The following should be etched on every wind turbine and displayed conspicuously on every website: "For best results place me 30 feet above and 250 feet away (and avoid contact with migrating birds, bats, and nimbys)."

—KEVIN SHEA

Overview

- Two basic requirements for a good wind generator site:
  Good _____ and low _____
- The energy available in a wind stream is proportional to _____.
- Wind potential is somewhat _____.
- The tower should be _____ higher than any barrier within a _____ radius.
- _____will have a significant effect on wind energy.
- Roof-mounted turbines have a problem with _____.

This chapter provides a toolbox of methods and formulas to assess your site for wind power capacity, as well as social, economic, and environmental factors. Some math will be necessary, but we will attempt to be gentle. If you don't think this is necessary, if efficiency and effectiveness are not your goals, you probably are just into a new Russian roulette model (don't know when it will fire off), physics-defying, hot-rod wind turbine with the extra-sleek flaming blades. In that case, please ignore this chapter and move on to the appendix, "How to Sell Your Wind Crap on Craigslist[1] and Handle Debt Into Your 70s."

FIGURE 5-1   This 80 kW Zond wind turbine in Waverly, Iowa, was replaced with a bigger 900 kW turbine in 2001. *Waverly Light and Power/DOE/NREL.*

Before you do your assessment for wind energy, we hope you verified that your site has legal allowances (Figure 5-1). If you are not sure before you start, this chapter will still help you, but you might be looking in the wrong place. The easiest way to do this is to go to your local town building permit office and ask for all information on permits for wind turbines or appurtenant structures. You can also read Chapter 7 on permits or variances before staking your lot.

Get Prepared

We recommend that you find a quiet place to read. To maximize your efficiency of learning, we recommend reading it several times using the QR^3 method—Question, Rush, Read, Review. First, create your questions that you want answered. Then rush through the chapter trying to answer those questions. After that, read it thoroughly. Finally, review the chapter one more time to confirm your comprehension.

If that wasn't enough to show you how important assessing is, please note that the overview box is missing words. You can fill them out before or after you read the chapter. And if you read it now, we will throw in some homework for you at the end of the chapter. You will work in this chapter, but we guarantee you will also have ample opportunity for fun.

How to Save Time on This Chapter

Follow these simple rules:

General rule #1: Ensure you have good average wind speed in your region.   This is the most important factor affecting success of a wind system, and early in this chapter you will learn what regions generally have good wind energy and will gain some observation methods to get you going. But if you find that your site has adequate *average* wind speed, not just gusts here and there, small wind turbines have a major advantage over large utility wind turbines in that you won't need a serious preinstallation evaluation. You can begin to get an idea of your wind resource by checking your local weather history or global wind maps. And although some online services charge, here in the United States there is a quick, free way to evaluate your site. Go to the Choose Renewables website.

Resource

Choose Renewables
www.chooserenewables.com

Simply type in your address, and you will get some idea if you may be a candidate, and you will get information on your incentives.

General rule #2: Avoid high air turbulence, which will slow your turbine down and damage it over time.   Setting up a tower at least 30 feet above all obstacles within a 500-foot perimeter is a must (even more is usually better). Being on a flat open hilltop is usually ideal for optimum wind levels.

General rule #3: Make it real high.   Beyond clearance, let us tell you the best solution: if possible, make your tower as high as you can afford. Balance this cost with cost for electric grid access and transmission, and social friction on visibility (Chapter 7). If you live near an airport, the Federal Aviation Administration (FAA) or other governing body usually has restrictions on towers higher than 200 feet, but other than that, today's robust technology can sustain sufficient energy for your home if you place it on a high tower.

General rule #4: Don't install a toy turbine.   If you follow all three previous rules, you can now purchase a turbine that has a good track record to perform for many years without falling apart ... or a turbine advertised with a "low wind cut-in speed," which supposedly permits you to grab low energy speeds, yet tampers down when substantial energy is available to protect the system. If you want toys, that is okay; we love toys too. Just don't put them in high winds, unless you like broken toys. And beware that toys probably will not generate meaningful amounts of electricity or pay for themselves.

The rest of this chapter deals with more specific scenarios and details about assessing your site.

Site Location Guidelines

Proper site selection is very important to the performance and longevity of your wind turbine. A poorly sited wind generator will probably bring nothing but poor performance, maintenance issues, higher costs, and frustration. The owners we interviewed who would fall into this class were unanimous in saying they wished they had taken more care during the siting process, and that they would not have built if they had known all the facts.

There are two basic requirements for a good wind generator site:

1. **Good average wind speed.** The single most important element to maximize performance of your wind system is average wind speed. For your wind generator to produce energy, the cut-in speed and the average wind speed at your site should be equal or close.
2. **Low wind turbulence.** The lower the turbulence, the more efficient the rotor, the less stress your wind generator will sustain, the more energy it will produce, and the longer it will last.

In order to correctly site and size a wind electric system, it is helpful to have the following information about your location:

- **Average annual wind speed.** As we showed in Chapter 3, a small increase in average wind speed results in a large increase in energy output of a wind generator. For example, an increase in wind speed of 10 percent (9 mph to 10 mph; 4 m/s to 4.5 m/s) results in roughly a 30 percent increase in power available. Therefore, the better the location, the better the performance. And almost universally, wind speeds increase with height. This chapter will cover various methods to determine average annual wind speed.
- **Prevailing wind directions.** *Prevailing winds* are technically winds that blow predominantly with the highest speed from a single general direction over a region. They are usually driven by global patterns. In general, low and high latitudes tend to have easterly prevailing winds, while the mid-latitudes tend to have westerly winds. Wind roses, covered later in this chapter, are tools used to determine the direction of the prevailing wind (Figure 5-2).

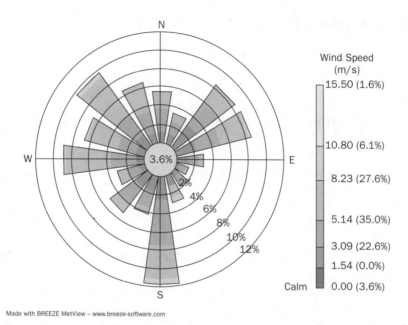

Figure 5-2   This wind rose shows the prevailing wind directions at LaGuardia Airport in New York City. *Breeze Software/Wikimedia Commons.*

- **Topography.** Although hills are better than valleys, there are also circumstances where the highest land available may not be the best place for your wind generator. High land may be awkward to get to, too far away from where you need the power, or may expose your wind generator to potentially damaging turbulent conditions.
- **Barriers.** Barriers (buildings, trees, etc.) produce wakes or eddies that may extend far downwind and high into the air. Such turbulence can rob your turbine of energy, or even damage it. The upcoming section on wind shear (first introduced in Chapter 3) will further cover effects and solutions of barriers.
- **Surface roughness.** As we mentioned in Chapter 3, the smoother the terrain, the less friction encountered by moving air, and the faster the wind speeds. We already introduced the wind shear exponent, but we'll build on that later in this chapter and provide calculating tools.

Power Up!   *In order to prevent wind speed impedance:*

What should your wind turbine not be near?

The ground, a roof, buildings, cliffs, foliage, large signs, plane routes (tee-hee), other wind turbines.

Basic Principles of Wind Resource Evaluation

We can't stress it enough: wind resource evaluation is a critical element in project-ing turbine performance at a given site. Here's the relationship we covered in Chapter 3:

> The energy available in a wind stream is proportional to the cube of its speed.

This means that doubling the wind speed (V) increases the available energy (E) by a factor of eight.

$$2V = P^3$$

Of course, a wind resource is seldom a steady, consistent flow. But many sites definitely have better average wind speeds than others. In general, according to the American Wind Energy Association (AWEA), annual average wind speeds of 11 mph (5 m/s) are recommended for grid-connected applications. For off-grid uses like battery charging, water pumping, and monitoring devices, annual average wind speeds of 7 to 9 mph (3 to 4 m/s) may be adequate.

In addition to average wind speed, another useful measure is *wind power density*, which indicates how much energy is available to a wind turbine at a given site, measured in watts per square meter (W/m²). We shall expand on this concept, with more factors that can affect the annual energy production of your turbine.

Know Your (Wind) Class

In the United States, the National Renewable Energy Laboratory (NREL) classifies wind power density into seven ascending classes, which help give an idea of the feasibility of a location for wind power. Table 5-1 shows how the classes correspond to wind speed measurements at three standard heights, 33 feet (10 m), 100 feet (30 m) and 164 feet (50 m). As stated previously, wind speed generally increases with height above ground.

Given the current state of turbine technology, class 2 areas and above can be suitable for small wind projects.[2] Class 1 areas are generally not okay for grid con-nection, but may supply supplemental energy, or make for conversations at your annual barbeque.

For countries that don't recognize this classification system, it is easy enough to use the wind speed averages shown in the table to see the corresponding power class and density to get a general idea based on global wind maps.

TABLE 5-1   Wind Power Classes

	Classes of Wind Power Density at 10, 30, and 50 m*					
	10 m (33 ft)		30 m (100 ft)		50 m (164 ft)	
Wind Power Class	Wind Power Density (W/m²)	Speed† m/s (mph)	Wind Power Density (W/m²)	Speed† m/s (mph)	Wind Power Density (W/m²)	Speed† m/s (mph)
1	50–100	3.5 (7.8)–4.4 (9.8)	80–160	4.1 (9.2)–5.1 (12.5)	100–200	4.4 (9.9)–5.6 (12.5)
2	100–150	4.4 (9.8)–5.1 (11.5)	160–240	5.1 (12.5)–5.9 (13.2)	200–300	5.6 (12.5)–6.4 (14.3)
3	150–200	5.1 (11.5)–5.6 (12.5)	240–320	5.9 (13.2)–6.5 (14.6)	300–400	6.4 (14.3)–7.0 (15.7)
4	200–250	5.6 (12.5)–6.0 (13.4)	320–400	6.5 (14.6)–7.0 (15.7)	400–500	7.0 (15.7)–7.5 (16.8)
5	250–300	6.0 (13.4)–6.4 (14.3)	400–480	7.0 (15.7)–7.4 (16.6)	500–600	7.5 (16.8)–8.0 (17.9)
6	300–400	6.4 (14.3–7.0 (15.7)	480–640	7.4 (16.6)–8.2 (18.4)	600–800	8.0 (17.9)–8.8 (19.7)
7	>400	>7.0 (15.7)	640–1,600	8.2 (18.4)–11.1 (24.7)	>800	>8.8 (19.7)

* Vertical extrapolation of wind speed based on the 1/7 power law.

† Mean wind speed is based on the Rayleigh speed distribution of equivalent wind power density. Wind speed is for standard sea-level conditions. To maintain the same power density, speed increases 3%/1000 m (5%/5000 ft) of elevation (from the Battelle Wind Energy Resource Atlas).

Maps Are Not Gospel

Although wind experts recommend conducting a preliminary site assessment with wind maps, these maps are not particularly accurate for local sites. In fact, there are many cases in which maps are woefully inadequate, especially in suburban and urban terrain.[3] Here are some points to keep in mind:

- If the height of the turbine is less than 33 feet (10 meters), which isn't recommended, a correction will need to be made. At 5 meters, the wind speed will be roughly 10 to 20 percent lower.
- The table is more accurate for flat, open land than it is for rough terrain. If you have high surface roughness, revise the wind speeds down.

- The table does not include any local thermally driven winds, such as sea breezes or mountain/valley breezes. If these are present, add 0.5 to 1.0 m/s.
- If the site is on a hilltop, adjust the average wind speed up. If it is in a valley, don't put a wind turbine there.

Wind Shear

Available wind maps or climate data may have brought you to an average wind speed at the height approximating your tower. But you will likely need to make adjustments based on hyper-local topography and wind shear. We introduced the notion of wind shear in Chapter 3, but we left room for a more comprehensive explanation in this chapter.

Anyone who has lost his or her hat from an unexpected gust on a city street has experienced wind shear. Wind shear simply refers to a large change in wind speed or direction over a short distance—either horizontal or vertical. Wind shear is closely related to turbulence, and is often the cause of it. It's important to be aware of because wind shear can be a factor in estimating wind turbine energy production.[4]

There are many causes of wind shear, including thunderstorms, fronts, temperature inversions, and obstructions on the ground, such as mountains, buildings, trees, and even sailboats. We won't get into the precise details, since this isn't a book about the weather. The important thing to know is that one of the results of wind shear is that wind velocity varies with height above the Earth's surface. As we've pointed out before, as the height increases, the wind speed usually increases. This results in vertical wind profiles like a multilayer cake moving across the Earth, each layer putting an increasing level of force on anything it strikes. (That layering is sometimes called the wind gradient.)

One small point: the increase in wind speed with height technically only holds true for the height above the *effective* ground level. That is, the wind rushing over amber waves of grain really sees the tops of the plants, not the soil on which they grow, as the effective ground level.[5]

So, since there are stratified wind speeds, your turbine may be receiving different wind power densities as the blades spin. The wind energy would probably be less nearest to the ground and more at the top of blade travel, and this does affect the turbine operation.[6] Even low-level wind shear can bend parts of a turbine, changing the blade angle to something less efficient at harvesting energy. Such forces can also put a lot of stress on the machinery.[7] The only consolation for small wind turbine owners is that these forces are going to be less than they are for large systems, where there is a greater distance between blade tips. (Although it's also true that larger swept areas mean much more energy can be harvested from the wind.)

As we've suggested, wind shear also tends to result in greater air turbulence, which is characterized by eddies and randomized air flow. *Mechanical turbulence* is common near the ground, and forms as wind blows over or around obstructions.

You can observe the same essential phenomenon easier in a rushing stream, where eddies that form in water make whitewater rafting exciting. Take that concept of bouncing around on a raft with swirls of water, and apply it to the invisible eddies that are oscillating your rotor. The faster the wind, the stronger the turbulence, just like the faster a moving stream, the stronger the eddies. Turbulence is one of the main reasons for failure in turbines. Don't underestimate the power of the wind, just as you wouldn't want to underestimate the power of moving water on a rafting expedition. Of course, to compensate for the extraordinary stress requires heavy-duty components, although that can reduce the performance. In the long run, it is better to find a way to mitigate wind shear.

Mitigating Wind Shear

When we often talk about obstructions, many people presume that they can be altered. However, it's worth remembering that the main sources of wind shear result from large-scale jet streams, tropical atmospheric circulation patterns, cold-core low pressure systems, geologic features, El Niño or La Niña events, and so on.[8] Unfortunately, these types of macro-effects are usually outside the range of your control.

However, there are a few things you can do to minimize wind shear locally:

1. **Move the turbine.** Knowing the cube factor effect of wind speed, sometimes moving the wind turbine a few meters up, away, or to the side of an obstacle, can make a substantial difference on performance. Many people like to have their turbine close to their home, but if the house is knocking out your wind, they make for a bad couple. It would be better if the turbine were matched with your dog's house.

2. **(Re)move the obstacle.** Cut that tall grass or remove that wayward tree. We understand if you are a farmer with cash crops or a homeowner who enjoys your shady forest, but we can offer some compromises. If you rely on your plantings, consider selecting (or possibly replacing) tall choices with shorter-growing vegetables or dwarf fruit trees. They are abundant, and make for easy picking. For the tree huggers who can't part with a forest view, select or replace your existing trees with dwarf deciduous trees (Figure 5-3). They can produce shade in summer, and when the leaves drop, the barren branches theoretically reduce the wind shear during the winter months, when the highest wind density typically occurs. If you can't bring the height of the trees down, perhaps you have the ability to bring the level of the ground down by making terraces of your terrain, or even a valley. That way, even if the trees do grow, they would be lower than if on a plain. Or if you can opt to sacrifice some forest or field for a pond, you have reduced the wind shear coefficient to 0.1, while providing water for your plants, home for nature, and beauty for you.

FIGURE 5-3   If you must have trees near your turbine, choose dwarf varieties. The shorter the obstacles and the farther away from your tower, the better wind your generator will see. *Kevin Shea.*

Kevin lives on a wildlife reserve protected by the national government. If he cuts down a tree, it better not be more than seven inches in diameter or a threatened species. If he wanted to clear the area, he would need to arrange to replant with a significant number of timber trees to replace it. And so he selected trees that would meet his height limit while making a beautiful addition to the property.

Wind Shear Exponent

Tech Stuff   *As mentioned in Chapter 3, wind shear can be expressed as an exponent in calculations to fine-tune understanding of what wind power density is realistically available at your site. Specifically,* wind shear exponent, α, *factors in wind shear when determining how wind speed varies with height. One way of calculating this is from the following equation, described by German engineer Jens-Peter Molly in his book* Windenergie *(Muller, 1978):*

$$\alpha = 1/ln(z/z_0)$$

where z_0 is the surface roughness (e.g., forest, buildings, personal statue, etc.) length in meters, z is the reference height, and the ln *is the natural logarithm that gives you the time needed to reach the aimed wind speed.*

As Table 5-2 suggests, if the wind comes across a fallow crop field or body of water, you do not have to reach as high for greater wind speeds as you would in a forest or suburb, where obstructions tend to cause greater drag and turbulence. This fact is represented in the higher wind shear exponents for the rougher terrain. For wind turbines on boats or anchored to the seafloor, the reduced wind shear over water means shorter and less expensive towers: a boon to the offshore wind industry.[9]

We provide a list of common wind shear exponents in Table 5-2, which you may be able to use to estimate conditions on your site.[10]

TABLE 5-2   Common Values for the Wind Shear Exponent (α), Which Varies with Terrain

Common Terrains and the Wind Shear Exponent (α)	
Terrain	**Wind Shear Exponent α**
Ice	0.07
Snow on the flat ground	0.09
Calm seas	0.09
Coast with on-shore winds	0.11
Snow-covered crop stubble	0.12
Completely open terrain with a smooth surface (e.g., concrete runways in airports, mowed grass, etc.)	0.14
Short-grass prairie; open agricultural area without fences and hedgerows and very scattered buildings; only softly rounded hills	0.16
Row crops, tall-grass prairie; agricultural land with some houses and 8-meter (26-foot) tall sheltering hedgerows with a distance of approx. 1,250 meters	0.19
Hedges; agricultural land with some houses and 8-meter (26-foot) tall sheltering hedgerows with a distance of approx. 500 meters	0.21
Scattered trees and hedges	0.24
Hilly, mountainous terrain; agricultural land with many houses, shrubs, and plants, or 8-meter (26-foot) tall sheltering hedgerows with a distance of approx. 250 meters	0.25
Trees, hedges, a few buildings	0.29
Suburbs, villages, small towns, agricultural land with many or tall sheltering hedgerows, forests, and very rough and uneven terrain	0.31
Larger cities with tall buildings	0.39
Woodlands	0.43
Very large cities with tall buildings and skyscrapers	0.54

Relative to a reference height of 10 meters (33 feet)

Adapted from *Wind Turbine Technology* (ASME Press, 2009) edited by David A. Spera; *Windenergie: Theorie*, Anwendung Messung by Jens-Peter Molly; Paul Gipe's *Wind Energy Basics* (Chelsea Green Publishing, 2009); and European Commission's "Urban Wind Resource Assessment in the UK."

How Does This Work for You?

Wind shear exponent is a starting ground for evaluating what terrain you want in order to minimize your wind shear, and thus maximize your wind power density.

And the answer seems to be that ice is the new "Green"! You will look to build your home near Superman's Fortress of Solitude or Santa's workshop. Hello, frigid Alaska, Canada, and Siberia! Get your property now! Or, for the timid, or those of us living in the temperate zones, turn your lawn sprinklers on in the winter just before the big chill. This chart also tells you to avoid anything with trees and nearby buildings, including your home.

So what you can reasonably take from this is that all obstructions decrease wind power density, but effects are based on the *roughness* of the terrain and *distance* from the obstruction. Every meter or foot you raise the tower, the wind shear generally has less negative effects on the performance of your turbine. It would be best if it could be calculated for your site and at the height of your tower. For this we follow up with the power law equation.

Power Law Equation: Determining Wind Speed at Different Heights. Information you get from charts and even local wind speed data might be measured at heights different from where you might feasibly be able to install your turbine. But this will not stop us from getting what we want with what we know. If we know the wind shear exponent and the average wind speed at a particular height, we can calculate what the wind speed should be at any other elevation.

The approximate change of speed with height for different types of topography can be calculated from what is called the *power law equation*. The power law equation is given as:

$$v \, / \, vo = (h \, / \, ho)\alpha$$

where *v* is the unknown wind speed at our tower height *h* above ground, *vo* is the known wind speed from the local wind charts at a second height *ho*, and the exponent α is the wind shear exponent.

Example: Wind Velocity at a Site with Several Buildings

Tech Stuff *Let us consider a hypothetical site. We have the following information:*

1. *Your regional wind speed map shows an average wind speed of 5.4 m/s (12 mph) at 30 meters (99 feet) above ground level. This is the reference height.*
2. *We plan to install a turbine on a 10-meter (33-feet) tower (this is quite short, and is rarely optimal, but we're using it as an example here).*
3. *The area surrounding the turbine consists of high grass. From the wind shear exponent table, we see that high grass corresponds to a wind shear exponent of 0.18. Nice!*

The wind velocity at level 10 m *can be estimated by our formula as follows:*

Meters/second	Mph
$v = (h / ho)^{\alpha}\ vo$	$v = (h / ho)^{\alpha}\ vo$
$v = ((10\ m) / (30\ m))0.18$	$v = ((33\ ft) / (99\ ft))0.18$
(5.4 m/s) (1)	(12 mph) (1)
v = 4.4 m/s	v = 9.8 mph

So, using our formula, we see that we lose about 2 mph (1 m/s) of average wind speed, which amounts to some good energy loss, by going from a height of 99 feet (30 meters) down to 33 feet (10 meters). However, this speed is the cut-in speed for many small wind turbines, so we could still consider this a favorable site. However, if the same turbine were placed at the same height within the middle of Manhattan, near skyscrapers with a wind shear exponent of .54, it would have halved the predicted wind power density, meaning it would not likely meet the cut-in speed for it to produce any energy.

INTERCONNECT

Small Wind Goes to College

College of the Atlantic (COA) in picturesque Bar Harbor, Maine, is often cited as one of America's greenest learning institutions. The small school was founded in 1969 with a deep commitment to the environment, and the only major offered is something called "human ecology," which is essentially an interdisciplinary study of humankind's relation to, and impact on, the natural world. In 2007, College of the Atlantic also became the first school in the United States to be effectively carbon-neutral.

In mid-2009, College of the Atlantic received a grant to install a small wind turbine as part of a course called "A Practicum in Wind Power." Like Kevin, the college chose a Skystream 3.7, rated at 2.4 kW (Figure 5-4).

Bar Harbor's restrictive zoning prevented the school from installing a turbine at the main campus. However, College of the Atlantic also owns an organic teaching farm in a more rural part of Mount Desert Island, and that seemed like the perfect location to set up a wind collector, due to the ample ocean breezes. In fact, the small turbine is the island's first, even though Mount Desert is the sixth largest island in the contiguous United States.

Today, the turbine sits on a 40-foot tower, where it powers the office building at Beech Hill Farm. College of the Atlantic lecturer Anna Demeo, who led the wind power class, told us via e-mail that the system will soon be set up to post energy production data to a public website. She also told us that it has been an invaluable teaching aid. Students get hands-on training in renewable energy theory and practice, and the surrounding community is invited to come and ask questions.

FIGURE 5-4   The Skystream 3.7 wind turbine is a popular choice for homes and small applications, including for college classes. *Southwest Windpower/DOE/NREL.*

"There was a lot of enthusiasm around installing the turbine," Demeo added. "We use the performance numbers from it to teach students about calculating payback and [return on investment] ROI as well as understanding wind power calculations in the 'Physics and Math of Sustainable Energy' class at COA." Demeo said the turbine has also helped spur other renewable energy projects, helping the college secure grants to add solar panels, a wood pellet boiler, and additional insulation at the farm house.

As far as actual energy production, Demeo told us "the turbine's performance has been disappointing." She explained, "We knew going into the project that the height might be a problem, but the local ordinance restricted the tower to 40 feet. There was talk at the time that the ordinance might change, but then we found out that the conservation easement on the farm would not allow more than a 40-foot tower anyway." "Even with less than impressive power production, I would say that the wind turbine has been a great learning experience for all involved," Demeo said.

Ways of the Wind: Assessing Average Annual Wind Speed

The average wind speed needs to be measured at the proposed wind generator's hub height. It's crucial to get an accurate number, because the relationship of energy potential to wind speed is cubed, or eight to one. You'll likely have only a 3 to 6 mph average wind speed just above your rooftop, but you might have a 10 mph average on a 100-foot tower that's 40 feet above the tallest trees. The difference in energy potential between 5 mph ($5^3 = 125$) and 10 mph ($10^3 = 1,000$) is significant.

So how do you obtain this hub-height wind speed at your property? Unfortunately, it's not usually easy to come by.

There are multiple ways to approach this value. It is really about how particular you need to be on your site, how accurate you desire the numbers, your time constraints, and your budget. We can only recommend that if you are off the grid, you should seek the most accurate information about your site, because every watt counts when your batteries are starving to be charged up after using your Jacuzzi on the rooftop of your container home in Nicaragua.

Not for Your Eyes Only

Without access to objective data, using more subjective resources becomes better than nothing for the initial assessment. Never underestimate the ability of your senses to evaluate your area. Your eyes notice the wind damage to trees, you can feel the force of a breeze on your cheek, you can taste salty sand blown into your mouth from a beach nearby, you can smell the dump five miles away, or hear cries of children carried from the nearby school.

Also, your ears are good for listening to others. Interviewing long-time locals for anecdotal comments about the winds in the area is very low on the list of quality methods, but not entirely useless, since it may deter you from trying to capture wind energy at a poor site. Ask questions. At the least, it can indicate an approximate direction and wind density of prevailing winds.

In this section, let's take action to objectively know your property a little so you can get closer to deciding on that critical spot for your turbine.

Beaufort Wind Force Scale

Let's start outside with detecting instantaneous wind speed at your site. That could be wind gusts—after all, it is often from gusts that people get the crazy idea for having a wind turbine in the first place.

The *Beaufort Scale* (Table 5-3) is an attempt to describe wind speed ranges based on some general observations. It doesn't record prevailing wind direction or frequency of the speed, so it is nothing more than a litmus test that can suggest if further analysis is warranted.

If you don't have trees or water, you are not necessarily out of luck. This scale has been modified since it was developed in the 1850s, and certainly you should be able to make some observations on your site. We added one based on the reported wind speed from the local weather station as a marker. See if you can find it.

Now, you don't necessarily need to stand there like a monolith to detect wind. If it is a large property, take a slow walk on every inch, and bring a lighter. See if the gusts can blow it out as you stand in different locations. Even bring a ladder if you want to get a little higher than your hedges. Or have a barbeque and invite some friends, and then watch the pattern of the smoke or see if sand from the yard gets blown onto your cheeseburgers, or lifts your canopy tent off your property.

TABLE **5-3**   The Beaufort Scale

Wind Speed Range m/s (mph)	Wind Effect
< 0.3 (< 1)	Smoke rises vertically.
0.3–1.5 (1–3)	Direction of wind shown by smoke drift, but not by wind vanes.
1.6–3.4 (4–7)	Wind felt on face; leaves rustle; ordinary wind vane moved by wind.
3.4–5.4 (8–12)	Leaves and twigs in constant motion; wind extends a light flag.
5.5–7.9 (13–18)	Raises dust, loose paper; small branches are moved.
8.0–10.7 (19–24)	Fresh breeze; small trees in leaf begin to sway; branches of a moderate size move; crested wavelets form on inland waters.
10.8–13.8 (25–31)	Strong breeze; large branches in motion; whistling heard in power lines; umbrella use is difficult; empty plastic garbage cans tip over; canopy tent dismounts and blows onto neighbor's lawn.
13.9–17.1 (31–38)	Whole trees in motion; effort needed to walk against the wind; moderate amounts of airborne spray from breaking wave crests.
17.2–20.7 (39–46)	Some twigs broken from trees; cars veer on road; progress on foot is seriously impeded; breaking wave crests with considerable airborne spray and well-marked streaks of foam blown along wind direction.
20.8–24.4 (47–54)	Some branches break off trees, and some small trees blow over; construction/temporary signs and barricades blow over.

Don't laugh too hard—Kevin lost three canopy tents on his property from wind. (Remember: never underestimate the wind, and if you find the tents, please return them.)

If you live in a temperate climate zone, while having some winter fun note where the snow drifts are and how high they go. Kevin's house in Long Island frequently gets buried in snow on two sides, while the rest of the property—and fortunately, the solar panels—get covered only by an inch or two. That gives some indication of wind direction and power density, at least for that time period.

We have located a wind speed calculator for Android, so consider giving that a try.

Resource

Wind Speed Calculator, Android
www.nrel.gov/gis/mapsearch

Griggs-Putnam Index of Deformity

Now let's try to get a sense of your average wind speed. The secret may lie in your trees or in the trees of the nearby area.

The *Griggs-Putnam Index* is a way of estimating the prevailing wind speed by observing the growth patterns of trees (Figure 5-5). It is based on the concept that strong winds often deform trees and shrubs, typically in a pattern called *flagging*. Being able to read this can suggest a record of the local wind speeds during the vegetation's life.

The effect shows up best on coniferous evergreens, which don't generally lose their leaves, meaning their resistance to the wind flow remains relatively constant during the year. Deciduous trees shed their leaves in the winter and thus change the exposed area tremendously. If average wind speed is too high, deciduous trees simply cannot survive, since their leaves get too dried out or blown off. However, it isn't always true that a lack of deciduous trees is as an indication of high winds, since there can be other factors at play.

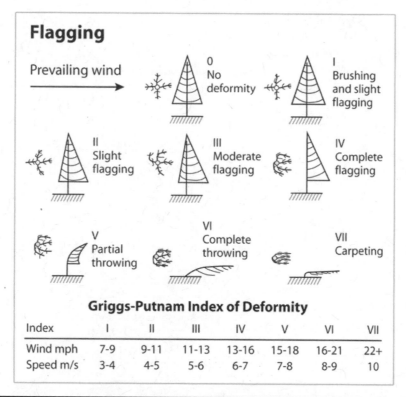

Index	I	II	III	IV	V	VI	VII
Wind mph	7-9	9-11	11-13	13-16	15-18	16-21	22+
Speed m/s	3-4	4-5	5-6	6-7	7-8	8-9	10

FIGURE 5-5   The Griggs-Putnam Index provides a way of estimating the prevailing wind speed by observing the growth patterns of trees. *Wind Power for Home & Business* by Paul Gipe (Chelsea Green, 1992).

Another point to consider is that absence of flagging does not necessarily mean an absence of wind, because flagging only occurs when winds blow consistently from one direction. Even Class 5 high-wind sites will not show flagging if the wind direction changes a lot. (The changes of wind direction can be best pictured in the section on wind roses.)

Data-Driven Pyramid of Wind Assessment

We can't stress this enough: use local data as much as possible. Possible sources include other wind energy users, weather bureaus, airports, newspaper historical weather data, and local weather enthusiasts who have their own monitoring stations and keep tabs on other local data. Mapping the data you have, plus characterizing similar sites, may give you some idea of your resource.

Figure 5-6 is a pyramid that shows you ways to assess your site from most general to most specific. We will look at all these methods:

1. Global wind maps
2. Regional maps
3. Climatic and meteorological data
4. Wind energy roses
5. Professional wind analysis services options
6. Anemometers

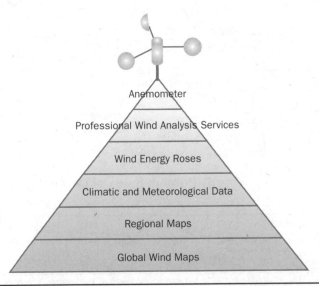

FIGURE 5-6   When considering your potential wind site, try to gather as much data as possible. Move from the most general to the most specific. *Kevin Shea.*

Global Wind Maps

First we start with the big picture, literally. Figure 5-7 shows a global average wind speed map, thanks to a team of scientists at NASA's Langley Research Center (LaRC). It is derived from ten years of geostationary Earth-orbiting satellite data. Although less accurate due to the indirect means of measurement at the time, satellite data are currently the only wind data with true global extent.[11]

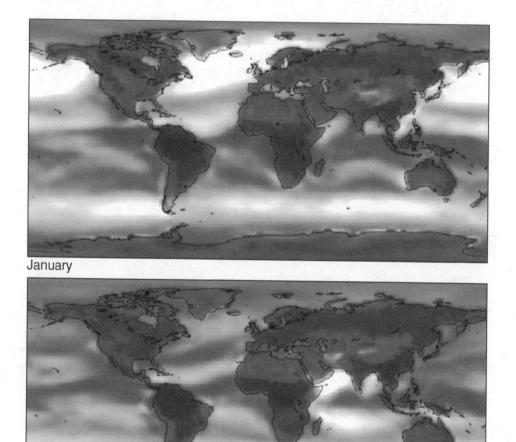

January

July

Wind Speed (Meters/Sec)

0   7   14

FIGURE 5-7   This global average wind speed map shows the largest wind resources (lightest areas) above the oceans and midcontinental plains. *NASA.*

Roughly 7 m/s (15.6 mph) and faster, shown as medium gray to white, are economically worth exploiting today, even in higher-cost locations for large wind farms. We see that the largest wind resources are above the oceans and midcontinental plains. This doesn't necessarily suggest that you should build an island out of recycled-plastic bottles, add a wind turbine, and then float off of Newfoundland. In many areas, especially on land, the 5 to 6 m/s (13.4 mph) areas, displayed in dark gray, are economically viable for small wind turbines.

In the United States, the South, with a few exceptions, has the lowest potential, but even there some coastal or hilltop areas may be feasible, even profitable, for small-scale wind. On the other hand, the American Midwest and the hurricane alley states have high-quality wind volume and good potential for both large-scale and small-scale wind installations. Parts of the Rocky Mountain states, West Coast, and Northeast also have favorable wind patterns, so that means small-scale wind may be possible near many densely populated cities (so long as you have no obstructions), as well as rural farms and exurbia.

The coastal oceans are of special interest because they have strong winds and are close to most of the world's population and electricity use. Internationally, many places that will benefit include much of Canada, Greenland, Iceland, Western Europe, southern Saudi Arabia, India, and southern edges of South America, Africa, and Australia.

One important fact that can be deduced from our global site assessment: Wind potential is somewhat regionalized.

Resources

NASA, Global Wind Map
1.usa.gov/visibleearth

Global Wind Map
bit.ly/windpowerworldmap

Regional Wind Maps

So let us discover more about how your region measures up. Governments, nonprofit groups, and commercial wind energy companies have all created large-scale wind maps. They may not have enough resolution to provide details on your specific site, but they may suggest first-cut feasibility for small wind turbines.

These maps are typically generated by data collected at various heights, often ranging from 50 meters (164 feet) to 100 meters (328 feet). Such data was most likely collected in open areas, and therefore must be adjusted significantly for wind shear in urban, highly forested, or built up areas.

Still, they may be able to give you a general idea of the local resource.

Resources

Full World Interactive Map
1.usa.gov/nrelmapsearch

World of Wind Atlases
bit.ly/WindAtlasDK

U.S. Regional and
State Map Links
1.usa.gov/regionalwindmaps

Wind Powering America
bit.ly/whereiswind

U.S. 50-Meter Wind Map
1.usa.gov/usawindmap

Wind Powering Maps
bit.ly/windpowermaps

The National Renewable Energy Laboratory website includes wind maps for 25 countries and all 50 U.S. states (Figures 5-8 and 5-9). Created in cooperation with the U.S. Geographic Information System (GIS) and the United Nations Environment

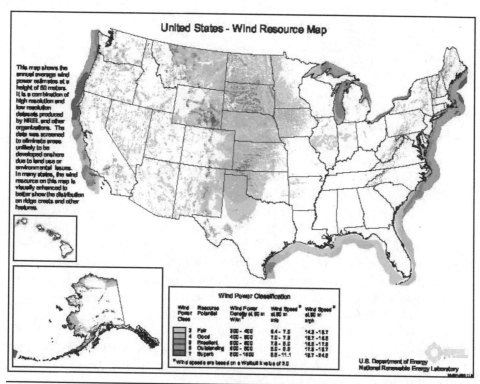

FIGURE 5-8   This regional wind resource map of the United States can give a general idea of large-scale trends, but it can't reliably predict what's going on at your site. *NREL.*

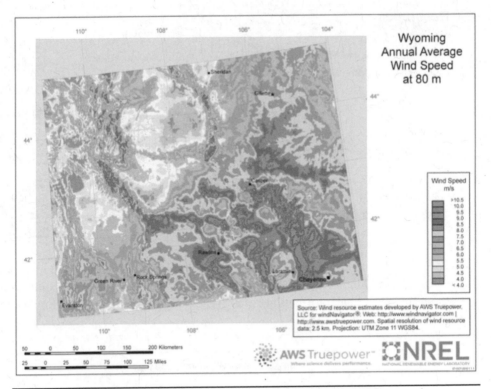

FIGURE 5-9   A closer view of the wind resource in the state of Wyoming. *NREL.*

Programme, these intuitive wind maps were produced at 50 meters elevation. They can help you quickly screen out less promising areas.

We include an image of the United Arab Emirates (Figure 5-10) to demonstrate that as we enter a post-oil era, there are feasible renewable resources not far from the existing petrol infrastructure. The same could be said about the state of Texas, which already has many wind farms.

If you are interested in installing in any other countries or looking into systematic and comprehensive collections of regional wind climates around the world, this is the place. The World of Wind Atlases site is established and maintained by members of the Wind Energy Division at Risø DTU in Roskilde, Denmark (a country that has long been at the forefront of wind energy). The site maintains major national and regional wind atlas studies and other comprehensive wind investigations and databases.

The most comprehensive free regional maps for the United States are currently located online at the Wind Powering America website. The maps are generated from average wind speed data from weather stations at 80 meters and 100 meters, as well as from previous projects.

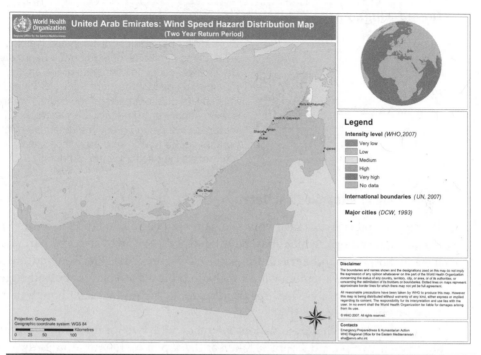

FIGURE 5-10 This regional map of the United Arab Emirates shows that even places long associated with fossil fuels may have wind resources to exploit. *World Health Organization*.

You are so lucky to be in the Midwest, particularly in Iowa and Oklahoma! Check out the wind power maps using the provided link to view down to the *county* wind maps. Yeehah!

Canada Wind Atlas

bit.ly/CaWindMaps

This map shows the Canadian territory sliced into 65 tiles, with calculated data on mean wind speed and energy for three different heights, as well as three geographic fields (roughness length, topography, land/water mask). We included a link to an interactive, color-coded map of numerous wind statistics for any one-square-kilometer area from as low as 10 meters (30 feet) in the province of Ontario. Cool, eh?

bit.ly/OntarioWindMaps

United Kingdom

Renewable UK
bit.ly/bweaWindMaps

Blokes! The Department of Trade and Industry wind speed database, offered on the Renewable UK website, contains estimates of the annual mean wind speed throughout the United Kingdom. This covers different parts of the country within a 1-km square at 10 meters, 25 meters, or 45 meters above ground level. The database even permits you to type in your ordinance survey grid reference to get specific sites. Cheers!

Climatic and Meteorological Data

To get more precise data near your site, you may be able to tap in to local weather observation stations, which may be established at nearby airports, participating schools, hospitals, and agricultural services (Figure 5-11). Precise wind measurements are made by anemometers, which are typically 10 meters above the ground level, often at open sites, and free from surrounding obstacles (although there are some exceptions).

In most cases, wind observations are made with two- or three-minute averages of three- or five-second samples, taken at the top of each hour around the clock.[12] So, these are not gusts. However, there are some locations where the observing program is limited to a shorter period than 24 hours.

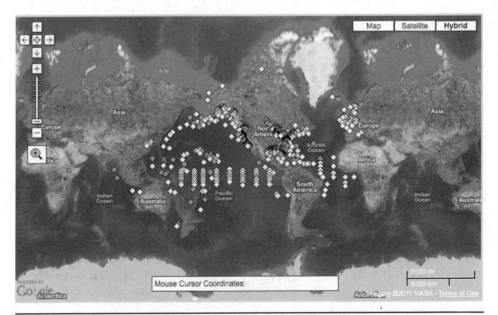

FIGURE 5-11   If you are on the coast, you may be able to take advantage of NOAA's buoy weather observation program, either by checking the website or dialing in for real-time data. If you live inland, there are comparable programs. *NASA/NOAA.*

True, many wind turbine installers include anemometers with their products, but this data is not always available to the agencies that generate the wind maps. If you know of any wind turbines in the local area, you might want to visit the owners to find out what wind data you can obtain from them. (You should also try to get a sense of how you like their kit.)

Resources

NOAA, Wind Energy Data
1.usa.gov/
NOAAWindData

NOAA, U.S. Wind History
1.usa.gov/
USwindhistory

Accuweather
accuweather.com

NOAA, Graphical
Forecasts
1.usa.gov/
NOAAgraphicalforecasts

Weather Underground
bit.ly/
WundergroundHistory

XCWeather
bit.ly/xcWeather

This book can provide a starter for your search into your community. The National Oceanic and Atmospheric Administration (NOAA) and Accuweather websites provide access to public wind data from as early as 1901, and you may be lucky enough to live quite near a test site (Table 5-4). You may be pleasantly surprised.

TABLE 5-4   The National Oceanic and Atmospheric Administration publishes current and historic data, all the way back to 1901 in some cases, for many sites. You may have to do some calculations and interpretations to make wind data relevant to your site. *NOAA.*

Florida		Jan	Feb	Mar	Apr	May	Jun	July	Aug	Sep	Oct	Nov	Dec	Ann
Daytona	DIR	N	N	ESE	ESE	ESE	E	ESE	ESE	ENE	ENE	NW	WNW	ESE
Beach	SPD	9	10	10	10	9	8	7	7	8	9	9	9	9
	PGU	52	58	77	49	69	67	67	68	48	56	47	43	77

Note that data from NOAA is usually in "raw" form, which means you will have to calculate average and maximum wind speeds, direction, and duration on your own. Once you find a weather station near you, the numbers will start to make sense. If you are spreadsheet savvy, you can enter the data and see immediately if your location has potential.

The table provides the prevailing wind directions (DIR) given in compass points, the mean wind speeds (SPD), peak gusts (PGU), and annual average wind speeds (ANN) in miles per hour (mph). One downside to the tables is that the data might not be current, unless you pay for the recent reports. This means recently installed obstructions, such as a skyscraper, or terrain modification, such as the reforestation of an area, may throw off that accuracy.

Weather Underground (www.wunderground.com), one of the earliest Internet weather websites, holds the world's largest network of personal weather stations (almost 19,000 in the United States and more than 13,000 across the rest of the world). Their latest online product is the WunderMap, an interactive weather map that allows users to choose from a number of different layers, including wind, that are plotted on top of a dynamic map interface (Figure 5-12).

In addition, the link takes you to an online tool that displays an easy-to-understand, linear weather graph of the average wind speed and direction. Simply select the range, and a chart filled with many data points fills the graph. At a glance, you can see the frequency of a particular wind speed or the wind speed distribution—and the numbers don't sugar-coat the truth the way a wind sales rep might. You

FIGURE 5-12   Weather Underground's WunderMap allows users to choose from a number of different layers, including wind. Weather Underground links tens of thousands of personal weather stations around the world. *Weather Underground, wunderground.com.*

can quickly step away knowing that areas close to your cabin have a 10- or 20-year history of varied speed ranges and the frequency of those ranges.

Finally, the UK-based XCWeather boasts current wind observation maps for Great Britain, Germany, France, Spain, and Italy, as well as the United States. The maps display average wind direction with colored regions, plus "gust" arrows, colored based on intensity (Figure 5-13). They even provide animation for wind activity up to the 24-hour range.

We are getting closer to your doorstep in terms of knowing what size power plant you will become. The obvious limit of this approach in grabbing climate data is that it only demonstrates the history from a weather station, which even if it is a block away, could be significantly skewed based on the differences between the topography of each site, the height of the anemometer used, and the validity of the observation methodology. In some cases, these weather stations were sited on airport tarmac just a few feet above the ground. Information on the particulars of a weather station is not always provided, which could be an issue, unless your site is the airport itself.

Wind Energy Roses

Although linear graphs of average wind speed are helpful, a better tool integrates wind speed *and* frequency of wind blowing from various directions. When a president's limo and vehicle train enter a town, various efforts are typically made ahead of time to ensure that nothing will impede their path of direction to their destination. The same should be done for the very important prevailing (VIP) wind. When

FIGURE 5-13   XCWeather's website shows wind observations for several European countries. The maps display average wind direction with colored regions, plus "gust" arrows. *XCWeather.*

you know *where* most of the wind energy is coming from, you can make sure that path to your wind turbine is free of obstructions.

Now, experiencing a few overturned flower pots on your property, watching a wind vane, or reading a passage in the farmer's almanac do *not* make for an empirical test. Instead, we use a rose.

A *wind energy rose* (often shortened as wind rose) is presented in circular format, with spokes or "arms" of differing lengths that give a summarized view of how wind speed and direction are typically distributed at a particular location (Figure 5-14). A wind rose can compactly display wind speed, direction, and frequency information for any selected period for which measurements have been taken. Annual wind roses represent wind measurements taken over a year, while seasonal wind roses depict data from a shorter period.

A wind rose is derived from a compass rose, and both usually have north pointing upward on the graphic. The subsequent arms represent the rest of the 16 primary directions, but in the case of the wind rose, they indicate the direction the wind is blowing *from* (e.g. N, NNE, NE, ENE, E, etc.). In a wind rose, the longer the arm, the more frequent the wind is from that direction. Summing the frequencies, shown at the tips of the directional arms, may not necessarily add up to 100 percent, however, since "calm" periods are not included.

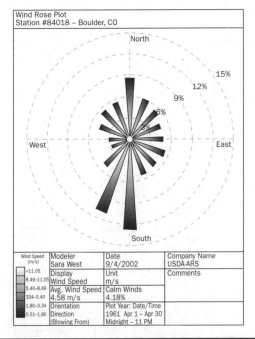

FIGURE 5-14   A wind rose compactly shows how wind speed and direction are distributed at a site. The longer the arm, the more frequent the wind from that direction. Color coding shows the amount of time the winds blew at that speed range. *USDA Natural Resources Conservation Service.*

The concentric circles drawn from the center of the wind rose represent the percent frequency of wind occurrences from each direction. Each circle represents a different frequency, from zero at the center to increasing frequencies at the outer circles. They help to show at a glance the frequencies within each wind speed category. The values of the concentric circles in this example are in 2-percent increments, but can differ between wind roses since they are scaled to best fit the observed data. In Australia, the percentage of calm conditions is often represented by the size of the center circle—the bigger the circle, the higher the frequency of calm conditions.[13]

The different colors (or in some cases the thickness) of each section of the arm can represent the wind speed range category. For example, a red section (in the black and white photo, it shows as the darkest grey, or is thicker than the slower range) could indicate a wind speed of 10 to 20 mph. As a result, each sectioned arm not only graphically represents the percentage of time that winds blow from a particular direction; it also displays the percentage of time that winds blow at certain speed ranges.

A quick analysis of our sample wind energy rose shows that significant wind energy arrives from the northwesterly region much of the time. In fact, the three arms around the northwest direction (NNW, WNW, and W) comprise roughly 25 percent of all hourly wind directions. This is quickly calculated by taking the sum of the frequencies (9.39 + 4.7 + 8.8 = 22.9%) of each of these directions. On the other hand, the winds from NNE to ESE are neither strong nor as frequent in any particular direction. Adding each percentage, they amount to 25 percent of reportable wind, yet most of the time the speed is too low to provide much energy.

These wind roses also provide details on promising speeds from different directions. Although the wind doesn't blow nearly as much from the southerly direction, the last section of the corresponding arms shows that the highest wind speed range, 10 to 20 mph, occurs as frequently as from northwesterly regions. With a glance, you can see at least two areas where the VIP might be arriving from, so it is best to site the turbine to exploit that wind energy.

To calculate the typical amount of time in *hours* that the wind blows from a particular direction and at certain speeds, just multiply the respective frequency by the appropriate amount of time. In our example, there are 520 days × 24 hours/day in the time period assessed, or 12,480 hours. From the wind rose we can calculate from the concentric circles and sectioned arms that winds blow from the northwest at speeds between 5 and 10 mph 4 percent of the time. This represents 0.04 × 12,480 = 499.2, or about 500 hours (21 days' worth) of typically having winds from the west northwest at these speeds at this specific location.

We couldn't readily find enough useful free sites that display wind roses outside of the United States, Australia, and Ontario. If you find one, send it to our Facebook pages. In the next section we can show you some websites that might offer services for your area.

Within the United States, the Idaho National Laboratory is gracious enough to provide a do-it-yourself, enter-the-location, and right-before-your-very-eyes calculate and display a wind power rose to show your family and friends, and then put on the refrigerator. Also included with the software is a program to help you create a wind rose from your own observations. It is designed to combine validated wind resource data from either climate data mentioned earlier or from data from your anemometer to calculate average wind speed and then display it in the wind rose map.

The bonus is that it includes sample power curves from some of the best-performing small wind turbines, with estimated annual energy production and capacity factor. After Chapter 8 on wind turbines, you can paste the power curve numbers, and you will know how your site performs with the machine of your choice before you put any money down (unless you got the data from a professional service or anemometer).

Resources

U.S. Wind Rose
1.usa.gov/USwindrose

Idaho National Laboratory (INL)
1.usa.gov/windrosegenerator

Australia
bit.ly/auswindrose

Ontario, Wind Rose
bit.ly/ONwindrose

Professional Wind Analysis Services

What was once used exclusively for large wind farm developments is now becoming increasingly accessible to community projects, and even individual installers, for a small fee. Getting a professional wind mapping assessment is much quicker, safer, and easier than using an anemometer, and it may bring similar results.

These assessments are typically online-based services that analyze existing data and modeling from climate history, nearby anemometers, and other sources, and they can be a fairly accurate way to estimate real-world values. Most services tend to use an interactive Google map that permits you to locate and place your wind turbine precisely on your property, with a suggested hub height. They include online tools providing annual mean wind speed, monthly mean wind speed, annual wind rose, and wind speed distribution for your specific location.

These services are made possible by a dataset created through applied weather prediction models, similar to national weather forecast services, as well as publicly available observations from meteorological towers worldwide.

A number of producers of wind analysis models have argued that the difference in annual mean wind speed data between their tools and actual on-site measurements is less than 1 m/s. Modeling is increasingly being proven effective, even in complex terrain and wind regimes. Not bad for an online instantaneous query. Some of them, like Seattle-based 3TIER (Figure 5-15), permit multiple locations or multiple hub heights (20 meters, 50 meters, 80 meters), so you can compare several prospects, or set one on your neighbor's property, so you can verify that you have more potential wind energy than the Joneses. Nice!

Oh, the price to assess your wind resource in moments, from your rocking chair? Could be as little as $150 for a full ten-page report with graphics and explanations. If you wish to make a career out of assessing other properties, you may consider leasing the software for $500 monthly, or buy it outright for several thousand dollars and earn revenue while educating your neighbors.

Resources

3TIER
bit.ly/3tiertool

WindStream
bit.ly/auWindstream

WindNavigator
www.windnavigator.com

Wind Atlas and Application Program
www.wasp.dk

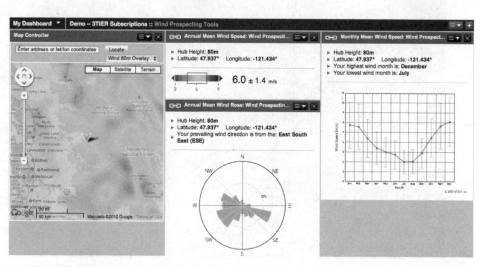

FIGURE 5-15 Services like 3TIER can provide quite detailed analysis of wind data and trends for specific sites, often for a modest fee. Remember that the results are going to be affected by your precise surface roughness and any obstacles. *3TIER Wind Prospecting Tools, 3TIER Inc.*

Anemometers

If you plan on investing significant money in wind power, it may be worth your while to actually measure the wind speed at your proposed hub height for a year or more. This is what commercial wind farmers do, and it's the most direct way to get a handle on the average wind speed on your specific site.

Note that we suggest doing a measurement for a year or more; ideally, you would measure for several years, or at least a year and a half. This is because winds can be quite variable, even year to year. Your area might be affected by El Niño or other periodic macro-level events, although ideally the research you would have already done would have provided some insight into those kinds of trends. It's most important to avoid relying on data from a single season, because winds tend to vary through the year, with highest speeds most often seen during the winter.

Wind speed is measured with a device called an *anemometer*, the most basic of which hasn't changed much since Dr. John Thomas Romney Robinson invented it in 1846, though there have been a few incremental improvements. Robinson's design had four hemispherical cups arranged at right angles. He observed that the prevailing wind direction would catch the cups, causing the device to rotate around a vertical shaft (making it a simple vertical axis wind turbine). By counting the number of revolutions per second, Robinson could work out the wind speed.

Today's anemometers usually have three cups instead of four, since they are more responsive (Figure 5-16). They may have a tag added to one cup as well in

FIGURE 5-16 The most direct way to measure wind is with an anemometer, like the rotating three-cup design pictured in the center. To assess a site, you should take measurements for a year or longer. *Brian Clark Howard.*

order to measure the wind direction by observing where that cup speeds up. There are many other ways to measure the wind, including with lasers, sound waves, a resistive wire, or even a ping pong ball attached to a string. Also common is the horizontal axis anemometer, sometimes called a "windmill" or "propeller" ane-mometer, or an *aerovane* (Figure 5-17). This device has a small rotor, like a tiny horizontal axis wind turbine, and a tail that looks like a weathervane. As a result, an aerovane can display both the wind speed and direction.

The biggest drawbacks to using anemometers are time and cost. Understandably, people are often not thrilled when they hear they may have to wait a year or more before they can start shopping for their new renewable-energy toys. Commercial anemometers are expensive, and weather stations are very expensive. Affordable ones can be found with a quick Google search for "average wind speed anemom-eter," with prices that range from $30 to $2,000, depending on the durability, qual-ity of results, application, and bells and whistles, not including the tower.

Homebrew methods of setting up an anemometer may drop the cost to between $1,000 and $5,000, depending on the height evaluated. A professional setup will likely cost three to five times that. Obviously, that's a lot of money to invest in something that may end up telling you not to get a wind system, which is why a lot of small installations are made without this step.

FIGURE 5-17 An aerovane is a type of "windmill" or "propeller" anemometer that measures both wind speed and direction. *Brian Clark Howard*.

Once operating, you should only need to check on the anemometer setup occasionally for guy wire integrity and to ensure that the cups are spinning when the wind is blowing. The cheap ones can have an accuracy of plus/minus 5 percent when wind reaches a speed above 3 mph, but to ensure that your equipment can handle the environment, check for feedback from previous customers. Not all anemometers are made for hostile environments. Also, get a clue of how high you want your final wind power tower to be before you buy the tower for the anemometer. And it is usually best to install a higher tower, because if you want to go lower, you can simply lower the anemometer. If you have to calculate for hub heights higher than your anemometer, you should expect increasingly inaccurate results.

What are some pointers on anemometers that you might not find in the installation guide? First, when is the best time of the year to start assessment with one? We can reasonably suggest a month when it is easy to install in order to mitigate risk and minimize effort. It is a different type of experience when you have to slosh through the Mississippi mud fields or pick through a foot of frozen earth to install anchors.

Second, leave it alone for the time you are measuring. You don't want to risk moving it a few feet and affecting the readings. And last, if for some reason in the first five months the system fails for more than a week but less than a month, you can average all the recorded months and get a fairly accurate average. However, if it fails within the sixth to the twelfth month of an 18-month period, we would suggest that you extend the testing to the same month(s) of the following year to ensure you are not losing data on a month that might have higher wind energy.

To aid in understanding how this works, simply recall what was discussed in the section on wind roses—namely that certain months typically display higher frequency and intensity of wind power energy. If Kevin were testing on his property on Long Island and the system failed during the cold winter winds of January, his main energy-producing month, he would monitor the following January to achieve a more accurate assessment before he flagged his conquest with confidence.

Bringing It All Together

In the end, it's best to use all available methods at your disposal to assess your wind resource, take an average of the results, and then round down. It's much better to predict lower wind speeds and be pleasantly surprised than to predict unrealistically high wind speeds and be disappointed.

Remember, the more time you spend and the more data you assess, the more accurate picture you'll have. Still, you may be surprised that many system owners we talked to didn't spend as much time as we'd recommend at this stage, and few used an anemometer. Some are disappointed with their system results, some are satisfied, and some are very satisfied, though it's also true that in many cases energy produced isn't the only value of the system. Only you can decide what balance of up-front cost versus risk is right for you.

We asked small wind installer, author, and trainer Dan Fink for advice on the matter. "The stuff available from NREL is really good, and can give you a decent ballpark figure," Fink told us. "Micro-siting makes a huge difference, much more than for solar power. Everyone knows that you don't put your solar panels in the shade. But a wind resource measured by someone 10 miles away could be completely different than what you have, because of ridges, valleys, trees, buildings, and so on. That's where people get confused if wind will work for them."

Tower Types and Site Evaluation

It's important to remember that when you are siting a wind turbine, you are siting for the tower as well. So, we provide here a brief section on the relationship of your site to your tower type.

As we mentioned earlier, there are three main types of towers: tilt-up towers, fixed guyed towers, and freestanding towers (Figure 5-18). Let's take a look at how your site might accommodate each one.

To understand how it affects your site decision, we need to first look at the general requirements for all towers. Your site needs at least a good-sized patch of open space, road access (unless you own or can afford to rent a helicopter), adequate soil, and—sometimes—level ground. We define open space here as any area that is not only void of obstacles that prevent wind, but also void of any geological feature that would prevent you from getting your turbine and trucks to the site. Rocky terrain, steep inclines, thick forests, ponds, and muddy roads are a few examples of features that impede passage of a wind turbine. It adds to your cost, and is some-

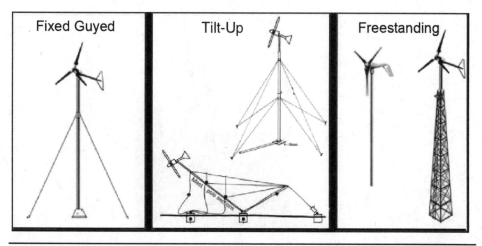

FIGURE 5-18   There are three main types of turbine towers: fixed guyed, tilt-up, and freestanding. *Kevin Shea.*

times the major factor in determining the feasibility of a project, at least until hover-crafting wind turbines are invented.

Once you are at the site, consider the soil type. If your ground is bedrock, making foundations may be relatively straightforward. However, if your alligator-infested swamplands have good wind, then creating a substantial base to hold your turbine in place may be your biggest challenge.

Fortunately, you can put a tower on an incline, as long as you can get through the installation. You can often use the slope to your advantage, and this can be cost saving.

Most geographical features that present challenges, with the exception of active volcanoes, are resolvable, but may require different approaches, and lead you to the tower type of choice.

Tilt-Up Towers

Rather self-explanatory, a tilt-up tower permits the turbine to tilt down to the ground via a hinged mechanism near the base (Figure 5-19). Tilt-up towers *require large open spaces* without obstructions, at least on one side, because you need to lower or raise it (hopefully only occasionally). They are usually made out of tubular steel.

Ensure either that you own the property that the tower is about to descend to or that you have legal permission to bring this heavy steel structure over your neighbor's white picket fence. This tower type also requires a totally clear cookie cut-out on the ground of a size at least the length of the tower and width of the rotor diameter.

FIGURE 5-19   Tilt-up towers offer easy maintenance, but they need a wide clearance around the tower. *Warren Gretz/DOE/NREL.*

Figure 5-20 Make sure your turbine's resting spot is over level ground. *Warren Gretz/DOE/NREL.*

If it requires manual assistance with a heavy-duty truck, ensure you have access to the area on the opposite site. You don't need to have level ground, and it can help if you tilt down the tower into the incline. Also, make sure the wind turbine's resting spot is not over a land depression like a gulley (or a swimming pool), because that would defeat the purpose of tilting the tower for easy access (Figure 5-20).

Fixed Guyed Towers

Like the tilt-up tower, your site doesn't need to be flat to accommodate a fixed guyed tower (Figure 5-21). Guyed towers are lightweight poles or latticed steel fixed in a concrete base, supported by at least three cables or guy wires, attached to the tower and mounted to anchors in the ground some distance from the tower.

As its name suggests, this tower type typically doesn't tilt down, but frequently it might have a hinged base, and for installation is raised using a small truck with a winch and a pivot bar, while the side guy wires assure the tower doesn't swing to the sides as it is placed in its final position. The cleared area required for install appears to be in the form of a Christmas tree, where its Christmas star (or angel) would be at the position of the rotor on the tower as it lies on the ground.

Although the installed tower itself might not take up as much of a ground footprint as the tilt-up tower during its lifetime, the paths of the tension wires always need to be clear of swaying and falling branches. In addition, care must be taken to avoid siting near roads or electric lines. In the case of a compromised cable (snap), it becomes evident that a recoiling and dropping line can cause its own havoc. Even for maintenance, the site should permit access for a bucket truck to raise and lower the turbine if major overhaul or blade replacement is required.

Figure 5-21   A Kansas family proudly shows off their Breezy 5.5 wind turbine (aka "the Sunflower"), held up with guy wires. *Timothy McCall/Prairie Turbines.*

Freestanding Towers

Freestanding towers are generally heavy steel structures, and they may have either a *lattice* or *monopole* design (Figure 5-22). They can maintain the same width through their entire height, like a typical flagpole, or they can start wide at the base and become narrower at the top, like the Eiffel Tower.

Freestanding towers have a relatively *small footprint*, but need a place to excavate an extensive hole for a deep, or wide and deep, concrete footer. If your site is on top of bedrock, using rock anchors may suffice, even for large towers. Not all stones are equal in terms of strength of anchors, so find out what you've got and talk to an expert about it.

Like the fixed guy tower, you will probably need road access and space for a crane to lift the heavy tower. A Toyota Hilux is not likely going to raise a freestanding tower; more likely it will catapult the car. After the tower is up, most maintenance will require climbing the tower, so make sure you have the gear ready.

Figure 5-22   This Northwind 100A turbine in Colorado is mounted on a 98-foot (30-meter), free-standing monopole tower. The 100-kW turbine has a rotor diameter of 62 feet (19 meters). *Lee Jay Fingersh/DOE/NREL.*

Consider Maintenance

You'll need to make scheduled maintenance on your wind generator at least once a year, and possibly more frequently. Are you willing to safely climb, or hire someone else to climb, the tower (Figure 5-23)? If so, then the vegetation could be allowed to grow around the tower. (Don't let it grow too high, or did you not read the wind shear section?) However, if you are thinking of hiring an expensive crane, then you need to ensure that you have an access road to the tower. The next decision is aesthetics.

Siting for Roof-Mount Micro-Turbines

Like many people, you may be excited over the prospect of having your very own small turbine *on* your home. And plenty of YouTube videos show some progress in the field, often with fancy computer-generated graphics and high production values. A roof seems like a shortcut to getting higher into the wind, and it seems like it could save space and the expense of a tower. For some, the roof may seem like the only option because they have so little space, especially in suburban or urban areas.

Figure 5-23 If you have a high tower, chances are good you are going to have to climb it at least once a year for maintenance, if not more often. Make sure you can ascend and descend safely, with proper safety gear. *Rosalie Bay Resort*.

As it turns out, rooftop wind turbines are quite controversial in the small wind industry, with many long-time pros condemning them en masse and claiming that they hurt the credibility of the whole field. We'll share the facts that we have available, and you can decide whether it is worth it to you.

In a nutshell, our current recommendation is to avoid roof mounts for the vast majority of situations, unless you want an expensive experiment or want to serve as some startup company's guinea pig. That's not to say roof-mounted turbines will never work, but there are currently very few examples of successful installations, and most of the technology is quite new and largely untested in the real world (wind tunnels can only tell you so much).

It's important to note that mounting a turbine on your roof can be dangerous and difficult, and it can endanger what is probably your most valuable possession besides you and your family—the building. Turbines do produce vibrations, which can be enough to cause structural damage and even roof failure.

If it is your home, you certainly don't want a heavy wind generator crashing into your bedroom while you sleep! In a less extreme scenario, the vibrations can be annoying, and can require expensive dampening equipment and a lot of tweaking. On a lighter note, if you can afford two turbines placed on either side of your bedroom window, it could be the most realistic flight simulator. "Honey, I think we are flying to your favorite country today!"

Probably the biggest drawback to roof mounts is that they tend to be in high-turbulence zones (Figure 5-24). The building itself tends to disrupt the wind, often in complicated ways, and many homes are close to trees, utility poles, and other buildings. Further, although your home may have two or three stories, that isn't really very high when it comes to wind turbines. Even if the wind weren't affected by the home, a roof mount would have a hard time getting a turbine into good flow.

Since rooftops create power-robbing turbulence, that disrupted air also tends to place wear and tear on the blades and the generator assembly, which can shorten a wind turbine's life. Turbine models that supposedly contain mechanisms for turbulence reduction most likely experience added vibration as a result, which would require more dampening, and that is additional cost to you.

To reduce rooftop turbulence, a wind turbine could theoretically be raised well above the roofline on a tower (Figure 5-25). But the engineering required for this can be formidable. Most likely, you're not going to want a big tower sticking way up off the top of your house, like something from a Monty Python sketch. That has a good chance of causing angry neighbors, and it sounds expensive.

Some supporters of roof turbines have pointed out that, in isolated areas, a pitched roof tends to cause wind to speed up. That is generally good, except for the presence of wind shear. Even turbines designed specifically to take advantage of the acceleration of wind across buildings have shown very poor energy perfor-

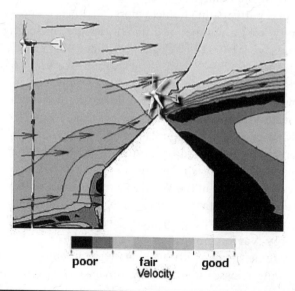

poor   fair   good
Velocity

FIGURE 5-24   Roofs tend to experience heavy turbulence, as do any obstacles to the wind. *Kevin Shea*.

FIGURE 5-25   A roof-mounted micro-turbine kicks in power when the sun doesn't shine on a farm in Cobden, Illinois. *Advanced Energy Solutions/DOE/NREL.*

mance in real situations, with payback periods literally in the thousands of years (more on this in Chapter 12).

In 2007, British researchers conducted a year-long study on the performance of 23 roof-mounted turbines at 30 sites in the United Kingdom. Called the Warwick Microwind Trial, the project was carried out by a consulting firm called Encraft, with funding by the Pilkington Energy Efficiency Trust, Warwick District Council, BRE Trust, and participating homeowners.

The study used a number of different designs, including 14 Ampair 600 systems, 5 Windsave WS1000s, 1 Eclectic Stealthgen 400, 1 Swift 1.5kW, and 3 Zephyr AirDolphin systems (23 + 1 spare). The locations were spread across several towns and included three tower blocks, a number of gable end installations in different contexts, a steel-framed building, and a timber-framed "ecohouse."

If you want to view the detailed results of the Warwick trials yourself, browse to www.warwickwindtrials.org.uk. But the "short answer" is that many small wind experts interpret the study's results as reinforcing their beliefs that rooftops are poor places to harvest wind energy.

Every turbine tested in the Warwick study produced less than 500 watt-hours (½ kWh) per day. To put that in perspective, the average U.S. home uses 30 kWh per day. That means, at best, one of the rooftop turbines could provide 1/60th of a home's energy. Not too impressive.

Further, some of the study turbines were unable to generate more than the "parasitic" power losses used by their inverters, meaning they consumed more electricity than they generated. This was largely due to very low wind speeds. At some sites, the amount of energy that had been predicted to be produced was more than 1,000 percent higher than what was actually generated.[14] A number of the study's roof turbines were turned off during the course of the trials because of technical problems or complaints over noise.[15]

In fact, the study authors noted a number of problems, including long waiting periods for equipment and difficulty in sourcing all the needed supplies. Remember when we mentioned that the rooftop turbine sector is largely unproven and immature? The researchers also concluded that having a good wind resource is the single most important factor that determines whether a wind energy project is going to be successful. (That should sound familiar.)

Perhaps not surprisingly, manufacturers of rooftop wind turbines are critical of the Warwick study. Some argue that the scientists used test sites in an area that historically has relatively low winds, and that were really not suitable for any type of wind-collecting device. Some said the trials failed to take into account their particular models and high-tech designs, although it's worth noting that several manufacturers were represented.

Small-wind owner, installer, and author Ian Woofenden pulls no punches when it comes to roof mounts. "They might make a profit for the companies that sell them, but if your goal is renewable energy, roof mounts don't work," Woofenden told us in a phone interview.

"The best use of roof space is solar panels," he added. "For wind energy, there is nothing that will work at that level. There is low wind and high turbulence, a one-two punch knockout."

Woofenden calls roof-mounted turbines a "crazy waste of resources." He said people often forget about the cubic relationship of energy and wind speed, and argued, "It's about kWh." He added, "If you move 20 feet above a roof you get a significantly larger resource, though that is still in a turbulence zone. If people don't understand energy then they're ripe for being taken advantage of."

Another small-wind expert, Mike Bergey, president of Bergey Windpower Co., said, "The plain facts are that roofs weren't designed for wind turbines and the wind over roofs is hard on a wind turbine. It's the hardest and worst place to put a wind turbine."

Southwest Windpower Vice President Andy Kruse was more sanguine: "There are conditions where a rooftop-mounted wind generator can work. However, like solar, an improperly mounted wind generator can be worthless or worse, cause damage to the roof. A wind generator must always be mounted on the upwind side of the structure and if at all possible more than 10 feet above the roof. Structural engineering must be completed on any building where a wind generator above 1kw is installed."[16]

For his part, Stephen Crosher of Quietrevolution, a manufacturer of small vertical axis wind turbines based in the United Kingdom, told us via e-mail that his products can work well on a roof if the right conditions are in place. "The building needs to be several stories higher than the surrounding buildings," Crosher wrote.

"The mast used on roof tops should be at least half as high as the depth of the building, e.g., if the building in plan is 8 m[eters] × 8 m[eters] the mast should be at least 4 m[meter] tall. Quietrevolution advises a six-meter mast as this is appropriate for most roof tops."

Crosher added that underlying topography needs to be considered. "A tall building in a dip is unlikely to work well, while a smaller building on a hill top is likely to perform," he explained. "Lastly, the choice of site should not be shadowed from the more dominant wind directions. In the UK this is all directions from south through west and the north east."

As mentioned previously, a small number of companies are working to address the difficulty of siting building-mounted turbines by using software and other computer models to predict wind resources more accurately in these environments. They are also using exotic designs or, frequently, vertical axis wind turbines (VAWTs), both of which we cover in detail in a later chapter (see Figure 5-26). Some of these companies have won high-profile design awards for their efforts, and roof-mounted turbines enjoy considerable media buzz and attention from the public.

Power Up! *"Breakthroughs" are frequently announced by makers of rooftop turbines, many of whom also boast that their devices have "very low cut-in speeds" or are "better at turbulence," although you should already be skeptical about the importance of these claims (we'll address them at length when we get to VAWTs). It's perhaps most important to note that actual installations of roof-mounted and VAWTs can often be counted in the tens or maybe hundreds. In comparison, more "traditional" propeller-style small wind turbines have been installed by the hundreds of thousands, and they have decades of proven track records.*

If you have a lot of disposable income, there's maybe nothing wrong with ordering the latest and greatest roof turbine and impressing all your friends and neighbors, although be aware that some people may accuse you of discrediting the small wind industry and turning away potential customers for them if you have problems or poor results.

The bottom line is that if you are on a budget and care about producing kilowatt-hours, putting a turbine on your roof these days is at best a considerable risk. Ten years from now, who knows?

Resources

Domestic Roof-Mounted
Wind Turbines Study
bit.ly/midwalesmicrowind

Urban Micro Turbine Article
cnet.co/CNETurbanmicro

FIGURE 5-26 This Savonius VAWT sits on a roof at Ben-Gurion University of the Negev in Israel. It's a demonstration project that produces little energy. *Brian Clark Howard*.

Social Friction Factor

Just when you get a site that you know would be perfect for lassoing wind energy and riding that horse into the sunset, you want to share your excitement with the neighbors and tell everyone to put one on their property.

Take off your thinking cap and put on some comfortable clothing. You might also want to grab a tissue box. We need to balance the science we've covered so far with a short section on belief and emotions. In Kevin's neighborhood, he tends to loosely call any resistance from the public to a wind project the Not In My Back Yard (NIMBY) policy. The fact is:

Your neighbors, whether the turbine can be seen, heard, and/or felt, or the nearby public, have an interest in the safety of the birds, bats, fish, or howler monkeys around you, as well as other factors you might not have even thought of, so all may have a stake in the success of your project. In Chapter 7 we shall discuss in detail winning over your neighbors, overcoming zoning hurdles, and working with the power authority and your government. It can seem overwhelming, but it can be done.

Here, we just alert you to the fact that your location may have an effect on your neighbors' emotional comfort or property value (Figure 5-27). Any site evaluation should include what the neighbors—Status-Quo Stan, Hyper-hearing Henderson, Viewshed Vicky, Seizure-stricken Sally, Paulie the Politician, Bird and Bat-loving Bobby—think of your project.

What? Outrageous you say! Just wait for the details in the ensuing chapters. Some jurisdictions do not yet have sufficient legislation specific to small wind turbines, and some that do issue permits have mechanisms that delay the process. Currently, it is still common for building ordinances to require consent from neighbors within a certain distance from the turbine, or a majority vote from town boards. In addition, they may require extensive and expensive testing, similar to what's applied to the big guns: noise pollution, shadow effect, avian mortality, epileptic seizure or syndrome effects, and so on.

Your only recourse if the permit ordinance doesn't address your innovative wind collector is to assess the situation for social temperature. Unlike the scientific method we use to evaluate the site for wind energy, the hot air chatter and cold stares from the community are not easily measurable. Your recourse is usually the old-fashioned way: tell your neighbors and town officials your plans and see if they applaud, accept, or abhor your ideas. And when your survey, formal or informal, results in a very hot-to-handle situation, you need to decide if the challenge is worth it.

However, the good news is things are changing. Small wind turbine performance and safety standards have been established from various wind energy associations. The certification process requires wind power systems to adhere to sound limits, strict installation protocols, and built-in safety components. Such quality control measures provide town officials substantial assurances, such that there is

Figure 5-27   This Bergey 10 kW Excel sits atop a freestanding lattice tower in Kansas. *Warren Gretz/DOE/NREL.*

no longer such a foreboding case-by-case challenge of assessment. Of course, if you are within a progressive jurisdiction where you can already get a wind turbine permit, you will be able to bypass some of the hurdles of paperwork and the Facebook surveys. If you get defriended because of your insistence on a wind turbine, let us know.

Summary

Before you start shopping for turbines, it's critical to make sure one will work on your site. Above all else, do you have enough wind to make the project worthwhile?

Luckily, a number of tools are available to help you answer that question. You can check out global and local wind maps, and examine data from nearby weather stations or other wind energy producers. You can talk to neighbors and even get clues from the landscape, including how trees may be deformed by the wind.

We recommend ordering a custom wind assessment from a reputable company online. For a modest fee, you can often get quite detailed estimates of the wind resource on your property. If you're really serious about the technology, you may also want to get an anemometer and take your own measurements. Unfortunately, setting one up can cost a good amount of money, and you really should do the monitoring for at least a year and a half.

Once you're sure you've got enough wind, you also need to make sure your site can support a tower, both for installation and maintenance. And finally, what will the neighbors think? Get them on board early, or risk stress and headaches later.

To help you start thinking about these decisions, we put together Table 5-5.

TABLE 5-5   Design Approach to Small Wind Turbine System

Approach	Cost	Advantages	Disadvantages
Certified manufactured system	$$$$	• Tested and certified system • Detailed installation manual • Zoning approval easier • Warranty • Insurable • Established support	• May require certified installer
Manufactured turbine/tower; supplier-approved balance of system (BOS)	$$$	• Tested and certified system • Detailed installation manual • Zoning approval easier • Warranty • Insurable • Established support	• Possible hassle to purchase from other suppliers

(continued)

TABLE 5-5 Design Approach to Small Wind Turbine System *(continued)*

Approach	Cost	Advantages	Disadvantages
Manufactured turbine; Supplier-approved tower kit/BOS	$$	• Tested and certified turbine • Detailed installation manual • Turbine warranty • Insurable • Established support	• Moderate risk • Warranty void if tower kit is not manufacturer-approved
DIY turbine certified tower kit/BOS	$$	• Stimulating work • Custom-design	• High risk of failure/injury • More science and math required • Zoning approval issues • Insurance issues • No warranty • No formal tech support network
DIY turbine and DIY tower; BOS from suppliers	$	• Stimulating work • Custom design • Innovation possible	• Higher risk of failure/injury • More science and math required • Zoning approval issues • Insurance issues • No warranty • No formal tech support network

CHAPTER 6

How to Finance
Small Wind Power

The two most abundant forms of power on Earth are solar and wind, and they're getting cheaper and cheaper....

—Ed Begley, Jr.

Overview

- Wind power is a long-term investment.
- Wind speed is the most important factor in determining if a project is financially feasible.
- Determining feasibility is not simple; there are more than a dozen factors that could affect the bottom line.
- No size for a durable turbine is too small to earn a return eventually.
- As you increase the size, the cost goes up, but its feasibility rises.
- Add a healthy "stuff happens" factor to installed cost estimates.

Well, you made it through a lot of considerations when it comes to small wind turbines, and hopefully you have determined that you have enough wind energy at your particular location. Congratulations! You may feel ready to click that "pay now" button to buy that shiny new turbine. But before you do that, you want to know financially what you are getting into.

Let's place all the items on the table, and go over all the pieces that affect your *final price.* Then, let's examine a few more questions that can ground you in your decision.

This chapter discusses the long-term costs of harnessing your wind, and includes the various agencies and programs available to assist home and farm owners in getting financing for their systems. We'll also cover tax credits and incentives from federal, state, and local governments, as well as incentives from utilities. You may be able to save two-thirds of your bill even before the first revolution of your turbine blades. In your reading process, we think this chapter is right where it belongs, just before the chapter on selecting your system.

Let's be honest with ourselves by taking a quiz. If you answer NO to any of these questions, then you may not be a candidate for a small wind system (Figure 6-1), and this book should be used as a resource for community wind projects, to make town code changes, or to become the wind geek at the next social:

- Does your site location have high average wind speed? (12 mph (5.4 m/s) or greater preferred)
- Does your governing body or power utility have any wind incentives?
- Does your power authority have utility rates at or above the U.S. equivalent of $0.10/kilowatt-hour?

Now, you may also still read on if you are like many people, in that we make very few decisions based primarily on financial considerations. If you don't believe us, think about some other big decisions you have made, such as your home, vehicle, or education. Further, did you ask what the payback is on your tool shop, roof, pool, refrigerator, or boat? Sometimes, energy decisions are not so different.

Putting it bluntly, small wind author, advocate, and trainer Ian Woofenden told us, "Most things we buy have no return at all, and in fact almost everything has a negative return. I think the payback return question is sort of out of line. To those who ask me about it for wind turbines, I say, 'cite me one thing you bought that you thought about ROI [return on investment].' We subject renewables to that but not our other choices."

FIGURE 6-1   Bison roam in front of a 10 kW Bergey wind turbine in East Glacier, Montana. *Northwest Seed/DOE/NREL.*

When pressed to estimate a reasonable payback period for a quality small wind system, Woofenden said, "It depends on the resource, utility rate, incentives, and reliability, but I think between 10 and 20 years is a reasonable range."

Let's take a closer look.

Cost Breakdowns for a Small Wind System

Price ranges for small wind turbines—and even for a single model—vary widely due to the numerous factors affecting installation (not including marketing hype). Still, costs for a well-sited small turbine tend to gravitate between:

- $3 to $6 per watt from the estimated power rating
  or
- $3,000 to $6,000 for every kilowatt of generating capacity
  or
- $0.15 to $0.20 per kilowatt-hour of annual energy output for the expected life of the entire system

For example, a Southwest Windpower Skystream 3.7, rated at 2,400 watts, would be calculated at a market price of $6,200. That is just the turbine.

However, as discussed, the manufacturer's power ratings are not always trustworthy. Instead, if we calculated for annual energy output (AEO) at average wind speed of 12 mph (5.4 m/s), we're talking about 3,800 kWh/year. That's roughly one-third of the average annual energy use of a typical American home (11,000 kWh). We'd have the full system for the modest price range of $11,400 to $15,200.

Electricity Costs Last a Lifetime

Have you ever been, or are you, in debt, in which you pay every day for something that you use every day? If the answer is no, think again. When you buy any electrical device, the cost doesn't stop at the cash register. It is all about perspective, so let's consider something that nearly every home has—a television.

Let's take two 55-inch high-definition (HD) televisions with comparable features and the same ticket prices ($1,600). However, one is plasma, and the other is light-emitting diode (LED). Stamps on the sides of the units indicate the number of watts consumed per hour, the average hours used nationally, and the approximate annual cost based on rates in the local region.

As you can see from Table 6-1, the LED TV uses only 100 watts an hour, for an average annual energy cost of $22. The plasma TV, which uses 324 watts an hour, requires $71 a year. So over the course of ten years, the total price of the LED TV is $1,820, while the average cost of the plasma TV is $2,310.

TABLE 6-1   Lifetime Cost Comparison of Two New Televisions[1]

Television	Technology	Price	Watts	Annual Energy Cost (avg 5 hrs/day)	Lifetime Cost* (10 years)
Toshiba Regza 55UL605U	LED	$1,600	100	$22	$1,820
Panasonic TC-P54G10	Plasma	$1,600	324	$71	$2,310

*For households that pay somewhere near the average retail cost for energy—12 cents per kilowatt per hour in 2010—and that watch near the average amount of TV—about 5 hours per day

If items were sold with the lifetime energy cost included, so long as the features weren't appreciably different, people would probably purchase the energy-competitive price. In fact, if lifetime costs were placed on all sorts of things that require energy, people would have a better idea of the true cost of their purchases.

To give another example, more than 90 percent of the total cost of owning an incandescent light bulb is due to the energy costs to keep it lit, not the initial purchase price of the bulb. If more consumers were aware of this fact, we would see faster adoption of more efficient technologies like compact fluorescent lamps (CFLs) and LEDs.

Permit us to suggest that wind power is a paradigm shift with energy, because the lifetime cost is up-front. Imagine that your power authority moved its conventional utility to every home to produce energy. Imagine having to pay for the fuel and having to deal with the noise, dirt, and air pollution of a coal or gas generator in your garage (in fact, many of our grandparents had to do just that). Then imagine discovering wind power. Wow, no fuel cost!

What Is It Worth to You?

Is your goal to save the Earth, save money, generate money, find peace of mind, or something else? You need to know this before you think about financing your system.

Kevin remembers watching the wind blow furniture across his lawn and wondering if he could tap that free resource to eliminate his electric bill. Here are his answers to the aforementioned questions: 1) The wind maps (Chapter 4) stated that he was getting high average wind speed; 2) he had, and still has, relatively high utility rates; 3) he was too early to receive financial incentives for wind. Kevin saw this attractive new technology called the Skystream wind turbine, he knew a certified installer, and didn't delay the thinking process further. Sadly, he didn't look at the rest of the equation—finances—until more recently. Why not? Because he determined it was worth it for the asking price. And it just so happens that every windy day on Long Island, he is pleased with his decision to cough up the up-front cost for the free resource he receives now.

But you need to ask yourself what you are willing to pay.

For example, Surfin' Sal has a boat in Nicaragua and no power source. Is it worth $3,000 to him to be assured of energy on his boat without the need to dock frequently?

Is it worth it to flower growers in Holland who pay exorbitant electric prices to spend $1 million to eliminate an equivalent of $145,000 in annual expenditures?

Farmer Fanny in Toronto has virtually no electric bills, but hears that the government is going to pay above retail price for wind energy. The bigger the turbine, the more energy, meaning a bigger paycheck.

Then there is Joe Green, who just finished defacing a gas station sign because he is angry about oil spills, and is eager to start a distributed wind project within his co-op development. He is leasing an area at the clubhouse to install a 100 kW system in return for waived fees and a legacy.

Then there is Resourceful Rahim in Nairobi. He is unemployed and rides his bike five miles to buy gas for his community's 3 kW generator. He hasn't any electric bill, because he has no public utility service to his home. He is willing to trek through unsafe regions on his way to the dump to find parts to build a turbine to produce light at night to study.

So, know what you are willing to sacrifice. We placed an unfinished sentence with tagged words below it in Table 6-2. You can fill in the check box for what applies to you. And the cost that you are willing to pay for it.

TABLE 6-2   What a Wind Turbine System Means to Me

· Energy when I want it	· Technology lust	· Money earned
· Integration with PV system	· Independence	· Money saved
· Cool lawn ornament	· Personal legacy	· Carbon offset
· Year-round sailing	· School science experiment	· Status symbol
· Night lighting for education	· LEED certification	· Security light for village
· An annoyance to my neighbors	· Marketing	· An air-conditioned cabin
· Improve viewshed	· Knowledge building	· An efficient pool heater
· Year-round irrigation	· Skill-building	

And is worth investing: $_____
(maximum money amount)

If you aren't sure what your limit is, that is okay. Once we get through this chapter, you will have a ballpark figure. The rest of the chapter also covers how you can shrink the price tag for small wind (Figure 6-2).

FIGURE 6-2   In rural Douglas County, Colorado, one of two adjacent 450-watt Winco wind turbines is tilted down to the ground for maintenance. *Jim Green/DOE/NREL.*

Factors That Affect Final Cost

Cost and return periods can vary due to the following factors, ranked in approximate order of importance.

Average Wind Speed

What goes up (wind speed) must come down (cost per kilowatt-hours). As stated previously, wind speed is the biggest variable determining the financial feasibility of a wind turbine. A 10 percent increase in average wind speed results in a 33 percent increase in power available to the wind turbine's rotor. Chapter 4 on wind assessment should still be fresh in your head.

Cost of Electricity Offset

The second biggest variable is your local energy cost per kilowatt-hour. The higher the cost of electricity, the more favorable the offset. Installations tend to be most cost-effective in regions where the cost of utility-provided electricity exceeds $0.10 per kWh or utility bills average at least $150 monthly. With utility costs less than that, you're talking about intangible returns like "awesome turbine dude!" or sizing your turbine smaller for power-specific applications such as water pumps or recreational vehicle lights and appliances.

Government/Utility Incentives and Net-Metering Policies

Net metering means that your utility credits you for the electricity generated and banked by the grid for later retrieval. If your area provides net metering, the price the utility pays for your surplus energy varies from the retail price (sweet!) to wholesale price (ho-hum), to, in some places, nothing! Some governments and utilities also offer rebate checks or credits.

In lieu of initial payments for the installation of the equipment, many countries have programs called feed-in tariffs that pay the homeowner at a higher-than-average electric rate for their clean energy production for a set time framework. This way, the return on investment is accelerated based on production. Your strategy for sizing your wind system will depend to a great extent on the way your utility treats net excess generation. Ideally, you want the meter to spin backward only if the compensation is adequate to improve your return on investment.

Rising Cost of Real Prices of Electricity

Rising electric prices particularly affect the financials of on-the-grid systems. Note that the future value of your investment (and money) is impacted by the inflation rate and the rising cost of electricity. So-called real prices of electricity take into account the fluctuation of the inflation rate during a period of time.

Once your system is paid for, what is the cost for wind? Wind is FREE, FREE, FREE! On the other hand, most analysts expect the price of oil, raw materials, and electricity to soar upward. If that happens, so will the worth of your energy, and your return on investment. You'll pass the whole gang of wind naysayers, and you'll soon take the lead.

"Except when you don't. Because, sometimes, you won't," Dr. Seuss wrote in his book, *Oh! The Places You'll Go* (Random House Children's Books, 1990). The beloved author was balancing the hopeful future of a young person with unpredictable bumps in the road. In a similar way, know that no predictions of the price of electricity are assured, because prices are at best highly volatile. Analysts often publish timelines showing the rising of utility rates over time, which for renewable energy owners would seem oddly pleasing. Yet, that would be one-dimensional. We would need to compare the rise of the price of electricity to the inflation rate—the rise in the general level of prices of goods and services in an economy over a period of time.[2]

For example, when your bank is offering you less than 1 percent interest annual yield on your savings account, while the nation's inflation rate creeps up to 2 percent average for the year, you can easily calculate that money left in that savings account is losing 1 percent value. And inflation is continuously fluctuating. If a linear graph of the annual average inflation rate for the last ten years is superimposed over the presumed rate of electricity increase, you will find that there are times when "real" rates actually decrease, eroding any of the potential value of

FIGURE 6-3   This Breezy 5.5 by Prairie Turbines works well in the breezes of Managua, Nicaragua. *Timothy McCall/Prairie Turbines.*

each kilowatt-hour produced. Depending on what country you are in, the difference can be dramatic. A wind turbine in Zimbabwe would not provide favorable financial numbers against inflation, which is 12,563 percent as of this writing. In contrast, most of North and South America (Figure 6-3), Europe, and Australia have been keeping inflation to less than 3 percent.[3]

All of these impact the present and future value of money and the value of your investment relative to other investment options, such as a savings account, the stock market or mutual funds, or even an addition to your house.

Application

Some government incentives are contingent on your application, whether residential, agricultural, commercial, industrial, or municipal.[4] Residential homeowners may have national governmental tax incentives for wind energy. For example, in the United States, you are likely to see a savings of 30 cents in tax deductions for every dollar spent. Further, the cost of financing a wind system through a home mortgage is currently tax-deductible. Meanwhile, a wind system purchased by a farm or small business might be eligible to have its capital expense depreciated over a number of years, and repair costs would also be deductible.

Cost of Equipment

Quality equipment costs. Of course, those costs vary by manufacturer and size. But here's a rule of thumb: the bigger the swept area, the higher the cost. Yet, the bigger the size, the lower the cost incrementals *per* additional foot length of the blade and/or the height of the tower.

Recently, there has been one other tidbit. For manufacturers that conduct American Wind Energy Association (AWEA's) new standards on performance, the additional costs for extensive testing will likely be carried over to the consumer.

However, we are currently seeing a dramatic decrease in the cost of wind energy system components. This price drop is partially due to more efficient manufacturing as well as overproduction during recent years, as the world slumped into an economic recession. Wind energy system surpluses allow the consumer to take advantage of dramatic price decreases.

Monitoring and Maintenance

Monitoring and maintaining your equipment is the only option for long-term investment, and this includes wind energy. Of course, this doesn't mean you have to stand by the machine all the time, watching it spin. Wireless monitoring is becoming commonplace, with associated costs ranging from $0.01 to $0.05 per kWh. Other calculation methods place total operation and maintenance (O & M) costs at roughly 1 percent of the retail cost of an installation, accrued annually.[5]

With more integrated, advanced small turbines and standardized tower practices, there is a likelihood that such costs will be reduced. But if maintenance is shortchanged any time during the 20- to 30-year lifetime, expect the likelihood of depreciating return.

Installation Cost

An installation service cost can be as high as at least a third of the price of your system (Table 6-3). Yet, this is where some people make an error of judgment. This is a 20- to 30-year investment in something that lives in the harsh elements 24/7. The price cannot be bargained down much without affecting quality of service.

For a truly micro-turbine this will not have as much effect, but when you are still negotiating the price on the lowest bid installer for a 33-meter do-it-yourself (DIY) tower kit, you are putting its success more on faith. The funds saved up-front are no consolation when you are forced to hire a lift so that you can scrape the rust off the high rungs of the tower because your more affordable subcontractor doesn't return your calls.

Also, the harsher your conditions, the more the total cost of installation will be. If you have 60 mph winds whipping gravel in a dramatic plume above your head, you should either relocate or invest in a bulletproof, "heavy metal" wind turbine. Invest well to produce well for the long term.

TABLE 6-3   Installation Costs for a Sample Small Wind System[6]

Core equipment (turbine, tower, inverter)	$26,400
Labor	4,000
Materials (concrete, steel, wire, etc.)	3,000
Services (permits, excavation, shipping)	2,400
Equipment rental (compactor, crane)	2,000
Total	$37,800

Insurance

To make insuring your wind energy system cost-effective, you will likely want to add it to your homeowner's, farm, or business policy. Homeowner's insurance specifically protects property owners from casualty (losses or damage to the home and its contents) and liability (damages to other people or property).

You'll want coverage for damage and repair due to "acts of God." The amount varies based on the size of the tower and proximity to other structures. There is more on this later in this chapter to help you make a wise choice, but for now, it is good to know that insurance on a galvanized steel bean stalk with a spinner on top that generates electricity may be added to your homeowner's insurance, with a price range from no cost to up to 1 percent of the total cost of the system.

Source of Money

Whether you withdraw savings from your account or borrow money from a bank (long- or short-term) will affect the return on your investment. The terms and rates may vary, but here's a general list of options:

1. You might find immediate funds in your savings account, but this may take away funds for emergencies. It is usually advisable to maintain at least five months of expenses in the bank.
2. Short-term loans are another option, but remember that they are usually at a higher interest, and are not tax-deductible.
3. Mortgage, home, or farm equity loans provide immediate funds at a lower interest and are tax-deductible.
4. Government-insured and some conventional loan programs can approve a larger mortgage payment based on the projected savings on monthly utility bills, or roll the costs of proposed improvements into the mortgage. These plans are often called green mortgages or energy-efficient mortgages. To find qualified lenders and a certified home energy rater in the United States, use this site: resnet.us/ consumer.[7] In the United Kingdom, go to: est.org.uk.

Related to this is *who* provides the money. We don't spend much time on who actually owns the wind system, but know that the source of money can significantly reduce the price. Private ownership is our main focus, but investor-owned systems with corporate financing have historically maintained a lower cost. And the least costly way to install a turbine is to obtain internal or project financing from the public utility, which usually means they would claim most, if not all, ownership (and operating and maintenance costs) of the generator.

Permitting Costs

If your local municipality supports legislation specific to wind turbines, permit costs can range from $0 to $1,000+. Some local governments give special consideration to "energy conservation devices" and "energy production accessories." In Riverhead, New York, for example, a permit for any "structure" can result in $1,500 or more depending on the square footage, but a wind turbine is only $150 dollars for a permit. That's a sizable discount that changes your bottom line for the better.

Zoning Jurisdiction

In some cases, municipalities may be familiar only with utility-sized wind turbines, and may try to apply existing legislation for that scale to small wind turbines. Local ordinances may, therefore, require costly additional studies and reports before a project can be considered.

For example, bird migration and audible sound studies are commonplace for larger-scale projects, and they can be conducted in no less than thousands of dollars. Your goal is to educate the town that small turbines deserve an exception. We advise you to use information in Chapters 1 and 2 to inform them of the differences.

Energy Usage Tax

Wherever you live, you can't be taxed on something you don't use. If there are taxes or tariffs on your energy consumption, the less you use, the lower your energy tax. Taxes have a profound effect on the economics of any investment. When you "store" your excess electrical generation "on the grid" in the form of a credit for later retrieval, you also offset the usage tax.

Raw Materials

Wind turbine systems are like any other product that requires many different elements, all of which can experience price volatility, such as copper, rare earth magnets, plastic composites, and steel. This can affect not only the size that you can afford, but the distance from the building or grid. Copper wiring currently can cost

$20,000 to $30,000 per quarter mile. The distance from the grid to your home and the home to your wind turbine can vary greatly. That same money could buy an entire wind energy system that will meet the electricity needs of an energy-efficient home.

Financial Incentives

Now that you have the list of factors that affect price, let's examine some of the specifics and provide some groundwork that you can further explore (Figure 6-4). By understanding the financial resources out there, you could cut thousands of dollars off your project budget.

One barrier that remains in the renewable energy industry is inconsistent financial incentives. As this industry continues to evolve, we can hope that the price of this technology comes down to a point where tax credits and utility rebates are not needed to make this investment affordable; but this day is not here yet. Currently, government incentives are one of the key forces driving this industry. Financial incentives are for many the "make it or break it" force behind the decision to purchase or not to purchase a renewable energy system. Some people actually move their project to different states just to take advantage of the benefits.

Many national and regional laws require utilities to include some amount of renewable sources of energy in their portfolios, and there are generally two ways they go about achieving this (Figure 6-5). Utilities can:

1. Purchase green energy from other generators, and then sell it to their customers at a premium.

Figure 6-4   This cup anemometer shows that the winds are strong. Now line up that financing!
Warren Gretz/DOE/NREL.

2. Build renewable energy generators, and then sell clean power directly to their customers.
3. Encourage their customers to invest in their own renewable energy systems through incentives like energy rebates.

All of these approaches help fulfill the utility's renewable energy standard, whatever the actual percentage may be. However, in many ways, the third option is the most economical decision for the customer.

In fact, well-sited small wind turbines can pay for themselves within 15 to 20 years, which is less than their serviceable lifetimes, if the right incentives are applied.

Worldwide, there are generally three incentives available to wind system owners:

1. Capital subsidies
2. Tax incentives
3. Low-interest loans

Let's take a closer look.

FIGURE 6-5 Many states and countries mandate some kind of renewable energy portfolio standard, which specifies that a certain amount of power must come from designated "clean" sources. *Wind Powering America/NREL.*

Capital Subsidies

In many jurisdictions around the world, those who install small wind energy systems may receive money toward the purchase price, in what's called a *subsidy*. The money may come from public treasuries, but the term subsidy may also refer to assistance granted by utilities with funds drawn from charges all electricity consumers pay for public benefits. Clearly, wind turbines have substantial up-front capital costs, but subsidies can lower the effective price.[8]

Capital subsidies take a number of forms, as the following sections show.

Grants and Rebates

On Kevin's first sales meeting with a successful greenhouse business, he recalls reciting the current federal and state benefits to the client, believing that the tax credits available would be enough to fund the project and bring the business to an easy solution for its high utility bills. However, one call to the accountant made it sound less appealing. They already had a laundry list of deductions from other tax credits that couldn't be combined; hence, the company would be unable to absorb the full value of a tax credit. This is a substantial limitation of tax credits.

In contrast, a direct payment in the form of a grant or rebate does not share the same problem. Anyone can benefit from those incentives, regardless of tax status. A grant is usually in the form of money paid directly to the recipient. With a rebate, the consumer must purchase the system first, and then file paperwork to receive money back, according to the value of the rebate. A number of countries and utilities offer such incentives.

Rebates are typically paid for with public benefit funds (PBFs) collected as a small fee on consumer utility bills.[9] So technically, you may have been paying for wind turbines for years, but it just wasn't on your property.

Direct payments can be made even more powerful through *cost sharing*, in which the government pays part of a wind system's costs directly, because the private investor would not pay taxes on the cost-shared portion. So that is a tax incentive on top of cash. Double dipping.

When our client was informed that if we commenced the project before a certain deadline set in the incentive program the federal government and local utility would cut a lump-sum check in lieu of a cumulative tax credit, they were more inclined to get the wind turbine on their property.

Production Incentives and Feed-In Tariffs

Some governments or utilities provide cash payments to wind power owners based on the number of kilowatt-hours they actually generate, rather than on installed price tag and its predicted capacity (Figure 6-6). To proponents of this strategy, including Paul Gipe, it's a good way to encourage consumers to invest in high-quality equipment on good sites, because they can only get paid based on production, not prediction.

FIGURE 6-6   The Northwind 100 kW turbine is designed to withstand especially cold climates. *Warren Gretz/DOE/NREL.*

As an example, in Washington state, the Chelan County Public Utility District's Sustainable Natural Alternative Power program (SNAP) has helped build two 10-kilowatt wind turbines and four solar projects to date. Funding for the program comes from voluntary customer contributions of $2.50 to $50 per month, which are promoted as supporting locally generated clean energy. Yet, it does not affect the electric rates of customers who do not want to participate. Producers are paid a percentage of the pool based on their percentage of the total renewable kilowatt-hours generated, up to $1.50 per kilowatt-hour. A SNAP starter program is available to other utilities.[10]

Perhaps the ideal incentive would be a combination of pre- and post-installation assistance. For example, Orcas Power & Light Cooperative Green Energy Program in Washington provides up-front payments for small wind turbine installations based on expected generation during the first year. Then, system owners also receive a production incentive based on output in subsequent years.

An increasingly popular type of production incentive is the *feed-in tariff* (also look for FIT, feed-in law, advanced renewable tariff,[11] or renewable energy payments). A feed-in tariff establishes a *set price* at which a utility purchases excess electricity from a renewable generator, such as your small wind system. This way, natural market forces can adjust the amount of energy produced (supply) and demand accordingly. Unlike up-front rebates and tax credits, this policy is based on output (cents/kilowatt-hour) rather than turbine size (kilwatt-hour), and therefore rewards real-world performance. And unlike most net metering systems, the feed-in tariffs are typically set well above average electricity prices.

Feed-in-tariffs are normally established with long-term contracts, with 15- to 25-year agreements common.[12] And unlike many renewable energy credits, FITs are often nondiscriminatory in terms of production, meaning they apply the same treatment to wind as they do to solar.[13]

By 2010, at least 63 jurisdictions, 50 countries, and 25 states/provinces had adopted FITs. Feed-in tariffs exist in most of the member states of the European Union, as well as in Australia, Brazil, Canada, Iran, Israel, Japan, Kenya, the Republic of Korea, South Africa, Taiwan, Thailand, Turkey, Ukraine, and in more than a dozen U.S. states.[14] The program is gaining momentum in China, India, and Mongolia.

Emilie Morehouse, the Small Wind Policy Manager for CanWEA, told us in an interview that there has been "interest" in feed-in tariffs for small wind at the provincial level in Canada. In fact, Nova Scotia recently announced a new FIT for small wind, which according to Morehouse will only be available on a community scale, and will only be offered to aboriginal groups, co-ops, or municipalities. As of this writing, a review board is working to develop the details.

If you want to know more about feed-in tariffs and see if your country or state has some mechanism set up, please visit:

Wind-Works.org
www.wind-works.org/articles/feed_laws.html

Wind author Paul Gipe has long been a strong supporter of feed-in tariffs, because he feels they are the best way to guarantee meaningful energy production out of public-subsidized systems. Gipe told us, "For small wind to be profitable you need to be paid a higher price [for the electricity you generate], so you need a feed-in tariff. Subsidies that just pay based on estimated production are a terrible way to develop the wind turbine industry because they invariably lead to development of products that are shoddy or poorly sited.

"Feed-in tariffs only pay for electricity that is produced, so a buyer must know when they install that the turbine will work and produce sufficient electricity to make a profit. If it doesn't they can go after the manufacturer. With basic tax credits there is a tendency for Americans to think it is a gift, and they will do something stupid financially to take advantage of the subsidy," said Gipe.

David Sharman of UK-based small turbine manufacturer Ampair told us, "FITs are good, probably the best, since they create well-aligned interests." He added, "Other [incentives] can also be helpful, but also unhelpful. There are many routes to heaven."

Renewable Energy Credits (RECs)

Another type of production subsidy for renewable energy is a *renewable energy credit* (REC), also known as a green tag or tradable renewable certificate (TRC), and in the United Kingdom as a renewables obligation certificate (ROC). These are tradable, nontangible energy commodities that represent proof that one megawatt-hour (1,000 kWh) of electricity was generated by a green energy provider (Figure 6-7), such as your wind turbine (for reference, remember that an average U.S. household uses about 920 kWh in a month).

A certifying agency gives each REC a unique identification number, while the associated "green energy" is fed into the grid. The REC itself can then be bartered or sold and traded on the open market, yet considered separate from the sale of commodity electricity.[15]

Prices of RECs depend on a number of factors, including supply and demand, the year the certificate was generated (called the "vintage"), the location of the turbine, and whether the REC is used by an entity for compliance under a scheme that requires a minimum of energy be provided by renewable sources. In the United States, several states have a so-called renewable portfolio standard (RPS) or renewable electricity standard (RES); in the United Kingdom it's known as the renewables obligation (RO).

Even the type of renewable energy created can make a difference. For example, in July 2010 Delaware and New Jersey paid as much as $665/REC for solar electricity,[16]

Figure 6-7 RECs are largely designed to support large-scale renewable energy projects, like this relatively new wind farm in rural Idaho. *Brian Clark Howard.*

while Canada offers $3/REC for 20-plus year contracts. In the United Kingdom, there's a ceiling price on ROCs (£36.99/ROC in 2010–11). While the value of RECs/ROCs fluctuates, most sellers are legally obligated to deliver energy continuously under penalty if they don't meet their obligations. The prices might not always be a fair market value, but if you have no other alternative, it will still improve your return.

The United States currently does not have a national registry of RECs, so you would need to have a third-party audit of your system to be eligible for certification. The Center for Resource Solutions maintains a list of auditors who meet the criteria to be listed on the program website: green-e.org/auditors.html.

In the United Kingdom, the RO system is a major supporter of green energy projects. It obligates UK suppliers of electricity to source an increasing proportion of their electricity from renewable sources.[17]

In a bid to make FITs more accessible to the developing world, the World Future Council has proposed setting up a global Renewable Energy Policy fund that would cover the feed-in tariff rate in order to avoid increasing electricity costs for low-income households. The global fund could portion out assistance based on the economic status of the country, with wealthier developed nations paying a portion of the freight.[18] So, to you, in East Africa, there is hope on the way. (By the way, this is a good time to remember that most people, especially in the developing world, tend to pay much higher rates for electricity than we do in the United States, especially as a percentage of income.)

Net Metering

We've mentioned *net metering* before, especially because it is pervasive in North America, and is a key part of the relationship between a distributed producer of energy and the grid. Net metering allows access on the customer side of the meter and specifies how the wind energy producer will be paid. Most net metering programs have project size caps, but these are normally well above what most small wind turbines can produce. All renewable technologies typically receive the same price for generation.

Under net metering, your wind generator feeds your surplus (unused at your site) energy into the grid, where it is "banked" with a utility for a particular period, usually a year. When generation is less than consumption, energy is provided by the grid to make up the difference, and the amount provided is subtracted from your balance of "banked" electricity. Thus, the price or tariff for generation is effectively the retail price of electricity.

Note that most net metering programs have a cap that restricts banking of electricity to even with the customer's total annual consumption—or, if you are lucky, to 10 or 20 percent above annual consumption. That is, if you consumed 10,000 kWh of electricity at your site that year but produced 14,000 kWh of electricity, an excess generation of 4,000 kWh, at best you would only receive payment for 1,000 kWh (110 percent) or 2,000 (120 percent), depending on your deal. As for the excess banked electricity beyond that, you provided it free to the utility.

If you think about it, it's not hard to see why the utility would want to structure the arrangement in this way. They aren't historically in the business of purchasing power from tiny producers. They normally buy at high volumes from big power plants and pay wholesale rates. Still, there are a number of benefits to distributed generation, as we covered in Chapter 1, so it's an evolving relationship.

You should also be aware that, sometimes, the net metered price is not the true retail rate. Some programs add fees, while some do credit according to the whole-sale rate (bummer), not the retail rate that the customer pays. So be sure to understand all the details before you sign.

Emilie Morehouse of CanWEA told us that Canada has net metering nation-wide, though the rate is set by each province. She said Nova Scotia has the best rate, at 13 cents per kWh. According to Morehouse, "Experience has shown that net metering programs don't really work to significantly deploy wind."

They certainly help, but net metering alone is usually not sufficient to make a grid-tied system financially profitable, unless some additional factors come into play.

Tax Incentives

Tax incentives have long been used in many countries to stimulate development of renewable energy markets and industries. Small wind energy is no exception. Figure 6-8 gives an idea of the relative availability of different incentives for renewable energy around the world.[19]

Let's take a closer look at the different instruments.

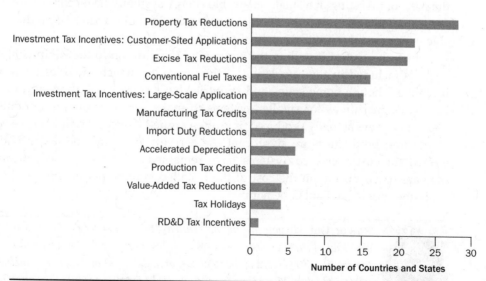

FIGURE 6-8 Relative abundance of renewable energy incentives around the world. *Center for Resource Solutions. International Tax Incentives for Renewable Energy: Lessons for Public Policy, http://www.resource-solutions.org/lib/librarypdfs/IntPolicy-Renewable_Tax_Incentives.pdf.*

Property Tax Reductions

Some people wonder if installing a wind turbine on their property would count as an "improvement," and therefore raise their property tax rate. That is possible, although luckily, many places have rules that prevent property taxes from being raised on a new renewable energy system.

For example, in the state of New York, there is a 15-year moratorium on property tax assessments for a solar installation, although no statewide waiver exists for a wind turbine. It would be under the discretion of your local government, hence your due diligence, to determine if property taxes should be affected.

Excise (Sales) Tax Reductions

A number of jurisdictions exempt renewable energy equipment purchases, including wind systems, from up to 100 percent of excise (sales) tax. Some countries that tax electricity consumption also provide an exemption for electricity produced by renewable technologies. A few places also offer tax rebates, which refund a specific share of an excise tax already paid. Consumers apply for tax rebates at the time of purchase of equipment and systems. Many U.S. states exempt all sales tax for the installation of the entire system if purchased by a certified installer.

Investment Tax Credits

Historically, so-called *investment tax credits* (ITCs) are one of the most popular incentives for renewable energy offered by state and federal governments. With an investment credit, tax deductions or credits are offered based on some fraction of the costs of investing in a small renewable energy system. The incentives are most often applied against income tax that is owed. Technically, deductions reduce taxable income, while credits directly offset taxes due.

Often, the cost of installing the equipment (in addition to the equipment cost itself) is included in the calculation of the tax credit, which provides a stronger incentive. Though some states and countries have considered applying production-based incentives for smaller systems to make sure meaningful amounts of clean energy are actually produced, it is often argued that the administrative costs of tracking production are significant. As a result, most income tax incentives offered for customer-sited systems have remained investment-based, meaning they are based purely on the cost of the installed system, not energy produced. (This does not make Paul Gipe happy.)

 Power Up! *If you are in the United States, you might have heard that small wind turbine systems (100kW or less) are eligible for a federal income tax credit of 30 percent of the total cost of installment of the system, including parts and labor. There is no maximum credit for systems installed after 2008.*

This ITC credit can be used in the first year of production, or it can be spread over a number of years. Systems must have been placed in service on or after January 1, 2008, or before December 31, 2016, which is when the current incentive expires. The home served by the system does *not* have to be the taxpayer's principal residence.

Corporate Investment Tax Incentives

Corporations sometimes qualify for corporate income tax deductions or credits based on investment in approved renewable energy technology—such as that series of 100 kW turbines behind the office park. As with individuals, deductions reduce taxable income, while credits offset taxes due.

Qualifications for companies vary from country to country, but usually governments require certification of performance and safety and set size minimums to qualify for a tax incentive.

Production Tax Incentives

Production tax incentives (PTCs) provide income tax deductions or credits based on a set rate per kilowatt-hour actually produced by wind-energy facilities. These incentives, like feed-in tariffs, are designed to encourage good siting and maintenance of systems.

In the United States, the federal wind energy PTC currently provides a credit equal to 2.1 cents per kilowatt-hour produced. It is available for the first ten years of a wind generator's operation, if it is running by the end of 2012.[20] Calculations show this credit could lower the life cycle cost of wind power by more than 25 percent. Publicly owned electric utilities and rural electric cooperatives are not eligible for the PTC.

When it comes to individuals, or any other groups that might qualify, they actually have a choice in the United States: elect for *either* the PTC or the ITC. Sorry, the law currently does not allow you to claim both types of credits. Note that for most home and farm owners, the ITC works out to a considerably better deal, which is why most people in the business spend their time talking about that, and essentially ignore production credits. The ITC is available immediately, instead of spread over ten years, and it isn't tied to production, meaning it isn't subject to the vagaries of the real world.

We aren't aware of any small wind owners who have opted for a PTC over an ITC, but if you hear of any, let us know.

Research, Development, and Manufacturing Tax Credits

In order to support development of clean energy, domestic manufacturing, and job growth, many countries have research and development tax credits that offer up to 150 percent[21] of money invested by a corporation into new, expanded, or re-equipped research or production facilities (Figure 6-9). This is usually determined

Figure 6-9   Many countries offer additional tax credits on research, development, and manufacturing of wind turbines in order to help build capacity in renewable energy. *Paul Anderson/Wikimedia Commons.*

based on a project's commercial viability, job creation, carbon emissions reductions, and other factors. China, India, the Czech Republic, Bangladesh, and Jamaica have recently made significant investments in wind energy.[22]

Unless you are looking to continue to build turbines in your shop, this doesn't necessarily apply to you now, although it's true that you may eventually have more options in terms of purchasing turbine equipment that's made locally. Morehouse, the Small Wind Policy Manager for CanWEA, told us that part of the reason why Nova Scotia recently passed a feed-in tariff plan for small wind was to help support and stimulate the province's turbine manufacturing industry, which is led by Seaforth Energy.

Accelerated Depreciation

For those of us who don't work in accounting or finance, the concept of depreciation can seem a bit mysterious, but it can be helpful, at least to those of us who run agricultural or commercial businesses. *Depreciation,* as you probably know, refers to the fact that assets decline in value over time as a result of wear and tear, obsolescence, and changing trends. The financial instrument of depreciation is a way to approximate that decline. It's a "non-cash" expense that reduces income as calculated for income tax purposes. If your wind turbine is a business asset, it could be depreciated over a 20- to 30-year period, the estimated life of the system.

In an accelerated depreciation incentive, agricultural businesses and investors in renewable energy facilities are permitted to depreciate equipment at a faster rate than typically allowed, often 15 years or less instead of 20 to 30. This substantially reduces stated income for income tax calculations during those years, and thereby reduces income taxes. In India, you can depreciate the entire value of the facility and equipment in the first year.[23]

Get your "non-cash" now, and get it while it lasts.

Value-Added Tax Reductions

Many countries have *value-added taxes* (VATs), instead of income, sales, or other taxes. This would include most of Europe, Australia, New Zealand, Russia, India, Japan, much of Latin America, and a good many other countries. A value-added tax is similar to a sales tax, in that the people who actually pay it are the end-level consumers. However, a VAT has a more complicated collection and accounting process, which proponents say takes the burden of deciding who is the ultimate consumer away from businesses.

In an area with a VAT, whenever a business or individual buys a good or service, they have to pay the stated tax. For end users of products, either individual or corporate, the process stops there. But if the buyer adds something to that purchase, by refining it into something else, for example, they will get a credit back from the government for the tax that was initially paid. In other words, they don't effectively pay taxes on things they buy in order to make other things.

Value-added tax can hurt renewable energy if the output of electricity is not taxed, while inputs, like capital expenditures on equipment, are taxed. In other words, a person who buys a small wind system is taxed as an end user of that equipment, instead of getting a credit for "adding value" by producing clean energy.

Typically, countries with VAT reduce the rate of the tax applied to the production of renewable energy and the domestic manufacturing of renewable energy parts, equipment, and systems. Alternatively, some countries collect the full tax on such activities, but refund a portion of it.

Import Duty Reductions

Some countries have reduced or eliminated duties on imported wind system equipment and materials, especially if the nation doesn't have a developed manufacturing and support infrastructure for the sector. Although reducing import duties may not have a huge impact on your final cost, it is not something to be overlooked. Import duties can vary by technology, depending upon the status of domestic manufacturing.

Taxes on Conventional Fuels

Most countries tax fossil fuels in some manner. In the United States, there is a lot of discussion about gasoline taxes, which are generally lower than in much of the world. Some countries, particularly in Europe, also tax consumption of fossil fuel–

based energy in the form of a *carbon tax*, which is typically assessed on emissions like carbon dioxide. If your country taxes only on the emissions side, wind-energy turbines are exempt from paying a tax. In other words, if your home or business is using clean wind power, you may be saving money on fossil fuel taxes that you don't have to pay, yet you maintain your same quality of life.

That's like your government saying, "Thanks for nothing" (no emissions, that is).

Tax Holidays

Particularly popular with consumers, tax holidays are time-limited reductions or eliminations of excise (sales), income, VAT, or property taxes. Time frames can be as short as a month and as long as a decade. In India, the government offers a ten-year tax holiday on income produced from wind systems.

INTERCONNECT

Dominica, the Caribbean's "Nature Island," Gets a Wind Turbine

One of the least developed islands in the Caribbean, Dominica is often described as a green paradise, teeming with lush forests, sparkling waterfalls, and spectacular diving opportunities. In 2008, "The Nature Island" also got its first wind turbine, a Norwin 225 (225 kW) (Figure 6-10).

Although a 225 kW turbine is typically considered utility scale, we thought this project was worth including because it shows how wind power can be accessible to a broad range of stakeholders, particularly in remote areas. Interestingly, the turbine's owners—Beverly Deikel and her partner Patrick Oscar—told us they originally wanted something smaller to power their new green resort, but say they were unable to entice any installers to come out to the island for anything less than 225 kW.

Along with some rooftop solar panels, the grid-connected turbine provides about 70 percent of Rosalie Bay Resort's energy needs. Open to guests in late 2010, Rosalie Bay Resort was founded by Oscar (he goes by his last name), who is from Dominica, and Deikel, who hails originally from Minnesota. The property occupies a gorgeous stretch of coast in the southeastern part of the island, and it is loaded with eco-friendly features, including an organic garden and advanced water treatment. One of the small resort's main goals is to preserve the area's nesting sites for endangered sea turtles, so local guards are hired to protect vulnerable baby turtles from poachers. Guests have the chance to pitch in or just watch the reptiles' march to the sea. Guests can also enjoy kayaking and tubing, nature walks, hikes into the surrounding mountains and to a boiling lake, or, of course, snorkeling and diving.

For a unique thrill, visitors can don a harness and ascend to the top of the wind turbine, where they are rewarded with spectacular panoramic views of the Caribbean and

FIGURE 6-10   In 2008, Rosalie Bay Resort in Dominica installed a 225 kW Norwin wind turbine, which powers 70 percent of the property's energy needs in combination with solar panels. *Rosalie Bay Resort.*

the rugged countryside. "I went up, and it is so beautiful," Deikel told us while she was in New York to kick off the resort's opening. "It's one of the few islands Christopher Columbus would still recognize," she added.

Rosalie's wind turbine is estimated to produce 586,000 kWh of clean energy per year, offsetting 160 metric tons of carbon dioxide emissions per year. According to Deikel, Dominica's government was helpful in getting the turbine approved and connected, although she said the utility was hesitant at first. The resort site had not previously had electricity or water (or much in the way of employment, though the resort has brought about 60 permanent jobs for locals).

"Some people don't like the look of it but most do," Deikel said of the turbine (Figure 6-11). "One neighbor doesn't like it but he doesn't like anything." Deikel said the neighbor asked them to paint it a different color to try to match the hills or the sky, but she doubted that would really make a difference to this "angry neighbor."

FIGURE 6-11   Resort guests can ascend the wind turbine for spectacular views of the Caribbean, and to learn about wind energy firsthand. *Rosalie Bay Resort.*

Rosalie Bay Resort's general manager, Jan Neel, told us, "Guests and locals alike are thrilled to see the turbine generating power. It's quite a landmark for the progress that Dominica is very proud of in the eco-sensitivity sector."

In January 2011, the Obama administration announced that Dominica will be receiving U.S. funding for a new large-scale wind farm, through the Energy & Climate Partnership of the Americas.

Low-Interest Loans

It's true that a wind turbine can call for a significant initial cash outlay. And even with all the available subsidies removing a big chunk, what remains might still make you uneasy. So, depending on the size of your system, you might want to consider a loan. Assuming you own some property, a standard home equity line of credit is definitely one option, since the terms are generally favorable.

Kevin remembers the day an astute client asked, "With all the federal and state tax incentives and local authority rebate cutting the cost by more than 40 percent, what would the loan amount be?" It turned out to be the initial total cost. That's because many incentives don't kick in until after it is built and producing. The installers are not likely going to defer the cost, so you are left with bringing the money to the table first.

For those who can relate, getting a big loan is not easy to swallow. And many people who are keen on eliminating their bills are the same people who might struggle to make the initial high monthly loan payments that would follow.

The good news is that if you are receiving a grant or rebate check a month or two following the install, you can quickly pay down the principal, which means that this payment is not going toward the interest accrued on your debt, but the actual amount you owe. So long as there is no clause that prevents prepayments, any additional payment amount to be applied toward the principal saves you money on the interest, and reduces the life of your loan.

After the recent global recession, financial institutions do not provide much incentive for saving your money in a bank, so that makes it doubly important to get yourself out of high-interest loans. Our most recent look at interest rates for certificates of deposit shows that banks don't break one percent unless you invest for at least five years. That is not even enough to adjust for inflation.

For those committed to lowering your bottom-line cost, don't stop with the lump-sum rebates. We suggest you send in two checks each month: one for the regular loan payment and the other an amount equal to your previous monthly electric bill, *clearly marked that it is to be applied to principal*. This can clarify your wishes to the bank beyond a doubt. Please remember to ensure that there is no prepayment clause that prevents you from being able to reduce the interest paid.

Well, that is nice, but how do I reduce the monthly payments? Well, that requires refinancing the loan. That is an option. Especially if you have just prepaid a substantial sum and the loan interest rates have reduced, you could seek a loan advisor on the steps to reduce your long-term debt.

Now, let's look at a few other types of loans.

Revolving Loan Funds

One potential source of funding worth exploring is a revolving loan fund for renewable energy projects. Banks in some areas offer the program, in which a loan is made to one person or business at a time to finance a new system. As repayments are made on the initial loan, funds become available for new loans to other folks. Hence, the money "revolves" from one person or business to another.

For example, Iowa's Alternate Energy Revolving Loan Program provides loan funds equal to 50 percent of the total financed cost of a project (up to $250,000), at 0 percent interest. Talk to installers in your community, others who have wind systems, and industry associations to find out more.

Related to this, it's possible that you may be able to tap into a local community development loan, or public investment loan, if you organize a group of people and propose a small-scale wind project for your town. There isn't a lot of money available for such projects, but some places do have funding to support green initiatives, so it doesn't hurt to ask and to talk to your fellow wind enthusiasts.

Green Loans

In many cases, lenders can approve a larger mortgage on a property to provide funding for energy efficiency and renewable energy projects, based on the projected savings on monthly utility bills. Such loans may be called *green*, *green business*, or *energy efficiency* loans, and they apply to all technologies that meet the lending bank's investment criteria. They typically provide a competitive rate, and will be calculated based on a wind-turbine's long-term return on investment.[24] Use this site to find qualified lenders and a certified home energy rater in your area.[25]

Resource

resnet.us/consumer

Does Small Wind Pay?

With all of these financial considerations, you may be feeling kind of side-tracked from all the constants we previously discussed, such as swept area, betz limit, and annual energy output. You also may want to put it all together: given all the factors that can affect the price of a system, the incentives that are available to you and the potential savings in energy, you may be asking if small wind will really "pay" for you. Undoubtedly, that is an important question, and one that receives a lot of media attention and discussion by the public, although as we have pointed out, some of the rhetoric can be blown out of proportion.

Consider this response from Mick Bergey of the well-respected manufacturer Bergey Wind Turbines: "You are already paying for it regardless if you participate or not." That is, even if you don't buy a wind system, the money you pay every month toward your electric bill is money that could go toward a wind turbine. So, if you decide not to become your own wind generator, at least you know you get the benefit of paying for one without actually having to install one. Congratulations! Yet, all facetiousness aside … we are sure you would like to make sound decisions regarding your energy solutions.

Before we get deeper into the question of how quickly your system might return your investment dollars, let's step back a moment and give some thought to that query itself.

Does a Small Wind System *Have* to Pay for Itself?

Although some pundits talk like return on investment (ROI) is *the* only variable worth considering when it comes to renewable energy systems, we know that the real world is more nuanced. Ultimately, buying a wind system is an individual choice, and like all consumer decisions, it must be made at the individual level, based on each person's unique circumstances, needs, and desires (Figure 6-12).

For an off-grid user who would be required to pay their power authority tens or even hundreds of thousands of dollars to install an underground electric line and transformer in order to provide power to a remote location, a wind turbine can readily appear to be a cost-effective solution. In fact, demand among the roughly 180,000 American families who are currently living off the grid has helped spur innovation in the small wind industry for decades.

What about farm owners with good wind sites and moderate distances from street-side power lines, or business property owners with high demand fees (explained shortly) due to their heavy energy requirements? In these cases, a small wind system may or may not pencil out in terms of ROI in a short period, depending on the specific circumstances.

What about the typical homeowner on the grid? Is small wind energy the least costly way of getting electricity? Well, no, not overnight at least. But the question arises, does it need to? Why do you think owning a wind turbine should be treated so different from all the other "necessary" purchases you have made in the past?

Consider typical big purchases people often make: a ready-to-walk-in house, a sleek new sports car, an iPad, a multimedia gaming computer, a backyard pool, a cruise trip, or a sailboat. True, a devil's advocate might say you can't really compare such items with a wind turbine, because the wind turbine *is designed to earn or save money*. But why must that statement be a fact?

One of us owns a small wind turbine, and it provides great gratification, even if on the surface it doesn't appear to provide an immediate positive cash flow. Isn't that like those other items we mentioned? It cost $10,000 out of pocket, but to this day Kevin remembers the joy of seeing it spinning during the installation process, even though there wasn't enough wind to meet the cut-in speed. Kevin reasoned that he is a homeowner, not a professional contractor, yet he chose to build his own

FIGURE 6-12   This grid-connected, 10 kW Bergey wind turbine offsets the energy use of a small business in Norman, Oklahoma. *Bergey Windpower/DOE/NREL.*

house. He is a consumer, not an application programmer for Apple or Google (yet), yet he bought a smart phone; he is a swimmer, not an Olympian, but will install a small lap pool. He realized that he didn't purchase any of these items for immediate cash flow. Perhaps in the long run, some of these things could earn some income—historically, at least, a home has been a relatively good asset—but for the most part, most things actually depreciate in value.

We ask, why must buying a sports car be only about "fun," while buying a wind turbine is only about "ROI"? Many people rarely drive their sports cars, and even if they do they seldom have a chance to really open them up, what with all the congestion and speed limits and everything. Sure, it can make you look cool … but then so can a wind turbine, which also has a significant image and "wow" factor.

Perhaps the devil's advocate is right: wind energy *is* different in a significant way, because it *appreciates*, like a rare collectible car, slowly over time. How? Renewable energy systems not only maintain their value (for the most part), they actually make money for the owner, both in terms of the value of the electricity generated and the taxes saved on not purchasing grid electricity. The same cannot be said of most other purchases.

And the true gift of wind energy, besides the fuel being free, is that it doesn't need to pay your whole bill to be economical. In fact, if you are not going to be compensated at least the retail rate for your surplus electricity, sometimes it is better to size the system down so it merely pays for your needs. Come to think of it, placing a micro-turbine in a yard with an awning and chairs, in lieu of a swimming pool, would over the long term be a more cost-effective solution—even providing respite on a hot day from the savings on running your air conditioner. A swimming pool has a high up-front cost, raises property tax, and requires constant maintenance and energy costs, yet it doesn't fully appreciate the property value over the long haul. (It may also be worth pointing out that swimming pools are much more dangerous than wind turbines, if you consider the number of people who drown in them every year.)

In Kevin's case, his wind system (1) did *not* raise his property tax (ask your local assessor), (2) has low maintenance, (3) has no energy costs, (4) appreciates the property value ("Do you want a home where this small turbine pays the equivalent of your cable bill every month?"), and (5) gives immediate gratification every time the wind blows.

That being said, there are also wind advocates who make a strong case for increasing the payback power of the technology. Author and industry consultant Paul Gipe told us: "For small wind energy to be used by many more people, for it to have a chance to reach a scale necessary to confront the crisis we face, it has to be profitable."

Gipe said he'd ultimately like to see "millions" of small turbines installed across the country. But in order for that to happen, he said people would need to be in a position to *make money* on them, not just *save money*. "To do that, small turbines must be much, much cheaper than what they are today, or owners must be paid a higher price for their electricity," he explained, alluding to the feed-in tariffs that he strongly supports. "Small turbines are far less productive and less reliable than large ones, and as a consequence, it's very difficult in the current situation to make small wind pay," said Gipe.

In any case, there are many ways to calculate value, so let's look at some more related factors.

Initial Cost

As you are probably aware, a wind turbine has a significant up-front cost. Seasoned wind installers tend to treat the price tag like the pot handle test—it might be too hot to touch. For the individual, small wind may at first appear to be a lost cause, if you look one-dimensionally at the initial price. True, if you compare the cost of hooking up to the grid to installing a wind system, hands down, it is cheaper up-front to stick with the grid ... unless you live more than a thousand feet from it (copper wire is really expensive). Heck, it might even be cheaper to buy a diesel generator or build a wood-fired steam engine. But initial price only determines if the wind system will cost more, not if it is a bargain.

The Concepts of Payback and ROI

For many people, the next logical consideration is to look at the payback period of the system. In renewable energy, *payback* is generally defined as the amount of time it takes for a system to pay for itself through energy savings (once you have already deducted any incentives from the cost of the system). In the world of finance, some investments have a short time span, making payback important, because you would want to know that you made your investment quickly so that you can turn around on another quick investment. But wind turbines are not stocks being sold short, or three- to five-year certificates of deposit. They last for 20 to 30 years.

Experience shows that the payback period for a small wind system can range from several years to several decades, depending on the cost of the system, incentives, the cost of the electricity you are offsetting, and the average annual wind speed the turbine "sees." Actually, the average wind speed is usually more critical to the payback period than anything else, including the initial installed cost.

In fact, if the only reason to install a small wind system were to make money, then wind turbines would be constructed only in the windiest areas. We know that this is not always the case. In fact, we are familiar with a number of projects that have been installed on urban rooftops that have payback periods far in excess of the presumed lifespan of the systems. Such installations are controversial in the industry, though some owners have argued that the other benefits they provide, such as press coverage and a green profile, justify their expense.

It's important to remember that the payback view alone gives no indication of the earnings a wind turbine will produce after it has paid for itself. Much of its return comes later—more like a cumulative interest on a lower interest bond. This is compounded by the reality that electric rates tend to go up, an average of 4.5 percent every year (of course, curbed by inflation). So then we are talking about a special "bond" with a cost of living and/or inflation control. Huh? Well, your wind turbine keeps producing electricity while the cost of electricity is going up, right? So the value of what you produce is going to rise. Now this long-term thinking is not a well-beaten path made by the average consumer. But we suppose you are not average and that you are ready to realize that *durable* wind machines, even many of those with seemingly low energy output, will pay for themselves—they are long-term investments.

Did you notice we slipped in "durable" there? It should be implied, but with the current small wind turbine boom, people sometimes presume that all wind turbines make the distance. But you can't sacrifice quality for price. The equation needs to be clarified: Initial high cost for a durable wind machine brings long-term positive cash flow; a low-quality turbine lowers initial cost at the risk of reducing longevity to the point that payback is never actualized. Dead turbines provide no cash flow.

Rather than payback, small wind expert Mick Sagrillo prefers the question: "What is the return on my renewable energy investment?" He rightly points out that no one asks what the payback is on a certificate of deposit. So, if renewable energy is pursued as an investment, it should be evaluated the same way that other investments are.

Working out ROI involves including the factors we've covered, such as the inflation rate, rising energy prices, any applicable depreciation, and other incentives (Figure 6-13). Sagrillo suggests starting a spreadsheet, and he recommends the two downloadable templates on Windustry's website (windustry.org/calculator/default.htm), namely Wind Powering America's "Wind Energy Payback Period Workbook" and "Bergey Wind Power Small Wind Project Calculator."

So let us take a look at the average U.S. residential home with a wind turbine adequate to supply typical annual energy needs, roughly 12,000 kilowatt-hours of electricity.

Residential Economics: Homer Hank

Homer Hank lives at sea level, where he installed a 10-meter tower and a 10k wind turbine that cost about $50,000 in total. The site has an average wind speed at hub height of 12 mph (about 5.4 m/s). The expected AEO is 12,000 kWh per year under these conditions. Nice setup!

Homer sees an average increase in electricity prices of 4.5 percent a year, and an annual inflation rate of 2.5 percent. That means 4.5 − 2.5 = 2%, for the real price increase.

FIGURE 6-13 In addition to straight ROI accounting on this small wind turbine at a supercenter in Colorado, Walmart would do well to consider new business attracted by the visible machine, as well as good publicity. *Brian Clark Howard.*

For his financing, Homer got an equity loan, and he put 20 percent down so he wouldn't have to pay mortgage insurance. Fortunately, he was able to get the loan for 4 percent interest for 20 years to pay for his wind system. He is in the 34 percent tax bracket.

To calculate his net present value, we need to know the *discount rate*—the interest rate used to discount future cash flows. At a minimum, the discount rate on a 20-year income stream will reflect the rate on a 20-year mortgage (currently in the United States this was lower than 4 percent).

With electricity used in Homer's home at a retail rate of $0.10/ kilowatt hour, and interest payments on his mortgage tax-deductible, his wind system will produce a net positive cash flow after ten years. This is *not* even taking into account government subsidies or other agency incentives.

Farm and Business Economics: Energy Demand

For commercial and some farm properties, the equation for earnings has an additional component. In many cases, such larger operations must pay not only for the energy they use, but also for the *demand* they put on the utility system.

One way this is handled is that many commercial accounts are given an allotted energy load. Yes, they get special rates, because they give the power authority plenty of revenue. But if their consumption of electricity exceeds what is allocated by their contract, they are charged a penalty and a "demand rate," sometimes multiple times the retail rate. This is common for businesses that have energy-hungry machines running around the clock, and it can be especially significant in populated regions, where the load is relatively high.

So, wouldn't wind power generation counter that excess? Not according to most power authority policies. As Paul Gipe states, "This charge is based on the maximum power drawn during the billing period in relation to their total energy consumption. This compensates the utility for maintaining generators online to provide power at the demand of the customer. By installing a wind turbine a business lowers its total consumption while hardly affecting its peak demand. Thus the wind system could actually *increase* the demand charge while lowering costs for the energy consumed."[26]

It has to do with how the power authority interprets your energy usage. During an interview with our local authority, they explained that they calculate average energy consumption from meter readings of the commercial account multiple times an hour during the day and consider the power entering the grid from the wind generator as a separate service. Even if a 100kW wind turbine owned by a printing service company produced enough energy for 25 New York homes (250,000 kWh), if the business's electricity consumption crossed the line into their demand zone at the time of the meter reading, it would be penalized.

Fortunately, the power authority was pleased to say they are working on updating their policies and infrastructure to better support wind energy. A simple thing like a smart meter would avoid this scenario. Perhaps by the time of publishing, you will see more wind turbines at Walmarts around the world.

Insurance: Protecting Your Investment

A wind system is a significant investment, and as such it is worth protecting. In addition, wind systems can pose some hazards, so you want to take care to reduce your exposure to liabilities. A time-tested method for meeting these goals is through insurance, which often ends up being the only ongoing fixed expense of the system.

We'll cover property and liability insurance and provide you with the kernels of information, but this is one area where it's a good idea to talk to a local experienced wind installer, such as Mick Sagrillo of Sagrillo Power & Light Co. in Forestville, Wisconsin. Sagrillo's knowledgeable and helpful advice columns on small wind systems have appeared in national magazines and in the AWEA "Windletter" for many years. We sum up his expertise in this section, but the footnotes lead you to his articles for further explanation.

Property Insurance

Here are some general guidelines for your wind system:

- Insure it as an appurtenant structure under existing homeowner's insurance.
- Insure from "acts of god," fire, theft, and vandalism.
- Install to National Electric Code standards.

The most cost-effective way to insure a wind system is under an existing home-owner's insurance policy on your house. This is far cheaper than purchasing a policy specifically on the wind system. However, this option is generally only available to you if your system is on the same property as your home. Otherwise, you will likely have to get a separate policy.

Fortunately, there is a precedent of insurance companies generally being supportive of renewable energy, including small wind turbines. Jason Cassee, director of marketing for personal insurance for Fireman's Fund Insurance in Novato, California, said, "Homes with wind systems and green features are more self-sustaining and energy efficient overall."[27]

"They offer benefits to homeowners from a financial perspective and to us from a risk-management perspective," Cassee added. It's worth noting that both State Farm and Allstate Insurance advertise that their offices are powered by renewable energy, so it seems appropriate that they, too, understand the risk-savings of such an investment.

If you have a grid-connected system, it should be comforting to know that insurance companies will often pay the expenses for inspection, reconnection, and permits needed to get the system back online after an outage, as well as the cost of replacement power until the system is replaced or repaired to the manufacturer's specifications. The homeowner is also reimbursed for any power-generation income lost while the system is down.

The best policy is honesty and complete information.   You don't want to deny your insurance agency crucial information, such as the existence of a wind turbine. Not only is this approach illegal and unethical, but the insurance company could ultimately deny future claims based on disclosure issues, especially with Google Earth and OpenStreets offering satellite photos of your property. On the other hand, it's true that simply informing an agent that you want to insure a "wind turbine," and expecting that they know what you are talking about, can be costly. Uninformed insurers may perceive greater risk and, in response, raise their rates, or simply deny coverage.

Appurtenant structure.   Instead, Sagrillo suggests that you say you wish to insure a "wind mill and tower," since both are terms everyone is familiar with. It's a good idea to also specifically ask for coverage of an *appurtenant structure* on your current homeowner's policy. This is a term used by the insurance industry to refer to any uninhabitable structure on your property, such as garages, silos, barns, sheds, towers, satellite dishes, and so on, although note that some companies place any farm structures in a different category.

According to Sagrillo, it is common for insurance companies to cover appurtenant structures for the total cost of materials and labor, meaning the entire installed cost of your system. However, as mentioned earlier, your system must be on the same premises as your house.

Cost.   Insuring a wind system as an appurtenant structure is relatively affordable, and shouldn't raise your premiums very much. If you want to raise your coverage, you should be looking at $2 to $3.50 extra per year per $1,000 of additional coverage. In general, rural properties have somewhat higher rates because they are usually farther from fire departments.

Note that you might qualify for a discount with some insurance companies if you can demonstrate that the equipment is UL-listed (Underwriter's Laboratories), meets the National Energy Code, is approved by an energy commission, meets national performance standards, and is fully warrantied by the installer and manufacturer. If you are an owner of a Leadership in Energy and Environmental Design (LEED)–certified home, Fireman's Fund offers a 5 percent discount off its premium.

What You Want in Insurance Coverage

Vandalism   Unfortunately, vandalism is one of the biggest concerns for wind turbine owners. Incidents are relatively rare, luckily, but the most frequently filed vandalism claim is from guns being fired at turbine blades or generators. Another risk is intentionally undone guy cables, which can cause a guyed tower to come crashing down. In either case, damage can be substantial.

Acts of God   As with homes, any wind system should include this coverage. While most wind towers are designed to withstand wind speeds greater than 120 mph, few can withstand a direct hit by a twister.

Lightning strikes   Hopefully, you took proper steps to ensure that your wind system is adequately protected against lighting, so you shouldn't have any problems. However, there's always the possibility of a freak accident, so you want to be covered.

Fire   If you want to scare yourself for a moment, Google "wind turbine fires." Go ahead and enjoy the fiery devastation, then take a breath and notice that you're looking only at utility-scale machines. When it comes to small wind systems, fire is quite rare. We hope you set your system up to at least the National Electric Code (NEC). Otherwise, if you try to make a claim, your policy may be canceled.[28]

Theft   While it may seem unlikely, there have been claims filed for wind turbine parts that have gone missing, both from the ground level and from atop towers. Bad form, Peter Pan!

Flood insurance   In the United States, flood insurance is administered nationally by Federal Emergency Management Agency (FEMA). The program is not without controversy, and flood insurance can be quite expensive for those situated in low-lying areas. If you live in a floodplain, it's a good idea to get an estimate before you start building a turbine. You don't want your turbine to end up as a boat propeller.

Liability Insurance

Points to consider:

- Incidents are extremely rare, but it's important to protect against damage to other people's property and personal injury.
- Cost tends to be inexpensive, typically an additional $6 to $50 dollars annually for up to $1,000,000 coverage on your homeowner's liability coverage.
- Your local utility may dictate level of liability.

Liability is definitely something worth thinking about when it comes to wind systems, even though accidents are rare and claims are exceedingly rare. Still, there is considerable risk, and Paul Gipe has cataloged at least 20 deaths that have occurred in the wind industry, including several with small home-based systems.

In July 2010, a three-year-old boy was killed in Ontario, Canada, after a pole fell off a small residential wind system that was under repair. According to the *Toronto Sun*, the boy had been playing beneath the 34-meter tower with other children, and their movement caused a holding strap to loosen. The boy became pinned by one pole and was struck by a second. He was immediately rushed to the hospital but reportedly died two hours later. A handful of small wind owners have also been killed by falls or collapsing equipment, and the utility-scale industry has seen electrocutions and amputations.

In order to protect a system, liability coverage needs to address two areas: property damage and personal injury or death. Possible accidents could include falls, falling objects, or electrocution.

Like insurance on the wind system itself, liability coverage is relatively inexpensive if associated with a homeowner's policy. In many localities, the basic homeowner's liability coverage is for $300,000. Increasing this coverage to $500,000 is typically only an additional $10 per year, according to Sagrillo. Spend up to $50 more a year for liability coverage of $1,000,000 or more.

If your system is connected to your local utility, accept the fact that liability insurance may be a requirement, and may have a specified level of coverage.[29]

Companies Offering Insurance for Renewable Energy

Lexington Insurance
(Program: Lexelite
Eco-Homeowner
bit.ly/insurelexelite
(617) 330-1100

Fireman's Fund
Königinstr.28,
80802 München
+49(0)8938004511
corporate.affairs@
allianz.de

Allstate Insurance
P.O. Box 12055
1819 Electric Rd.
S.W.
Roanoke, VA
24018
1-800-ALLSTATE

Warranties

If your turbine shuts down due to an impact from a UFO, contact your insurance agent (and the news). If it locks up on a sunny, breezy day and never turns back on, you will need to contact the installer or the manufacturer and remind them of your warranty.

Many people choose to forgo extended warranty coverage, believing their property insurance will cover things like prematurely worn bearings, only to be surprised when that claim is denied. It is better to rely on your warranty as much as possible. Only involve your insurance company when absolutely necessary—filing an insurance claim may result in higher premiums down the line, or possibly even denial of coverage if it happens multiple times.

Most wind power systems and their controllers come standard with a one-, two-, or five-year warranty on workmanship and parts (Figure 6-14). Inverters for wind-electric systems normally have two- or five-year warranties. And the certified installer should offer at least a six-month warranty on installation. However, the expected lifetime of most turbines is 20 to 30 years, so some thought has to go into who is receiving your funds and faith to keep your dream alive.

FIGURE 6-14   Your wind system is a significant investment, so you want to make sure it is well protected. *Stephen Wilcox/DOE/NREL.*

In Riverhead, Long Island, Kevin saw one of his pioneering neighbors install the first 10 kW wind turbine on the island. Later, he watched it fail consistently. When the turbine manufacturer went out of business, the warranty was no longer honored. The installer had left the state, so each time there was a problem, he had to travel hours back to Riverhead to conduct repairs. Eventually, the cost of "extraditing" the repair crew became so burdensome that the owner decided to dismantle the turbine and reuse the copper cable elsewhere.

You may want to stick with the original equipment manufacturer (OEM), sign a deal with a third-party provider, or perhaps venture into managing operations and maintenance in-house. On any deal, the devil is in the details. Carefully consider the terms of the agreement and the guarantees provided. Don't hesitate to ask questions.

The biggest drawback to manufacturer's extended warranties is higher cost. They can run as much as 80 percent more than third-party services. Of course, since the original manufacturer has access to the complete history of their machines, you will enjoy a better integrated service package. Moreover, if you have high ambitions to expand to a fleet of turbines, you may benefit from discussing discounted longer-term relationships. Or you may simply get discounts on parts.

Another advantage of using an OEM is that some are adding monitoring capabilities, so they can stay abreast of issues before they happen. Finally, it's sometimes of benefit in keeping lenders and insurers happy for you to consider OEMs.

In-House Management

"In-house" is another word for you or your company cutting the umbilical cord from the manufacturer and doing it yourself. Theoretically, it can save you thousands of dollars. But you should know what you're getting into.

You need to really understand all the parts of your system. Question other owners about that specific model's repair history, check out forums, and ask yourself if you are really ready to troubleshoot problems. Note that smaller turbines generally have fewer parts to fail and less maintenance. Larger turbines are generally sturdy, but upon opening up the nacelle, you might find an amusement park of contraptions that are not as familiar as your turbo diesel truck motor.

Note that if you received government incentives, the installers may be required to offer longer warranties. For example, in California, a ten-year warranty is mandatory, largely because the state wants to ensure that it gets a good return on its subsidies.[30]

Third-Party Warranties

Your local certified installer may offer an extended warranty for an added annual or biannual cost—a package maintenance and fix-up agreement. Ensure it doesn't overlap with what may be already included from the manufacturer.

You can also explore a third-party warranty. Coverage is often sold in blocks of five-year periods. Third-party warranties are often calculated on a maximum probable loss (MPL) basis. That means providers look at the weakest links of wind turbines, measure the probability of loss, and calculate the foreseeable costs.[31] If something goes wrong, they will cover up to the limit of the policy.

Consider a five-year warranty for a 10 kW wind system:

OEM Warranty Costs	5 years × $3,000 = $15,000 warranty bill
Third-Party Warranty Costs (with deductions)	$10,000 in limits, 5-year block = $3,000 warranty bill

In this example, the wind turbine owner would save an estimated $12,000 with a third-party warranty, reducing warranty costs by 80 percent. One possible drawback is that there is less of a guarantee that service provided through the third-party warranty will be of high quality, by properly trained technicians and with premium parts. Still, references from satisfied customers can give some insight into the type of long-term relationship you would be expecting.

On the surface, the choice between OEMs, third-party warranties, and self-maintenance may appear to be relatively simple, based on cost considerations. But before making this cost-critical decision, consider all the factors at play (Table 6-4). Remember to read all the details of any contract before you sign.

TABLE 6-4　Comparison of Warranty Options on a Small Wind Turbine

OEM	Third-Party	In-House
Benefits	**Benefits**	**Benefits**
Qualified technicians	Lower fixed costs	Lower fixed costs
Remote monitoring	remote monitoring (possible)	Remote monitoring (possible)
More price control	More direct control over O&M	More direct control over O&M
Integrated service package	Annual deductible structures	Local resources
Lower risks		
Best OEM parts pricing		
Disadvantages	**Disadvantages**	**Disadvantages**
Higher cost	Handling emergent issues	Rental or capital cost for maintenance equipment and labor
	No recalls	No recalls
	Costly upgrades	No free rides and no discounts

O&M stands for post-warranty operations and maintenance agreement.

Resource

Power Guard Insurance
19000 MacArthur Blvd., Suite 575, Irvine, CA 92612-1447
Tel: 1.877.556.0090
Fax: 1.949.253.9665
Email: info@powerguardins.com

Financial Resource Checklist

To make sure you take advantage of all incentives that may be available to you, please consider calling your local authorities and electric utility. You could also start with an Internet search by typing: +wind+[name of financial incentive] +[your city, local, state, or federal government].

If you are in the United States, check out dsire.org, which has a long list of incentives by location. Outside of the United States, you can use Paul Gipe's website windworks.org/articles/feed_laws.html, where there is an extensive listing of all countries and their incentive policies.

✔	Incentive Programs	Terms, Amounts, Comments
	Capital Subsidies	
	Grants and Rebates	
	Production Incentives	
	Feed-in Tariffs	
	Renewable Energy Credits	
	Net Metering	
	Tax Incentives	
	Property Tax Reductions	
	Sales Tax Reductions	
	Investment Tax Credits	
	Corporate Investment Tax Credits	
	Production Tax Incentives	
	Research, Development, and Manufacturing Tax Credits	
	Accelerated Depreciation	
	Value-Added Tax Reductions	
	Import Duty Reductions	
	Taxes on Conventional Fuels	
	Tax Holidays	
	Loans	
	Mortgage/Home Equity Loans	
	Revolving Loan Funds	
	Green Loans	
	Public Investment Loans	
	Miscellaneous	
	Income Taxes	
	Insurance	
	Warranties	

Handy Cost Calculators

Now that we have explained the financing behind small wind systems in some depth, let's talk about simplifying your work. Most of the following tools are designed for utility-based projects, so don't be alarmed by some of the costs that do not pertain to your project. But these calculators can help you get a sense for some of the variables involved.

New Roots Energy newrootsenergy.com 	For a three-month online subscription ($99), New Roots Energy can run calculations for you with just a few inputs and provide a financial analysis that includes: • Self-populating wind speeds using 3Tier wind layer data • Latitude/longitude positioning and report visualization maps • Calculation of federal, state, and utility incentives (United States) • Energy production calculation using manufacturer-verified power curves • Accurate financial and equipment configuration comparisons • Presentation-ready reports for end users, CFOs, and CPAs
Windpowering America Wind Energy Payback Period Workbook bit.ly/wpa_smallwind 	Specifically for small wind energy systems, this is a downloadable spreadsheet tool that can help quickly analyze the return on your investment. It provides four sample small turbines, namely Seaforth's AOC 15/50 (50 kW), Bergey's Excel-S/60 (10 kW), Jacob's Model 29-20 (20 kW), and Skystream's Whisper 500 (formerly known as Whisper H175). It provides one column for you to specify the turbine, site, and financials in three easy steps. If you follow the instructions, it can provide you a rate of return, a cash flow chart, power graph for your turbine, and the payback period. Note: You need Excel to be able to use all the macros within this easy-to-use worksheet.
Bergey Windpower Small Turbine Project Calculator bit.ly/bergeyfeasibility 	This is an Excel spreadsheet for the economic feasibility of a small turbine. It uses the Bergey 10 kW turbine as a default, but permits you to change financial data for any turbine. What is unique is that it allows you to include your federal and state incentives (if you are in the United States), loan information, and it will calculate the cash flow, including internal rate of return (IRR), monthly payments, and depreciation. It has charts and graphs that display project metrics in an easy-to-visualize format. It's one of the most thorough, free, and nicely displayed worksheets out there.

Wind Turbine Economic Feasibility Calculator bit.ly/PSUwindcalc 	Penn State University boasts a free application for a wind farm that can be sized down to the individual small turbine user. It is a very comprehensive, downloadable spreadsheet, designed for those who are well acquainted with regulatory, construction, and associated costs. If you would like to manually plug in all your costs, here is the download, and use the step-by-step instructions to get you through the forest.
Paul Gipe's Cash Flow Modeling for Community Wind Projects bit.ly/wind-works 	You will need to enter specific information about the type of turbine you are considering, the estimated annual average wind speed, electricity use and electric rates, and financing and income taxes. The program will estimate the cash flows for investing in a wind turbine and the rate of return on the cash investments.
Windustry Wind Project Calculator bit.lyWindustryCalc 	Designed by Alice Orrell and Brian Antonich, the Wind Project Calculator is a downloadable spreadsheet that is arguably the most comprehensive and easiest of the financial analysis tools to use, from the small-scale to utility scale. The instruction sheet is included on a worksheet to guide you through the process. Use the calculator to begin to understand what the return from a potential wind project could look like; to understand the effect of local, state, and federal taxes and incentives on project economics; and to do preliminary sensitivity analysis on how different project variables such as changes in capacity factor, financing terms, or ability to utilize various incentives can affect the returns of a project. Note that you will input some of the financials yourself, including information about power purchase rates, tax utilization, government grants and other incentives, yearly expenses, and information about project ownership.
National Renewable Energy Laboratory's Wind Energy Finance Application bit.ly/NRELturbinecon 	NREL outdid itself with this online wind turbine finance analysis tool. You'll need to set up a user name and password, but it's free. This seems to have the most intuitive design of the many tools here, and is quite comprehensive. You fill out forms with assumptions, capital costs, and operating expenses, and the analysis report is dynamic. You will need to convert your kilowatts into megawatts for it to work with your small turbines, and if some of the items they are asking for do not apply to your situation, you can simply type 0 in the text box to complete the analysis. If you get lost, click the help button.

Danish Wind Industry Association's Wind Energy Economics Calculator bit.ly/wp_turbinefeas	This is a nicely designed, quick online tool with limited input necessary. For those who need to get an idea of their real rate of return (compensating for inflation), it is a good start. You would input the cost of the turbine system installation, the life expectancy of the turbine, an estimated annual energy output of your turbine, the existing electric rates, and your estimated operations and management costs, and it immediately spits out your real rate of return. A few clicks and you get your economic feasibility estimate. Check out the rest of the site to get a more fundamental understanding of how the calculations are made.

Summary

A small wind system is a sizable investment, so it isn't something you want to choose on a whim. Fortunately, a number of incentives are available that can help make it more affordable, from tax credits to rebates and grants. And once you have your system, you'll want to make sure it is well protected with insurance and warranties.

When all is said and done, all the feasibility studies in the world can't predict without a doubt if a wind system will be right for you. There's always going to be some uncertainty and some risk. It's not a bad idea to take everything you have learned to your accountant, who may be able to give you some additional advice to help you realize your dream.

CHAPTER 7

Navigating Permits and Zoning

Given the fact that you can count on one hand the number of other towns in Maine that have a wind turbine ordinance—if there's any at all—I think it's unrealistic to expect us to get together an ordinance from scratch that covers all the safety issues.

–STEVE BENNETT[1]

Overview

- How do I determine if I am legally allowed to install a wind turbine?
- How do I secure the necessary permits?
- How do I request an exception to a restrictive zoning policy?
- How do I work with the utility?
- How do I make sure my system passes inspection?

Here is a test to see if you need to read this chapter:

Construct a 30-foot pole made of PVC tubing, and install it vertically directly on your most visible property line. Then raise your national flag positioned upside down, or some other highly provocative banner. In addition, play a looped recording of any audible sound of at least 60 decibels as measured from the nearest neighbors. Next, stand on your roof and angle a handheld mirror toward the sun, then flicker the reflection toward the most trafficked streets in your neighborhood.

If you do not receive any form of feedback from neighbors, passersby, or the municipality, you do not need to read this chapter. For the rest of us, read on before you buy your wind system equipment. If you already bought your turbine and other gear, keep it boxed up, and don't advertise your recent purchase just yet.

In the good ol' Ancient Near East, during Hammurabi's time, the first known building code specified:

If a builder builds a house for someone, and does not construct it properly, and the house which he built falls in and kills its owner, then that builder shall be put to death.

Things have changed a bit since then, but municipalities are not typically known to keep up rapidly with the technologies.

As you may know, a building permit is required in most jurisdictions for any new construction, for adding on to pre-existing structures, or for major renovations. Typically, the construction project must be inspected during building and after completion, and an official must certify that you met national, regional, and local building codes. If you are caught without a permit, you may be fined and your new structure could be demolished. Luckily, you aren't put to death, right?

This chapter discusses the unavoidable, and sometimes most difficult, part of the process: the paperwork and public relations. We will provide a survey of typical ordinances and will walk the reader through understanding local laws, as well as how to avoid or get around obstacles, so you can get your dream system up and running in as little time as possible (Figure 7-1).

Getting Started on Permits

First, let's find out what you need to do by calling up your local municipality and requesting all forms you'll need to get your project authorized. A good place to start is your town's website, or do it old-school and drop on by city hall.

Figure 7-1 The Bergey Excel 10 kW wind turbine is designed to supply most of a home's needs with an average wind speed of 12 mph. *Bergey Windpower/DOE/NREL.*

Step 1: Contact Your Municipality

"Hello, my name is [your name], and I live at [address], and we are planning to install an appurtenant structure on our [residential/commercial/agricultural/industrial] property. Would you please tell me what resources, such as zoning regulations and permits, pertain to this?"

Notice: There was no mention of a "wind turbine." At the beginning, it is best to be on the safe side and not specifically alert local town officials that you intend to build a 50 kW wind turbine in your vineyard, until such time that you and/or your contractor have read through the existing zoning ordinances so you are prepared to best make your case. Instead, we used the term *appurtenant structure*, which we introduced in Chapter 6. An appurtenant structure is usually defined as a non-dwelling additional structure—such as a garage, silo, barn, tower, or satellite dish—on the same premises as the main building, but of lesser value.

You may have online access to your local building codes, which makes your research even easier. You may also try the free website ecode360.com, which contains many local codes for the United States and Ontario, Canada. Some municipalities may have a small fee, but from your sofa, you can now look up, search, copy, and paste your code. Other countries might have something similar, so check it out. It makes it easier to complete step 2.

Step 2: Read the Code

Bunker down and read the building code. It will save you headaches later. In addition to references to appurtenant structures, you are looking specifically for info on an *accessory wind turbine*. However, if there are no such regulations, don't rush ahead with your project assuming that you are in the clear. There might not be code specific to obelisks made of Jell-O, either. That is a quick way to get mail that says something to the effect of "cease and desist," along with a code violation number and a hefty tariff, not to mention a less-than-empathetic building inspector.

It is likely to be most beneficial to find and highlight regulations that may be related to wind turbines, such as: 1) height limits, 2) noise limits, 3) structure usage or purpose, 4) setback distances, 5) visibility, 6) protected zones, and 7) aesthetics. If you are fortunate to obtain the online version of your code, use your search engine or browser to sift for these key words through the small novels of legal and technical jargon. If you are toying with putting a turbine in various locations and want to compare building codes between jurisdictions, you might want to create a comparison chart with each of the aforementioned key words at the top of the columns, and make a row for each corresponding area.

Kevin was recently asked to assist several towns with modifying the building code to accommodate wind turbines, and he used a similar comparison chart, which was helpful in the process. One thing he noticed is that certain towns had a detailed list of acceptable adjusted decibel levels for every hour of the day. They went to the extent of denoting frequencies, duration, and intensity, as if their regulations were being published in a peer-reviewed audiology journal.

 Power Up! *More and more local zoning regulations do specify "wind power turbines," "wind energy generation accessories," or even specify the type of wind turbine—wind farm, industrial, utility, residential. In this case, you have been rushed to the front of the line and provided special client access. Now, you just need to know if the legislation gives you the green light. Either way, the paperwork begins.*

Step 3: Complete Applications

You are about to install a rather sophisticated piece of equipment, and your local government may not be familiar with it. You are (probably) not a Doctor Frankenstein creating some abomination. However, the community has a right to know about this alleged "beast" in case they need to mitigate some unplanned incident. And we would like to share with you some tips, terms, and concepts that make this process a breeze. Hopefully, it will head off any pitchfork-wielding mobs.

As you may know, most places have some form of zoning rules, which specify which land can be used for what purpose—agricultural, industrial, residential, "mixed use," and so on. Zoning is often even more detailed than that, specifying maximum square footage, building height, number of parking spaces, lot size, and so on. The concept has been around since ancient times. If you live in a true wilderness area or a rare community without it, you may be able to skip over the concept of zoning. But some of the other aspects of this chapter may still apply to you, including national building codes.

Some salient features of zoning are discussed in the following sections.

Permitting Fees

As fun as they are, a wind turbine is not a toy (Figure 7-2). It is a device that requires knowledge of mechanical engineering, construction, electricity, safety procedures, and even psychology. So don't be surprised if you are required to submit a building permit, use permit, zoning permit, or "plot plan." Permitting requirements, procedures, and fees vary widely among counties, from waived fees to thousands of dollars. A range of $400 to $1,600 is typical.

Keep in mind that there also may be other fees for public notification, hearings, or even environmental impact studies. Find out what fees you face and offer to provide documentation that proves they are unnecessary. If a fee seems excessive, ask if you can get it reduced or waived. As we've mentioned before, oftentimes your best strategy is pointing out that small wind turbines avoid many of the issues that surround utility-scale machines.

Figure 7-2   In April 2010, the tenth Kansas Wind for Schools project was installed at Hope Street Academy in Topeka, with support from the DOE's Wind Powering America. *Ruth Douglas Miller/DOE/NREL.*

Conditional (Special) Use Permits

While reading the building code you might have noticed a reference to a *conditional* or *special-use* permit. Look at that, you are special!

Technically, a conditional or special-use permit allows a certain specified structure, as long as all the stated conditions are met. So if zoning rules specifically list your type of wind turbine (micro, home, farm, etc.) as an approved conditional or special use for your property, then you need only comply with the conditions, which usually cover minimum lot size, tower height, setbacks, and electrical code compliance. The manufacturer or dealer of the wind turbine system, and certainly the registered installer, should help with the documentation.

To receive a conditional permit, you may need a public hearing at a plan commission or city council. Of course, this means the decision whether or not to grant the permit can become politicized. If you are hoping to build a wind turbine in a residential neighborhood, the local zoning authority may want to review the potential impact on the area.

Use Variance

This is the step that we see as the walking-on-eggs challenge. If the zoning specifies that small wind turbines are not an allowed use at your site, you won't be able to apply for a conditional-use permit. Instead, you will need to apply for a *use variance*. Similarly, even if your zoning code doesn't explicitly bar a turbine, some other aspect of the regulation might effectively exclude it.

A use variance is simply a project-specific exemption from local zoning rules. Most zoning systems have a procedure for granting variances, which are often supposed to address some perceived hardship on the part of the property owner.

When it comes to wind turbines, height restrictions are the most common obstacle, since many zoning rules prohibit any structure higher than 35 feet, unless exemptions are already in place for wind energy systems. That height restriction is commonly on the books due to the historic fact that firemen were formerly unable to fight fires higher than 35 feet, given the limitations of wooden ladders and old water pumping technology. Today, that's not a problem, but residents often defend the height limit because they are afraid that taller structures would hurt the character of the neighborhood.

So without an exception, you may need a variance for a turbine tower. If you need to appear before a planning board, do some prep work to be able to provide answers to an audience of your peers. We hope you took public speaking in school.

Your application for a variance should cite the specific rule (e.g., Section 32.1: Height Limit). The catch is that you need to show adequate reason for the requested exception. You need to prepare an argument that your installed wind turbine (1) compensates for some unnecessary hardship, (2) is needed for reasonable use, (3) does not alter the essential character of the neighborhood, (4) and/or is the least intrusive solution.

Explain that the impact of your wind system will be minimal (Figure 7-3). Calmly point out other tall structures your neighbors already accept: water towers, cell phone towers, radio antennas, satellite dishes, and so on.

Figure 7-3 If you live in the wilds of the American West, you may not be subject to any zoning, meaning you can build as many small wind turbines (like this Ampair 600) as you like without a variance. *Ampair.*

Secure the Permit Before Buying Your Equipment

We definitely recommend that you secure the required permits *before* purchasing any wind system equipment. If you build first then ask for permission later, you may anger your zoning committee, giving them reason enough to deny your variance. Expect to pay a nonrefundable application fee, and possibly preparation fees if you have someone else help with your application.

Picture This

To apply for a variance, you may have to submit a detailed structural plan drafted by a licensed engineer, but sometimes documents from your turbine manufacturer or dealer may suffice. We included Kevin's wacky design for a wind turbine to be installed in Nicaragua, made with a free Google application called Sketch-up. It has many models from which you can select to build your own wind turbine design and place it on your site in the Google Earth map for all to see.

Resources

Sketch-up   Google Earth

Love Thy Neighbors

Your primary strategy should be one of diplomatic public relations. While it isn't supposed to be that way, a zoning application can easily morph into a popular referendum on an issue—say, wind power, renewable energy, or land use in general—or it could become a "trial" on an applicant's reputation and character. To be prepared, try to build good, open relationships with your neighbors long before you apply for permits.

Also be aware that some regulations require that you give notice of your plans to anyone living or working within 300 feet of your proposed site. We recommend notifying neighbors further than that as well, and before you even apply for a permit or variance (Figure 7-4). That courtesy will often head off full-blown opposition. A letter like the following suggested by American Wind Energy Association (AWEA) can help address many common concerns.

> Dear Neighbor,
>
> You may be interested to learn that I plan to install a small wind energy system on my property at [site address]. This modern, nonpolluting system will generate electricity solely for my own use, reducing my dependence on the local utility. Any excess generation will be supplied to the utility system, and I will receive compensation from my utility for this exchange at the rate of [$/kWh] (or, will not receive compensation from my utility).

FIGURE 7-4   Some of Kevin's neighbors are angry at his unique green dome house, which has taught him that it's a good idea to foster open communication before you start any major projects. *Kevin Shea.*

I plan to install a [turbine make and model] that will be mounted on a [height] foot/meter tower, set back [distance] feet/meters from the street, and [distance] feet/meters from the [north, south, west, east] property line. This wind turbine uses a blade [length] feet/meter in diameter, and has only [quantity] moving parts. It does not turn on until the wind reaches at least [number] mph (m/s). On calm quiet days, the wind turbine doesn't make any noise. When the rotor is spinning, the audible sounds will reach no higher than [number] decibels (dBa), at a distance of [number] feet (meters) from the wind turbine, equivalent to the sound level of a [conventional household item emitting sound levels which will barely be audible over standard ambient noises caused by the wind].

[Manufacturer] located in [city, state] has installed [number] of [turbine make and model]. They have been certified by [authorized certifying agency, e.g., AWEA] for meeting stringent performance and safety standards, and have a proven track record of producing energy quietly, cleanly, and safely. If you have any questions about the proposed installation, please feel free to contact me at [contact information].

Sincerely,

[Your name] (signed in blood)

Be aware that local ordinances sometimes attempt to issue stringent code that goes beyond the rules of the United States Federal Aviation Administration in terms of clearances for aircraft. If you find that to be the case, you might question the municipal officials on their authority on the issue, and for their reasoning.

INTERCONNECT

Going Off the Grid as a Matter of Principle

When Maynard Kaufman moved into a saltbox house in rural Bangor, Michigan ten years ago, he made a conscious decision to go off the grid (Figure 7-5). "I would do it again," Kaufman told us in an interview over the phone.

Kaufman could have connected to the grid for $10,000. Instead, he opted to spend $30,000 on a solar power system and $12,700 on two wind turbines. "It was totally a matter of conscience," Kaufman explained. Specifically, he told us he bought a 1-kilowatt PV array on a tracker and two Southwest Windpower 1-kilowatt-rated turbines.

Kaufman retired from teaching environmental sciences at Western Michigan University, and he has long been active in the local organic farming movement and with green and anti nuclear power groups. Calling his home Sunflower, Kaufman explains on his website (www.michiganlandtrust.org) that he wanted to live as free as possible from nuclear and fossil fuels. He also hoped to establish an ecovillage in the area dedicated to clean and self-sufficient living, but he told us the plans fell through during the recession.

Kaufman explains on his website that during the cloudy Michigan winters, he gets most of his energy from his Southwest Windpower Whisper H-80 wind generators, which are mounted on 4-inch pipe 66 feet high. "This so-called hybrid system, using both sun and wind, is an extremely necessary part of the total system," Kaufman wrote (Figure 7-6). He told us, "Because we use energy-efficient appliances that is a great plenty, but I do not bother to keep records."

In addition to producing its own renewable energy, the Sunflower house is loaded with passive design elements, including earth sheltering (Figure 7-7).

FIGURE 7-5   When Maynard Kaufman moved into this house in Bangor, Michigan ten years ago, he could have spent $10,000 to connect to the grid. Instead, he spent $30,000 on a solar power system and $12,700 on two wind turbines. *PJ Chmiel/Flickr.*

FIGURE 7-6   Kaufman, a retired professor of environmental sciences, said his decision to go with renewable energy "was totally a matter of conscience." His two Southwest Windpower Whisper H-80s are rated at 1 kW each and are mounted on 4-inch pipe 66 feet high. *PJ Chmiel/Flickr.*

FIGURE 7-7   "Sunflower," as Kaufman calls his home, also boasts high-efficiency appliances and passive design elements, including earth sheltering. *PJ Chmiel/Flickr.*

Kaufman told us, "When I built our house with renewable energy ten years ago there was not yet a market for surplus energy. If I did it now I would try to sell the surplus back to the energy utility and use the grid for storage instead of batteries. I replaced them after 9½ years at a bit over $4,000, double the price I paid for the first set. I would also include more circuits of 24-volt DC power in the house."

Taking a broader view, Kaufman said, "The general issue is the value of distributed versus centralized energy production. Production where it is used avoids line loss and is more efficient. Also, as the grid gets older it is more subject to breakdown and outages, a problem that home power avoids."

When asked for advice for others interested in home wind power, he said, "The main thing to recognize about wind turbines is that they are more difficult to set up than PV panels. Also, they are really used only in rural areas with plenty of space. Many people are interested in renewable energy installations, but are put off by the up front investment. Yet they buy vehicles that are even more expensive."

Step 4: The Hearing

If you are required to attend a public hearing, you need to be prepared. Ironically, despite the name, there often isn't a lot of listening at a public hearing, but we offer some helpful suggestions based on sound advice from small wind installer and author Mick Sagrillo.

Shoot the Breeze

In addition to an aforementioned formal letter, chat up your project with key neighbors well in advance of a public hearing. Answer questions in a friendly, nondefensive manner. Allay fears by distributing appropriate articles on wind turbines or equipment specifications whenever the opportunity arises.

Engage Officials

Remember that zoning officials live in your community and must answer to the voters. Also remember that discussions are probably not going to be limited to your zoning hearing. They will take place in the officials' religious buildings, down at the hardware store, and in the local bar or restaurant. Avoid meetings in bathrooms.

Work the Media

Get your local newspaper or news website to do a story on your proposed wind energy project. You can even write and send out your own press release. Include as many facts and benefits as you can. Sagrillo suggests staying away from any negative aspects of the project. Many communities are also now using some sort of social media, such as Twitter, Facebook, blogs, and so on. Join them and let them know what you are doing or give some wind turbine facts. Or start your own account, and invite others to join you.

Supporters Needed

When it comes time for your zoning hearing, pack the room with local supporters. Again, zoning hearings are not supposed to be popularity contests, but they often end up that way. If you find a vocal advocate—actor and activist Leonardo DiCaprio might be available—invite them to speak on your behalf. Also, seek out and invite anyone in your community or nearby who has successfully installed a wind turbine.

Hot-Air Rule

Presume that meetings could get out of hand. Unfortunately, local zoning committee members rarely have training in how to facilitate meetings, Sagrillo warns. There may be an especially vocal attendee who doesn't respect the rules. If the discussion devolves into a shouting match, try to appear calm while sweating or soiling in your suit. If things seem unmanageable, suggest that the meeting be continued with an independent neutral facilitator.

Instant Rebuttal

At the zoning hearing, you are the alleged expert, and the focal point. Control the flow of information. Sagrillo suggests responding immediately to all wild claims about wind generators and insisting that extraordinary claims require extraordinary proof. It isn't hard to find the websites that opponents usually grab their information from. Some are notorious for misquotes of valid resources, or they often transfer what is in the news about large wind turbines directly onto your small turbine. Often the sentence, "That applies only to larger utility wind turbines," could resolve most claims (Figure 7-8).

Just the Facts

Sagrillo suggests coming to the meeting with copies of fact sheets on the issues that are likely to be raised, addressing concerns about property values, noise, bird kills, TV reception, or any other topic that arose during your preapplication PR campaign. Remember that long involved articles will probably not get read, so use brief bulleted lists and a few paragraphs, with citations from independent authorities. We provide an authoritative quicksheet so you can confidently and concisely address some of the myths and misconceptions in a few choice words.

Don't Know? Get an Expert

If an important question comes up that you do not have the answer to, make like a lawyer on a TV show and ask for a recess. Ask to continue the meeting later so you

FIGURE 7-8 Oftentimes, your best defense at a hearing is to point out that someone's objection to wind power applies only to utility-scale turbines ... such as this 20-turbine installation in upstate New York. *Kevin Shea.*

can look into the issue. If the question seems important, ask an expert to the next meeting to support your position. For example, a cell phone technician can explain that towers do not cause TV or radio interference, Sagrillo suggests.

We Will Get Back to You on That

Sagrillo also suggests that you can avoid delaying tactics by opponents by suggesting "homework," with due dates for yourself or the zoning committee members. This will keep things moving, and may give you a chance to talk to committee members privately to better understand their primary concerns.

Resources

AWEA small wind
turbine permit
bit.ly/aweapermit

AWEA small
wind factsheet
bit.ly/aweafacts

Utility Approval

Without a doubt, no grid-connected small wind energy system can be erected, constructed, or installed until the utility gets on board with the plan. The project size and details must be approved by the utility, and all required interconnection paperwork must be submitted before construction can begin (Figure 7-9). Contact your utility and follow their instructions for approval.

FIGURE 7-9   Before your utility starts accepting power from your wind system, you need a net metering agreement in place. You also may have to prove that your equipment meets safety standards. *Brian Clark Howard.*

Step 5: Engineering Verifications and Inspections

For better or worse, engineering verification of small wind turbines is not currently routine, since the industry is relatively small and immature. However, there is a good chance your local town building inspector will pay you a visit once your project is complete, or even during construction.

Kevin remembers the day his town inspector came to check out his Skystream. The inspector claimed no experience in wind turbines, and just asked how high the turbine was and if it was approved by the power utility. He then provided his stamp of approval. To be prepared for an inspection visit, have your system's specifications and certifications in a handy place. If you have any concerns about what you might need, check with the manufacturer.

Be aware that although a wind turbine might meet certification by a nationally recognized testing laboratory, such as the Small Wind Certification Council (SWCC), building inspectors are not necessarily going to accept this without reservations. Remember, SWCC certification is primarily a verification of durability, function, power performance, and acoustic characteristics of small wind turbines. It does not ensure that the wind system necessarily meets the local electric code.

Underwriters Laboratories (UL), the gold standard in product certification, has announced that they are in the process of creating two new small wind turbine standards to address this issue:

- UL 6142 Small Wind Turbine Generating Systems
- UL 6171 Wind Turbine Converters and Interconnection Systems Equipment

Once complete and deployed in the field, a UL rating should help smooth over conversations with inspectors, who are likely to be familiar with the Underwriters Laboratories name.

Brief Checklist

Here is a handy checklist to help you get through the permitting process, which we modified from AWEA's checklist.

1. **Contact your county planning department or permitting agency.**
   - Find out if small wind energy systems are addressed by local ordinance and, if so, get a copy of the law. (If not, see item 2.)
   - Learn the relevant permitting procedures.
   - Ask what documents you'll need. Are you required to submit plans from a consulting engineer, or will plans from the turbine manufacturer or dealer suffice?
2. **Review the applicable standards and restrictions.**
   Small wind energy systems need to meet local rules, which may include:

· Minimum parcel size	· Allowable tower height	· Setback distances
· Noise levels	· Aesthetics	· Equipment certification
· Electric code compliance	· Building code compliance	· FAA requirements (USA)

- Contact your utility or local building department, which may refer you to the website of AWEA or another certifying agency for a list of certified small wind turbines.
- Standard drawings of system electrical components and an engineering analysis of the tower may be required to show compliance with building codes.
- Local ordinances sometimes attempt to issue stringent code that goes beyond the rules of the United States Federal Aviation Administration in terms of clearances for aircraft. If you find that to be the case, you might question the municipal officials on their authority on the issue, and ask for their reasoning.
- Other siting restrictions: Small wind energy systems may be subject to local restrictions that relate to coastal areas, scenic highway corridors, or other specially designated areas.

- Ensure your turbine will be in compliance with state and national codes. Your local municipal building office should advise you of this.

3. **For grid-connected systems:**
   - Get approval from your utility. You may need to show your permitting agency that you have, in fact, informed the utility of your intent to install an interconnected wind generator.
   - Apply for any rebates you may be eligible for.
   - Apply for your interconnection agreement. Fortunately, many utilities now have simplified, consumer-friendly interconnection agreements, which spell out the terms of net metering and technical rules. Utilities are often required to process these applications within one month.

4. **Notify your neighbors.**
   - Some communities require that you give notice of your plans to anyone living or working within 300 feet of your proposed site. We recommend notifying neighbors further than that as well.

5. **Comply with permitting requirements.**
   Permitting requirements, procedures, and fees vary widely among counties and countries.
   - Building permit, use permit, zoning permit, or plot plan fees can range from a free waiver to $1,600, or a certain percentage of the total expense to install.
   - Other costs for public notification, hearings, or environmental impact studies may range from a few hundred to several thousand dollars.
   - If a particular fee seems excessive or inappropriate for your situation, find out the basis for it. You may be able to avoid it or have it reduced!
   - Your system might need to be installed by a licensed local contractor.
   - Obtain a final inspection sign-off prior to claiming your rebate. Net metering provisions usually take effect when the permit is obtained or the wind turbine begins operation.

Summary

That wasn't so bad, was it? While the permitting process is rarely fun, it is an essential part of every construction project, and small wind systems are no exception. Making sure all your t's are crossed before you break ground will go a long way toward making sure you avoid major headaches later.

Plan ahead, read all the relevant rules for your jurisdiction, and work with your utility if you are going to connect to the grid. If you need to get a zoning variance from your town, we've provided some pointers on making the process go smoother (Figure 7-10).

Figure 7-10 Oftentimes, a good strategy when applying for a variance is pointing out other tall structures that the community already accepts, such as water towers (pictured) and cell phone antennas. *Nigel Chadwick/Wikimedia Commons.*

Don't forget, your installer and/or manufacturer is likely going to be an invaluable resource in helping you navigate the permissions process. It is certainly in their best interest to see you get the permit, and they typically have a lot of experience in getting things done. (One more reason to go with experienced providers with a proven track record.)

CHAPTER 8

Wind Turbines Demystified

You know, you're very beautiful. You're also very quiet. And I'm not used to girls being that quiet unless they're medicated. Normally I go out with girls who talk so much you could hook them up to a wind turbine *and they could power a small New Hampshire town.*

–Adrian Cronauer (Robin Williams) in *Good Morning Vietnam* (1987)

Overview

- What are the main parts of a wind turbine?
- What is the difference between upwind and downwind turbines?
- What are the advantages and disadvantages of vertical axis wind turbines?
- What certification programs and standards apply to small wind turbines?

As you are hopefully aware by now, the part that spins is only a small portion of a complete wind system, which is a fairly complex series of interconnected parts, all of which need to be functioning properly. Further, you must ensure that your site has a good wind resource, and that you can legally—and safely—put up a tower.

All that being said, let's take a closer look at what is arguably the star of the wind system: the turbine or wind energy generator.

Drag vs. Lift Turbines

First, let's examine how turbines actually work (Figure 8-1). On the most basic level, a turbine captures some of the kinetic energy from the wind that passes through it. After the encounter, the wind is slowed and the turbine is sped up.

239

Beyond that basic fact, turbines (or windmills) can operate in one of two basic ways. The oldest types are *drag-based*, meaning the blades are pushed solely by the force of *aerodynamic drag*. As wind hits the blades, it pushes them, and they spin. Simple, right?

This is the type of windmill that was likely invented in the Middle East (Chapter 1) and has been used for centuries. A cup anemometer (Figure 5-20) is actually a drag-based design, though it is usually used to measure wind speed, not produce electricity. A Savonius vertical axis wind turbine is also a drag-based design (more on this later, but see Figure 5-30 as an example).

Drag-based turbines are straightforward, but they can't produce much energy. Like an anemometer, a drag turbine's blades can never move faster than the wind because it is pushed by it. Further, for half the cycle of rotation, the back side of the

FIGURE 8-1   Students built this test wind turbine to learn about renewable energy at the Jacob Blaustein Institute for Desert Research in southern Israel. *Brian Clark Howard.*

blades face against the wind. This strict limit on rotation speed reduces the amount of spin that can be applied to the generator, and therefore, the amount of energy that can be produced.

However, you may have noticed that most current wind turbines have blades that are gently twisted or curved. In cross-section, they resemble an airplane wing or boat sail. The reason for that is so the blades can generate *lift*. Lift is an aerodynamic phenomenon that is surprisingly complex to explain, and possibly not even fully understood. The traditional "textbook" explanation of different airspeeds of flow causing pressure differences that "suck" a wing up has been found lacking (for one thing, it can't explain how airplanes are able to fly upside down without rapidly losing altitude). More detailed discussions of lift are beyond the scope of this book, but know that the result of lift is a force that is perpendicular to the oncoming flow direction. Drag, in contrast, is a force that is parallel to the flow direction.

The effect of.lift on a wind turbine is to give the blades an extra push, speeding their rotation up faster than the wind. This is significant, because higher speeds mean you can make more energy, and doubling the rotation speed quadruples the effect on the wind.

In general, the goal of wind turbine design over the past century or so has been to minimize drag and maximize lift, because that increases the efficiency and amount of power you can extract. Sure, high-drag designs still have their place. Anemometers, which only use drag, are obviously useful. You can also still find multiblade, old-fashioned windmills on rural ranches, since such designs provide high torque at low rpms, which is fine for pumping water. Those old windmills (as seen in Figure 1-6) use mostly drag, with some lift.

Anatomy of a Wind Turbine

Let's take a tour of a wind turbine's components (Figures 8-2 and 8-3) before we get into design choices. In this book, we will primarily focus on horizontal axis wind turbines (HAWTs), because they are by far the dominant choice in the industry, despite recent press for their rivals, vertical axis wind turbines (VAWTs). We do cover VAWTs later in this chapter, and also further explain our preference. But for now, let's get to HAWTs, which are also called "propeller-type" wind turbines.

A horizontal axis wind turbine is composed of several main parts, which the following sections describe.

Rotor

Located in front of the turbine, the *rotor* is often the most visible aspect. Like a fan or propeller, the rotor is made up of blades, though in the case of a wind turbine

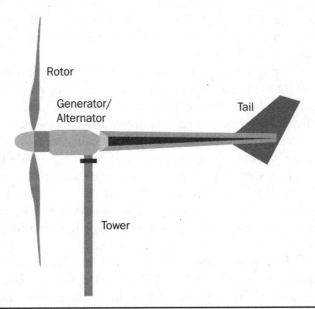

FIGURE 8-2   A small wind turbine is made up of a few key parts, all of which must be functioning properly. *U.S. Department of Energy.*

7.4 Exploded View of *AIR-X marine*

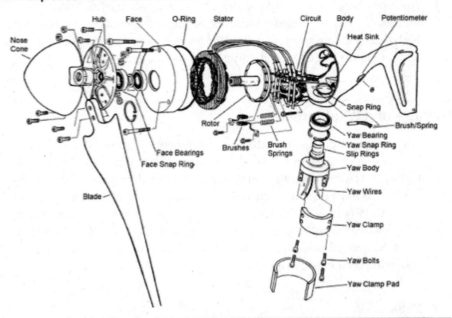

FIGURE 8-3   A cut-away view of a small wind turbine showing the major components. *Southwest Windpower.*

they are shaped to maximize aerodynamics for energy harvesting. Most wind turbine rotors have two or three blades, attached to a hub plate, which is often covered by a nose cone. The rotor is named for what it does—it rotates.

Early European windmills were made with what looks like extended wooden broomsticks and cloth sheets. This produced slow rotation, which worked okay for turning a grindstone or pumping water. If it got too windy, operators would roll up the canvas. The first wind electricity generators used wood blades, borrowing the airfoil design from the airplane industry. Wood is flexible and fairly light and durable, especially when coated with some good marine lead-based paint or (later) polyurethane tape. Wood blades are still used by homebrew crews, even up to a diameter of 100 feet, and wood is what William Kamkwamba used in Malawi. If you are going to make your own, look for straight, knot-free spruce or pine. Avoid hardwoods, as they are too heavy.

Before long, blades started to be made out of metal, which is extremely reliable and durable. The millions of farm and ranch windmills produced by Aermotor of Chicago used galvanized steel blades, starting back in 1888 and on through the 20th century (Figure 8-4). The blades last a long time, but they were heavy and spin slowly,

FIGURE 8-4 First introduced in 1888, Chicago-made Aermotor windmills have been called the "Cadillac of windmills" because of their rugged design and quality workmanship. *Brian Clark Howard.*

which isn't good for energy production. Aluminum is lighter and can be readily extruded, but over time it has a tendency to fail from fatigue. Plus, people eventually noticed that metal blades tend to interrupt television, satellite TV, and wireless Internet signals. Hence, no manufacturer today builds wind turbines with metal blades.

Nowadays, most blades are made from molded plastics, with fiberglass or carbon fiber composites. They may be hollow or have a foam core. These composite blades are relatively cheap and extremely rugged. They are easy to modify and then commence manufacturing. Some say plastics aren't "green," but for a component that will be used constantly for more time than a car, lasting up to 20 years, without having to periodically sand the surface and either wrap it with tape or coat with some paint or linseed oil, it comes close enough.

Blades are typically wider and more rigid near the rotor, where they need the most strength, and then taper off toward the tip. The front of the airfoil (from the leading edge to the thickest part) needs to be nicely rounded, and the back part is pretty much a straight line. Turbine blades are commonly molded with a fixed angle, ranging from three to nine degrees, so they can always present an angle of "attack" to the wind. In some models, they merely have the blades attached at the hub at an angle. Some homebrew builders like "The Dans" Bartmann and Fink take hand-shaped lengths of wood and simply mount them at an attack angle of five degrees, which is a pretty solid average.

The Dans, like many builders, also like to introduce a twist into the blade, often ten degrees at the base and only five degrees at the tip. "Having more pitch at the blade root improves startup and efficiency, and less pitch at the tips improves high-speed performance," they explain. "The wind hits different parts of the moving blade's leading edge at different angles, hence designing in some twist."

Naturally, there are slight differences in blades among manufacturers, including notches on the leading edge and varying flexibility and curvatures, but there isn't a lot of evidence that any particular design is significantly superior to others (although you may want to check out the section on WhalePower in Chapter 12).

It is certainly true that bigger blades can harvest more wind energy, but they can also start to get heavy. Since blade mass scales as the cube of the turbine radius, loading due to gravity becomes a constraining design factor. Every component has its stress limits, so this sets up a dance between performance and robustness that will need to be addressed in every new model. Regardless of size or orientation, it's important to balance the blades as well as possible. That minimizes vibration and wear and tear, and helps everything run smoothly.

People new to wind energy frequently ask about the optimal number of blades. Theoretically, the perfect wind turbine would have an infinite number of blades that are infinitely thin, so it could collect all the energy from the wind without blocking it. Haven't seen one of those around, have you?

As Charles Brush learned in the late 1800s (Chapter 1), having too many real blades impedes wind flow through the rotor, and that decreases the amount of energy that can be collected. Some engineers swear by one or two blades in order to allow greater airflow, but these designs are harder to balance and require more maintenance. They experience greater cyclical forces, which can cause "yaw chatter" and fatigue equipment. It is currently too costly for a small wind turbine to come installed with a yaw dampening system or teetering hub to minimize this problem.

Over time, the industry has tended to prefer three blades as a compromise of stability, reliability, and excellent energy collection, particularly due to greater longevity (Figure 8-5). Five-bladed small turbines are also somewhat popular, although most experts question if they offer any real benefits.

FIGURE 8-5 Most wind turbines today have three blades, since that seems to offer a good compromise of stability, quiet operation, and high energy production. *Brian Clark Howard.*

TABLE 8-1   A Quick Comparison

Number of Blades	1	2	3
Longevity	Most wear and tear due to yaw chatter; results in shortest life. But thicker blade brings strength, and additional teetering hub helps reduce stress to blade.	Wear and tear due to yaw chatter; could result in potentially shorter life. But thicker blades bring strength, and additional teetering hub helps reduce stress to blade.	Suffers least wear and tear; 20-year life span reasonable.
Efficiency	Potentially most energy extraction, yet in testing still captures ten percent less than two blades due to aerodynamic losses.	Ten percent more energy extraction than one blade due to high rotor speed.	Three percent more extraction than twin blade rotors.
Sound	Loud due to greater aerodynamic loading at higher rotational speed and yaw chatter as the nacelle rotates.	Depending on design, may be loud due to greater aerodynamic loading at higher rotational speed and yaw chatter as the nacelle rotates.	Most quiet due to lower rotor speed and less yaw chatter.
Installation	Easiest to install, although the single blade would be significantly heavier than one blade in a multiblade rotor.	Relatively easy to erect because the blades may be mounted to the nacelle while it is on the ground.	A three-bladed rotor must be lifted separately and mounted to the nacelle, which may require a crane.
Aesthetics	Least visually appealing.	Appealing; at the same rotational speed, the blade passage frequency of a two-bladed turbine is two-thirds of the value for a three-blade rotor, resulting in a shadow frequency of one-third less. The high rotational speed on a small two-bladed turbine has the drawback of making the rotation appear irregular, which disturbs some people. However, the impression disappears by increasing turbine size and decreasing rotational speed.	Arguably most appealing and elegant.

Cost	Expensive, due to one heavy-duty blade and a sophisticated teetering hub, and it might need more repairs.	Cheaper to build than a three-blade rotor, but the sophisticated teetering hub is expensive.	More expensive due to extra blade construction; in addition, during installation (see earlier) the separate lift is an additional cost. Yet, typically low maintenance (with polymer-based blades).

We could continue the table with an infinite number of blades, but significantly decreased efficiency with each additional blade makes the point futile. Further, as you close up the space between the blades, the wind is more apt to place exorbitant force on the rotor and tower. You wouldn't mount a brick wall up there, would you?

Power Up!   *Be wary of a manufacturer that boasts significant improvements based on any particular number of blades, especially higher numbers.*

Alternator (Generator)

In small wind turbines, the rotor must attach to a generator, which is located in the *nacelle,* or housing, behind the blades. For the generator, most wind turbines use an *alternator,* which converts mechanical motion into electricity (Figure 8-6). It's the same type of device found in your car for charging the battery. Inside an alternator is a set of stationary windings, known as the *stator,* and a set of rotating magnets, known as the rotor (yes, there is a rotor inside an alternator, in addition to the larger rotor of the turbine). The motion of the turbine's blades causes the magnets in the alternator to rotate. This induces an alternating current (AC) voltage in the stator windings.

Such alternators can have permanent magnets, or they can induce magnetic fields to form in coiled wires through electromagnetic induction. The former must interface with the utility through a synchronous inverter, while the latter has its own specialized controllers for utility interconnection. In permanent magnet alternators, sometimes the stator is placed around the rotating magnet, and sometimes a cylindrical can with strong earth magnets embedded rotates around a fixed stator.

Permanent magnet alternators are generally the cheapest and simplest to make, and they produce three-phase wild alternating current, or "wild AC," which means that the frequency and voltage vary with the wind speed. As rotor speed increases, so does the frequency. Wild AC cannot be used by standard appliances, and must

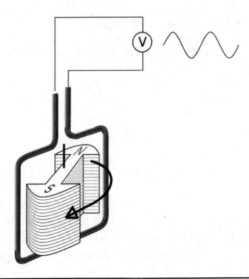

FIGURE 8-6   Diagram of an alternator, showing the rotating magnet (rotor) and the stationary wire winding (stator). The rotating magnetic field induces a current in the wire. *Egmason/ Wikimedia Commons.*

be rectified to direct current (DC) before it can be used for battery charging or sent to the grid.

You might think that a turbine would use a DC generator to produce direct current right off the bat, but very few do. DC generators are more expensive than AC alternators, and they require more maintenance, since they contain brushes that have to be replaced every so often.

People often ask if a car alternator will work for a small wind turbine. In fact, a number of homebrew turbines have been made with car alternators. They can work, but they are a poor choice and should be avoided as much as possible. Car alternators are designed to work at very high rpms (revolutions per minute), meaning they don't start producing electricity unless they are spun very fast. Unfortunately, a small wind turbine's blades are unlikely to achieve the speeds necessary under most wind conditions.

Sure, you could add a gearbox to increase the rotational speed applied to the alternator, but you are better off getting an alternator that is specifically designed to work at wind turbine speeds. Similarly, you could make a smaller turbine rotor that spins faster, but you are going to end up harvesting less energy from the wind if you do that. Finally, people who have made turbines with car alternators say they tend to be loud and temperamental, which may make for angrier neighbors.

Gearbox (Optional)

Most small wind turbines are *direct drive*, meaning the blades turn a shaft that connects directly to the alternator, without intermediate gears in between. Utility-scale wind turbines, of course, are larger and more complicated machines, and they usually have a *gearbox* to control the mechanical energy directed to the generator. A gearbox is similar to gears on a bicycle or in a car, and it makes it possible for the alternator to spin nearly six times faster than the rotor, thus increasing the electrical output. A few small wind turbines have been made with gearboxes, although that adds a level of complexity and makes the system larger, heavier, and more expensive. A gearbox also requires additional maintenance, and may need to be replaced a few times during the life of the system. Many installers say a gearbox shouldn't even be considered for anything less than 20 kW.

 Power Up! *Many VAWT and rooftop turbine manufacturers tout their designs as being "gearbox free," as if this were a significant advantage. What they fail to point out is that the vast majority of small wind turbines in the market are already gearbox free, since that device is primarily used on utility-scale machines.*

Tail

Like a wind vane, a horizontal axis wind turbine needs a tail (Figure 8-7) to keep it properly oriented into the wind (at least if it is an upwind design). The tail is connected to a boom behind the nacelle. When the wind changes direction, it pushes on one side of the tail, swinging the turbine around to face the wind. In *tail-furling* designs, the tail is also involved in protecting the turbine from high winds, and tail and boom length are designed carefully for weight and area.

You may have noticed that most small wind turbines do have tails, meaning they are *upwind* designs. However, there are numerous examples of successful downwind turbines as well. The distinction is simple: an *upwind* turbine always faces into the wind, with the blades in front of the nacelle, while a *downwind* turbine always faces away from the wind, with the blades to the rear of the nacelle.

Downwind turbines like the Skystream 3.7 or models from Scotland-based Proven can "find" the wind in different ways. In passive yaw systems, the rotor is designed to sit away and slightly to the side of the yaw bearing. So when the wind blows, it pushes it downwind. Another method is to use hinged blades, which fold with the wind like an umbrella collapsing from a gust. Another method is the active yaw, which is driven by an anemometer. If the system senses a change in the wind, an electric motor can rotate toward it.

For residential turbines, there is no definitive "winner" when it comes to upwind versus downwind, even if upwind is more common (Figure 8-8). Each style has its own advantages and disadvantages.

Figure 8-7   This turbine was hand built from scratch by Dan Fink and participants at one of his workshops. Note the tail and rotor. *Dan Fink.*

Upwind Turbines

- **Advantages:** Reduced tower shading, because the rotor is upwind of the tower, so the winds aren't blocked by it. The tail acts as a lever to position the turbine optimally into the wind, even if it changes frequently.
- **Disadvantages:** The nacelle must be longer to keep the blades positioned out away from the tower so they don't strike it. The blades must be rigid enough so they aren't blown back into the tower.

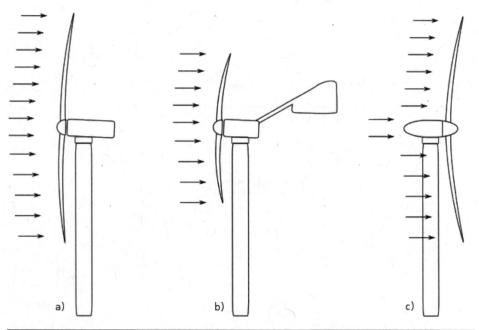

FIGURE 8-8   In this diagram, turbines a and b are upwind, while c is a downwind turbine. *Hanuman Wind/Wikimedia Commons.*

Downwind Turbines

- **Advantages:** Does not need a tail because the nacelle is designed to orient toward the wind. The blades don't have to be as rigid, because the wind blows them away from the tower. This means they can be made of cheaper material, and they can absorb some of the stress on the system in high winds.
- **Disadvantages:** Part of the wind is blocked by the tower, which can also cause turbulence. They tend to be noisier, and flexing of the blades can wear them out faster. In addition, if the wind dies down for a bit and then shifts 90 degrees from its previous direction, a downwind turbine can be "stuck" in a position that doesn't generate much energy. It can take a gust to realign it properly.

Yaw System

The yaw system is responsible for the orientation of the wind turbine, to make sure it is in place to maximize energy collection, or to face away from the wind to reduce overly high speeds during storm events. This system consists of a series of roller (yaw) bearings on a horizontal axis atop the tower (Figure 8-9). As wind direction changes, the yaw bearings permit the turbine to rotate on the tower.

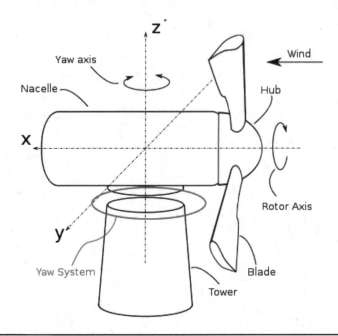

Figure 8-9 This simple schematic shows the basic elements of a yaw system, which swivels to keep a turbine properly oriented for energy production or governing during high winds. *Hanuman Wind/Wikimedia Commons.*

Slip Rings

Ever wonder how a turbine can safely yaw greater than 360 degrees without twisting or breaking the wires inside? It's due to an innovation called *slip rings*. These handy devices connect via wire to the alternator, and they are located at the junction between the nacelle and the tower.

The slip rings are a center core of rings of brass surrounded by copper metal brushes (Figure 8-10). This combination allows electricity to bridge over seamlessly to the wiring that runs down the tower, toward the balance of the system. Yet it allows the turbine to yaw endlessly in circles without breaking anything inside.

Slip rings can work in one of two different ways: (1) The center core ring can stay stationary while the brushes and housing rotate around it, or (2) the brushes and housing can stay stationary while the center core is allowed to rotate. If only our headphone wires did the same!

Tower

As we have mentioned, you need a tower to bring the turbine to the fuel, the good winds. All towers need a rugged design to handle the wind loads and harsh weather over the lifespan of the system. It is getting more common to see manufac-

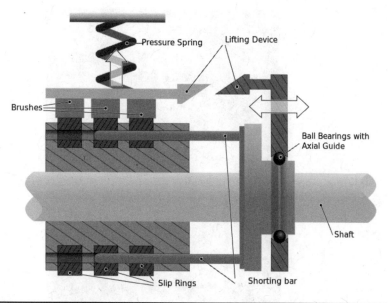

Pressure Spring

Lifting Device

Brushes

Ball Bearings with
Axial Guide

Shaft

Slip Rings

Shorting bar

FIGURE 8-10   A diagram of slip rings, which allow electricity to flow down the tower without breaking any wires as the turbine yaws. *Biezl/Theanphibian/Wikimedia Commons.*

turers' recommendations on engineered and certified towers that work well with their designs.

Turbine designs do vary, of course. Some variations may have benefits, while others seem to make no appreciable difference. Next, we'll take a look at one of the more fundamental design differences: the orientation of the rotation axis.

Vertical Axis Wind Turbines

We mentioned in Chapter 1 that some of the earliest wind-powered devices had a vertical axis, meaning the main shaft is perpendicular to the ground (Figure 8-11). This is in contrast to the vast majority of wind turbines and windmills that have been built over the past few centuries, most of which have had a horizontal axis, meaning a shaft that is parallel to the ground. To many leaders in the wind industry, horizontal axis wind turbines (HAWTs) have long been the best choice for any sensible, self-respecting wind prospector.

The HAWT is certainly a tried and true design, with a long history of refinement and improvements. But this hasn't stopped tinkerers and inventors from experimenting with vertical axis designs. In the past few years, there has been a renaissance in VAWTs, with manufacturers touting high-tech designs, and the media and public doling out enthusiastic support.

FIGURE 8-11   Vertical axis wind turbines have seen renewed interest in the past few years, though they remain controversial. This elegant model comes from UK-based Quietrevolution. *Quietrevolution.*

In a lot of ways, VAWTs can seem very appealing. They often look exotic, and they may seem less threatening than HAWTs, which have blades that spin faster, and which some people equate with giant buzz saws that knock birds out of the air and ruin property values. VAWT manufacturers have built slick websites loaded with dramatic animations and impressive claims about efficiency, versatility, and eco-friendliness. Journalists and bloggers, particularly those on the green and design beats, champion VAWTs, and have heaped awards on them.

To many professional small wind installers, VAWTs are too expensive and too unproven to be given a serious look. While some may admit that VAWTs may eventually prove useful, others question the need for them at all. Small wind owner, installer, and author Ian Woofenden asked us, "What's their advantage? We already have a design [HAWTs] that works well, has proven itself, and look at how many are currently in use versus VAWTs. The marketplace speaks loud and clear." Woofenden added, "Do vertical turbines have the potential to be better than horizontal? The clear answer from physics is no. There is only one real advantage: aesthetics, which is a very personal thing. But are they better in turbulence? No. And they're going to be less efficient."

To many in the utility wind industry, VAWTs are a curiosity, maybe even a toy. They're a has-been that was tested decades ago and found to be wanting. When we asked Morten Albaek, senior vice president of global marketing for Vestas, what he thought of the current resurgence in new VAWTs, he said he was skeptical. "Our conclusion is that the maturity of these types of inventions is very low," Albaek said.

To critics like Woofenden, new VAWTs are likely to be a waste of people's money, and may end up discrediting the whole industry. Some accuse VAWT manufacturers of making fraudulent claims, while the manufacturers fire back that their technology is brand new, and that opposition is based on outdated assumptions about past designs, not what they have developed in the lab today. More measured observers concede that VAWTs may one day play a bigger role in our renewable energy mix, providing real-world testing validates new and future designs.

Let's take a closer look at the advantages and disadvantages of vertical axis wind turbines, as well as what people are saying about them.

A Typical VAWT Pitch

Based on an Internet search, we called up Sean Taylor, who told us he had ordered a 12 kW vertical axis wind turbine for his company, 3rd St Hydroponics, in Oakland, California. Taylor and his wife Mary have been in the hydroponics business for about 30 years, and they operate a permaculture and greenhouse training center.

"We're making a zero-footprint farm across the street here," Taylor told us. "We've got LED lights, we use reclaimed graywater, and we've got aquaponics. We teach microbiology, chemistry, math, AutoCAD, design, flowering, propagation and much more."

For the past several years, Taylor has also sold VAWT turbines made by Aliso Viejo, California–based WePOWER. He has had a small 1.2 kW demonstration turbine set up inside his store, but he told us he expected a bigger turbine outside would attract more customers. "The city of Oakland has been real accommodating and is excited about the windmill," Taylor added.

When asked why he chose a vertical axis design instead of a horizontal design, Taylor echoed many of the technology's enthusiastic boosters. "VAWTs are so much more efficient than HAWTs," he told us. "They use a lot less wind to drive than a standard HAWT. They can capture the wind from any direction without having to adjust to sudden wind changes. They are 85 to 95 percent efficient." [We didn't get a chance to ask him what reference point he was using for that efficiency claim, but we know nothing can beat the Betz limit, 59.3 percent efficiency.]

Taylor told us his new WePOWER turbine would be installed on a 30-foot pole, with the top of the blades reaching 52 feet. "I'm right on the bay, and I have no obstructions. We have great wind here. I'll be the first all-green hydroponic store in the country, if not the world," he added.

Taylor's claims about the purported benefits of VAWTs are in no way unusual, and they have been repeated so frequently by manufacturers, salespeople, and the media that they are widely believed as "fact." However, it's also true that the critics of VAWTs aren't always up on the latest technology. Let's break some of these issues down so you'll be able to sort hype from reality when you read a blog post or visit a showroom.

Brief History of VAWTs

Although vertical axis windmills were used more than 2,000 years ago in the Middle East, horizontal designs have been dominant for a long time. Still, VAWTs have played a role, and a few major types have emerged over the years.

Savonius VAWTs

In 1922, Finnish engineer Sigurd J. Savonius invented a style of VAWT that uses aerofoils, or scoops, mounted around a vertical shaft. Typically, there are two or three scoops per section, and they work through drag (that is, the scoops experience less drag when moving with the wind and more when moving against it, which causes them to spin).

Because they only use drag, Savonius turbines can only harvest a limited amount of energy from the wind, so they are said to have low efficiency. However, the devices are simple and require low maintenance, so they have long been used for applications that don't require much power, such as marine buoys or bus ventilators (Figure 8-12). Most anemometers are actually Savonius wind turbines, although their function is measuring wind speed, not producing electricity.

Darrieus VAWTs

Sometimes described as looking like an "egg beater," the Darrieus VAWT was patented in France in 1925 by aeronautical engineer Georges Jean Marie Darrieus (he patented it in the United States in 1931). In addition to drag, the Darrieus design uses lift, so it can harvest more energy out of the winds. Theoretically, a Darrieus turbine can be just as efficient as a HAWT, if the wind speed is constant. However, in practice, these VAWTs have historically lagged behind in real-world energy production, due to a number of design challenges and limitations.

In addition to the primary "egg beater" shape, a Darrieus turbine can be made with straight vertical blades, which are usually attached to the shaft with horizontal struts. These types of Darrieus machines are called *giromills* (Figure 8-13). Giromills are easier to make, but the blades have to be stronger, and they tend to result in a more massive structure. WePOWER makes several examples.

If a Darrieus machine is built so each blade can rotate around its own vertical axis, it is called a *cycloturbine*. These designs provide more even torque and can self-start, but they are more complicated, heavier, and require additional sensors.

Darrieus turbine blades can also be canted into a helix, as they are in the designs by Quietrevolution and Helix Wind. These manufacturers claim the shape provides

FIGURE 8-12   A Savonius vertical axis wind turbine is based solely on drag, since breezes push the curved blades at the speed of the wind. These designs are not good for making energy. *Brian Clark Howard.*

FIGURE 8-13   Giromill Darrieus VAWTs have straight vertical blades, and they are easier to make. *Stahlkocher/Wikimedia Commons.*

near-constant torque and reduces stress on the bearings, and that it is better able to handle updrafts.

After Darrieus's invention, VAWTs remained in relative obscurity for decades. However, in the 1960s, two Canadian researchers at National Research Council Canada (NRC) independently came up with the same concept, at first thinking they had invented it. Upon research, they discovered Darrieus's patents, and they went on to improve on his design. Notably, they added a twisted, *Troposkein* shape to the blades. This helped decrease stresses on the hardware.

In the 1970s and 1980s, in direct response to the 1973 Arab oil embargo, the United States and Canada launched significant research and development efforts on wind power. They looked at both VAWTs and HAWTs, in both large and small iterations. A number of materials, systems, and designs were tested.

In 1975, Hydro-Quebec Research in Canada installed a 40 kW Darrieus VAWT. Two years later, a 230 kW version was built in the Magdalen Islands of Quebec. In 1984, work began on the world's largest VAWT, at a place called Cap-Chat on the Gaspé Peninsula in the St. Lawrence River (Figure 8-14). Called Project Eole, the massive Darrieus turbine is roughly 360 feet high and 200 feet across, and vaguely resembles the launching platform from the movie *Contact*.

The huge turbine was rated at nearly 4 MW, and reportedly generated 12,000 MWHs of electricity over the six years that it operated. Project Eole was a joint venture between Hydro-Quebec and the National Research Council of Canada,

FIGURE 8-14   The world's biggest Darrieus wind generator, Project Eole in Quebec (center), stands 360 feet tall and nearly 200 feet across. Although it only operated for a few years, due to bearing problems, it was rated at 4 MW. *Guillom/Wikimedia Commons.*

and it reportedly cost $65 million Canadian. The giant Darrieus turbine is still standing, although it was shut down in 1993 due to a bearing problem after repairs were deemed too costly. Today it serves as a park and educational attraction.

In the United States, NASA and the Department of Energy collaborated with Canadian researchers in the 1970s and 1980s, and Darrieus-type VAWTs were tested in sizes of 5, 10, and 17 meters. The largest Darrieus machine built on U.S. soil—34 meters high—was located at the USDA Agricultural Research Station in Amarillo, Texas.

A few companies spun off VAWT designs as commercial products, notably California-based FloWind, which produced a 17-meter Darrieus turbine. However, sales were slow, and FloWind eventually failed after a few VAWTs were installed in California and a few other places. At the Canadian Wind Energy conference in 1993, Lawrence A. Schienbein of Battelle, Pacific Northwest Laboratories, pointed out that more than 500 VAWTs had been placed into commercial operation, from five manufacturers. Still, that's a tiny subset of the wind industry, and VAWTs remained niche players until the recent ambitions of a new crop of manufacturers.

Purported Benefits (and Disadvantages) of VAWTs

There's no shortage of supporters of vertical axis wind turbines among the general public, and there are several start-ups making waves. So let's examine some of the common claims about VAWTs, and see how they measure up to HAWTs.

Convenience

There's no question that VAWTs seem to offer convenience, since the generator and gearbox are located near the ground, where they are easier to maintain. VAWTs don't require a tower per se, since the main axis supports the turbine, although newer models are being sold with (typically short) towers. As you should be well aware by now, the fact that VAWTs are usually placed near the ground is one of their biggest drawbacks, because that's not where the good wind is.

Some supporters have pointed out that vertical axis turbines can be packed closer together in groups, which would seem to be an advantage when it comes to wind farms. It's true that HAWTs tend to slow the air more, meaning installers usually have to separate them by at least ten times their width. However, it's also true that there have been very few large deployments of VAWTs. To date, wind farm developers have almost exclusively relied on HAWTs, as conventional wisdom has been that they offer a more consistent and effective technology, all things considered.

Forward-thinking architects, environmentalists, and technologists have recently argued that VAWTs are better suited to urban and rooftop applications. This argument is controversial in the wind world, and we'll explore it more later.

Energy Output and Efficiency of VAWTS

As we've tried to stress previously, the most important factor when it comes to wind turbines is energy output over time, in real-world conditions. This is the biggest problem with new turbines being developed and advertised today, whether of vertical or horizontal design: they often don't have much performance data available, especially from third parties. Manufacturer power curves and marketing claims will not convince us to invest tens of thousands of dollars in a new system. While it's theoretically possible for a new design to give impressive results, proving it can take time.

When it comes to VAWTs, historically they have not produced as much power as good HAWTs. Interestingly, a number of people we interviewed in the VAWT industry were quick to argue that it's unfair to directly compare the two types of turbines at all.

"Horizontal and vertical axis turbines are totally different devices," Howard Makler, president of VAWT-maker WePOWER, told us. "Their applications are totally different. If you're talking about large-grade wind, horizontal is the only thing you can use. They have to be erected up into the high winds. But vertical is very different: you've got no gears, and they are more suitable to urban situations. It does not produce as much energy, but you can't put a horizontal wind turbine on a commercial or residential street," said Makler.

There is some truth to what Makler is saying; however, it's also true that the primary reason to install a wind turbine is to *generate energy*. So power is an important factor, even if it isn't the only consideration.

In truth, the biggest drag on energy produced by VAWTs is their siting, which is normally lower than HAWTs. The lower the turbine, the less power it is going to generate, because the less access it has to wind. Yes, VAWTs could be theoretically placed on tall towers, but if they were, why not just get a HAWT? There is no evidence that a VAWT would do better than a time-tested HAWT up on a tower.

Historically, another drawback to VAWTs has been that they are more susceptible to damage from high winds. Placing them higher up would increase this risk.

Those critical of VAWTs often argue that the design is inherently less efficient because the blades must actually push *against* the wind during part of its rotation. Manufacturers counter that blade shape and spacing has minimized this issue. Similarly, original Darrieus turbines had to be manually pushed to start up, although manufacturers now say their designs have starting features.

Is it possible to get good energy production out of a vertical axis wind turbine? Yes, although it has to be a good design in a good wind resource. Figure 8-15 shows that the best Darrieus turbines can't quite match the efficiency of good HAWTs. Currently, you can get more kWh for less invested money with a HAWT.

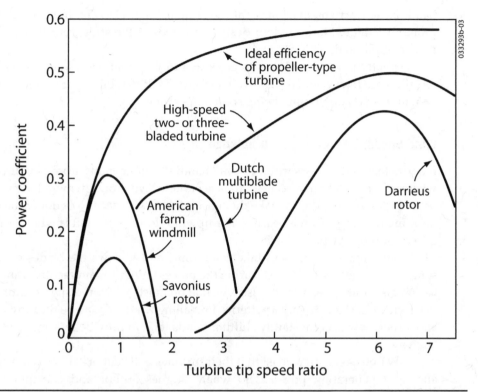

FIGURE 8-15 Although good Darrieus turbines can approach the efficiency of horizontal axis turbines, they still fall short. Savonius turbines, on the other hand, leave much to be desired when it comes to energy production. *NREL*.

Reliability and Fatigue

According to Ian Woofenden, "Efficiency is really a red herring [when it comes to turbine selection.] Size is important, but reliability is really important: a crashed wind turbine doesn't make any kWh."

So how reliable are VAWTs? Historically, their record is not particularly good, and the newest designs are too untested to say. In the judgment of Hugh Piggot, "Vertical axis turbines do have intrinsic fatigue problems, which along with the problems of starting and stopping, and the problems of putting them on proper towers have made them a poor choice for wind energy."

Classic Darrieus turbines experience considerable stress and strain, in large part because mechanical loads vary from zero to maximum operating values two or three times per cycle, depending on the number of blades. Twisting the blades and other techniques, such as making the whole thing helical, have reduced these stresses, although again, long-term reliability data is scarce.

Similarly, supporters of VAWTs sometimes claim that the systems require less maintenance. There is little evidence to support this in the real world. On one hand, basic Darrieus machines can be simpler than HAWTs, but then they are prone to fatigue. More modern designs can reduce the stress, but then they are more complicated.

Currently, your safest bet is to go with a turbine that has a long proven track record. Since few commercially available VAWTs fit that bill, your options are limited, unless you don't mind being someone's guinea pig.

Wind Speed, Direction, and Turbulence

It's true that vertical axis turbines can "handle" wind coming from any direction, whereas horizontal turbines need to face either upwind or downwind, depending on their design. In order to make them useful, HAWTs *yaw* into or out of the wind, meaning they swivel on a bearing. For upwind designs, the tail is what keeps the generator properly positioned.

It's true that yawing does take some time, and it slows the blades down, so some energy in the wind is lost during the process. VAWT manufacturers are quick to point this out. However, it's also true that on most good wind sites, the air tends to be streaming along regular patterns. Prevailing winds tend to be the most steady and strong, whereas constantly shifting winds tend to provide less energy overall anyway.

VAWT boosters often boast that their turbines will "cut in" at low wind speeds and start generating power. "So what," argues author and consultant Paul Gipe. "There is so little energy available at low wind speeds that it is essentially meaningless."

VAWTs are often advertised as being "good with turbulence." However, according to Gipe, turbulence has always been the enemy to power generation. In his view, turbine designers have always tried to minimize turbulence as much as possible.

Interestingly, the UK-based VAWT manufacturer Quietrevolution claims that, on some sites, its 6 kW qr5 VAWT has actually shown *increased* power produced with increased turbulence. Quietrevolution's commercial director, Stephen Crosher, told us via e-mail: "Turbulence has traditionally been the enemy; however, Quietrevolution's control system tracks the wind and the gusts and keeps the rotor at peak Cp [coefficient of performance] across its entire operating range." He added, "Our control system actually improves our power curve with increased turbulence as measured in the field, and we have presented these findings to the American Institute of Aeronautics and Astronautics."

Crosher did agree, "No wind turbine works well in low wind speed environments, and the qr5 is no different. We agree that strong winds are important, vital; however, there are many sites that have both high winds and high turbulence—e.g., a coastal site where the wind from the sea is strong and laminar, while the wind from the land is much more turbulent but can still be strong." Crosher said such sites are "ideally suited" to the qr5.

According to Crosher, on a "good site," a current qr5 turbine can generate up to 8,000 kWh per year, though 3,000 to 4,000 kWh per year is more common for what he called a "typical site." "By the end of 2011, we would anticipate being able to generate up to 10,000 kWh per annum on a good site from the same essential product, through the process of continual improvement in aerodynamics, electronics, and software control solutions," he added. For a typical site, he estimated it would be 4,000 to 5,000 kWh by the end of 2011.

Cost

The commercial VAWT industry is immature, so despite claims to the contrary, vertical turbines tend to be more expensive than horizontal ones, especially for the same rated power. Quietrevolution's qr5 costs £20,000 (the equivalent of $32,000 at the time of this writing), not including the foundation or installation costs, which push the total to £28,000 to £32,000 ($45,000 to $51,000).

That's fairly high for a 6 kW machine, for something that produces up to 8,000 kWh per year on "great" sites (with 15.4 mph average wind speeds), or 3,000 to 4,000 kWh on more typical sites (with 12 to 12.8 mph average winds, which are still relatively brisk), according to Crosher. In contrast, the 10 kW Bergey Excel horizontal turbine tends to cost $40,000 to $60,000 for the complete installation. For roughly the same amount of money, an Excel can produce about 16,000 kWh in a year, with 12.5 mph average winds, according to real-world data supplied by Ian Woofenden. That's four to five times the energy produced by the qr5 for the same amount of wind.

Crosher told us that, in general, a VAWT system is likely to cost 7.5 to 10 percent more than a HAWT when purchased. "However, a VAWT such as Quietrevolution's qr5 has only one moving part and is therefore likely to suffer far less fatigue than a HAWT due to both the number of moving parts and the turbulence on site," Crosher argued. "The additional 7.5 to 10 percent is likely to be more than offset in lower maintenance costs over the life of the product. This is proven in extreme locations. Where maintenance is the paramount concern, VAWTs are the technology of choice, for example by the German Antarctic Survey."

Crosher may be correct for some remote installations (although see Figure 3-6 for a HAWT doing fine in Antarctica), but it seems hard to believe that maintenance fees on a typical home or small business system would be more than the difference in power produced by a HAWT versus a VAWT, especially in cases involving multiples of four or five, as noted earlier. Further, the qr5 only comes with a two-year warranty, compared to the Bergey Excel's five-year warranty. That alone is a significant disadvantage, and perhaps underscores the relative immaturity of the design and company.

INTERCONNECT

One Missouri Dentist Loves His VAWT

Colin Malaker is a dentist in Columbia, Missouri, though he also has a small side business, Midwest Power Solutions, that sells renewable energy equipment. In addition to solar panels and solar thermal systems, Malaker sells small vertical axis wind systems made by WePOWER. (Malaker is not related to WePOWER's Howard Makler, despite the similarity in their names.) In August 2010, Malaker installed a Falcon 12 kW WePOWER VAWT in front of his practice, Sterling Dental Care, in Columbia (Figure 8-16).

FIGURE 8-16   Colin Malaker installed this Falcon 12 kW WePOWER vertical axis wind turbine next to his dental practice in Columbia, Missouri, in August 2010. He says his patients love it. *Colin Malaker.*

We spoke with Malaker on the phone, and he told us his turbine measures 57 feet from the ground to its top, including 40 feet of tower. The turbine projects just a few feet above the building, though it's only a few feet away. "[There are] trees around it, but it spins pretty good. The building kind of channels the wind toward it," Malaker said. He added that he had to get a variance from the town (Chapter 7), and that it's the first wind generator in the community.

When asked how much power the turbine has generated, Malaker said it had done 250 kWh over a few months. "So far it hasn't been super windy yet, and August was pretty still," he added. Malaker told us his site averages 11 to 15 mph winds (which is surprisingly high), and he estimates that, with the federal tax credit, his system will pay for itself in about 12 years, assuming average wind speeds.

The Falcon retails for about $30,000, including about $13,000 for installation. "Some of these turbines placed out in the country will pay back in about seven years, because the USDA has a 25 percent grant for small businesses in rural areas," Malaker added.

Asked if he was happy with his investment, Malaker told us, "Oh yeah. It draws people into my business. A lot of people come here just because there's a wind turbine. When you see a wind turbine at the Best Buy or a grocery store, it's not really helping their electricity cost, it's helping their image, it's marketing."

Malaker told us he expects only yearly maintenance on his WePOWER. "Since it's direct drive, it will just take 30 minutes, to check the bolts," he said. "Horizontal turbines break a lot, and they have to be geared down. This has a sealed magnet system, so you don't have to go in there and mess with anything. Your maintenance is a non-issue."

Malaker told us the turbine is extremely quiet, and that his neighbors were on board from the beginning. "They think it is a neat design," he added. "Everyone likes to watch it, and my patients think it's cool. It gets people driving down the street, and the businesses around here like that. Some are interested in getting one as well now."

To Malaker, WePOWER turbines currently make the most sense for businesses, who have the most to gain from good publicity and curb appeal, and farms, which benefit from additional incentives.

Sound, Vibrations, and Flicker

It's true that vertical axis turbines tend to be exceptionally quiet, in large part because they have lower tip speeds than horizontal designs. "Horizontal turbines have issues with noise, vibrations, and the flicker effect," Howard Makler of WePOWER told us. "Vertical axis [designs] are much more quiet and suitable for the urban environment. It's all about putting the right turbine in the right location."

However, as we have shown, small HAWTs are rarely "loud," vibration shouldn't be an issue with a well-made tower, and only large commercial turbines can cause the flicker effect. Putting a turbine in an urban environment or on a roof is rarely a good idea, regardless of the generator style.

Even so, very small VAWTs are starting to show up on rooftops and on utility and lighting poles in test settings, and on the properties of those who want to make a statement (Figure 1-10). These ultra-micro-turbines don't generate much power, and currently aren't very economical.

Still, it is tempting to imagine a future in which our built environment literally bristles with clean energy generators. If energy prices get high enough and micro-turbines get cheap enough, it could happen.

Effects on Birds and Bats

Supporters of VAWTs routinely claim that the design is intrinsically friendlier to birds and bats. However, there is scant evidence to support this belief.

We asked Howard Makler, president of WePOWER, why his company advertises that its VAWTs are safer to animals. "It's common sense," he told us. "Vertical turbines look like solid objects." Makler admitted that he wasn't aware of any empirical tests to back up that claim.

For their part, British VAWT manufacturer Quietrevolution chooses not to market their products as safer for birds or bats, according to Stephen Crosher. "We would agree the impact of all small wind turbines on birds and bats is incredibly small, if it exits at all," Crosher told us. "Climate change, pollution, habitat loss, loss of food sources, domestic pets, and windows in buildings are the real threats to birds and bats, and energy from the wind is attempting to address some of these issues without impacting on the others," he added.

On balance, we don't think there's enough evidence to warrant much weight being placed on the argument that VAWTs may be less harmful to birds or bats than HAWTs. At current levels of deployment, we think the risk to animals from all small turbines is exceedingly low. If you do live in a corridor where there could conceivably be an impact on an endangered species (Chapter 2), we don't think there is enough data to suggest that a VAWT would have a significantly lower impact.

Aesthetics

If you are mostly interested in how a turbine looks, you're probably best off getting a sculpture. You'll save a lot of money (well, depending on whom you hire to make the sculpture). Still, we recognize that aesthetics are important, especially in consumer choices.

To many people, VAWTs look better than HAWTs (Figure 8-17). We suspect that part of the appeal is their relative novelty, and that this effect could wear off over time if they were more widely adopted. Tech and design bloggers are clearly drawn to VAWTs, perhaps because they're "not your Grandma's wind turbines." It may be partly generational, because seasoned wind installers, on the other hand, often exhibit a visceral reaction against them. Perhaps remembering problems with the

FIGURE 8-17   Makers of VAWTs often tout their products as being quieter and more suitable for urban locations, although it's still true that the wind resource is rarely good in built-up areas. *Quietrevolution.*

technology in past decades, they assume VAWTs are doomed to be clunky, unreliable, and inefficient.

In any case, aesthetics are deeply subjective, and opinions tend to change quicker than most people like to admit. We interviewed some owners of VAWTs who are very happy with their purchase. A new turbine makes a bold statement in a time when there's seemingly a Prius and a solar panel on every block.

Is that worth something? Sure, as long as you know what you're getting into.

WePOWER Shows Momentum

In November 2008, 16 small WePOWER turbines and 64 solar panels were scheduled for installation to light up a sign in New York's Times Square. According to news reports, the $3 million advertisement for Ricoh was to save $12,000 to $15,000 per month in electricity and 18 tons of carbon a year. The sign's power system has a battery backup, but if there's no sun or wind for more than four days, the billboard goes dark (although it never really gets that dark in Times Square).

Interestingly, several Times Square signs were already "wind powered" through the purchase of renewable energy certificates, though the Ricoh sign brought the wind power directly to the swarms of tourists. We checked out the Ricoh sign in early 2010, and the solar panels were still there, but there was no sign of the turbines. Calls to WePOWER and Ricoh weren't returned.

According to WePOWER's president, Howard Makler, about 300 of his company's giromill VAWTs have been installed (Figure 8-18). "We probably have as many out there as anyone," he explained. A number have been shipped across the United States, including to Nevada, Utah, Wyoming, and Minnesota. Two have been sent to India, and three were bought in Hong Kong, according to Makler. The

FIGURE 8-18 According to WePOWER's president, Howard Makler, about 300 of his company's giromill VAWTs have been installed. *WePOWER.*

turbines are marketed for both residential and light commercial use and range from 600-watt to 12-kilowatt models. They all have a base about five feet in diameter.

Makler said most of his company's turbines get installed on 18- or 30-foot towers. Asked if that is high enough to provide sufficient wind, he said "of course." When pressed, he added, "We have some installed out in the desert, where winds are very high." When asked about installations in more developed areas, such as at a dental clinic in Columbia, Missouri, Makler responded, "Some have a lot of wind." He added, "It's a problem when people install turbines in places where they shouldn't."

Makler said his customers can see a payback period as short as 7 years, with around 15 years being his estimated average. "People are very satisfied with our turbines. The small wind industry is growing, and people are buying," added Makler.

Quietrevolution's High-Tech VAWTs

The 6 kW qr5 vertical axis wind turbine from Britain's Quietrevolution is a high-tech update on the Darrieus design. Standing five meters high with a diameter of three meters, the turbine is made with carbon fibers and incorporates advanced active controls. The three vertical blades have a helical twist of 120 degrees, so the wind pushes each blade around on both the windward and leeward sides of the turbine.

According to Stephen Crosher, Quietrevolution had sold about 140 qr5s by the end of 2010, for both ground and rooftop installations (Figure 8-19). He told us he expects to sell about 250 by the end of 2011, still mostly in the United Kingdom, although the company hopes to expand to more international sales in the near future. Quietrevolution also hopes to soon introduce a smaller, 2.5 kW model.

"We have a highly innovative product developed by some of the best young engineers in the [United Kingdom], that addresses many of the shortcomings of early VAWT designs," Crosher told us. "We have moved from prototype to market in about 18 months." The turbines are manufactured in South Wales.

Asked who his customers are, Crosher said anyone who uses electricity and is at a windy location. Currently, customers must also be connected to the grid. "Our customers could include infrastructure projects such as motorway corridors, rail lines, ports, and harbors, or they could be mounted on sea defenses or promenades," he explained.

qr5s are said to have low noise and low vibration. They use active controls to maximize power harvested from the wind, and come with remote monitoring and automatic fault reporting, as well as what Crosher calls "advanced safety systems that go far beyond international standards."

According to Crosher, the qr5 has an aerodynamic coefficient of performance (Cp) of 0.41 across the turbine's entire operating range, as measured by Canada's

Figure 8-19   Quietrevolution claims a coefficient of performance (Cp) of 0.41, thanks to active controls that keep the blades oriented to the wind. *VisMedia/Quietrevolution.*

National Research Council. "This is unique for a small wind system due to our patented gust tracking technology," he added. It's also near the top of the theoretical curve for Darrieus machines, according to the chart (Figure 8-15) prepared by NREL. The machine's electrical system efficiency is said to be 0.92, which is also high.

As we mentioned previously, the 6 kW qr5 currently produces up to 8,000 kWh per year on "good" sites, with 15.4 mph average wind speeds, or 3,000 to 4,000 kWh on "typical" sites, with 12 to 12.8 mph average winds, according to Crosher (8,000 kWh is enough for the electrical needs of a 10- to 15-person office). The qr5's swept area is 13.6 meters squared, which Quietrevolution says is roughly equivalent to a HAWT diameter of 4.2 meters.

The manufacturer suggests the qr5 should be installed only on sites with annual average wind speeds of 11 mph (5 m/s) or higher. Although Quietrevolution claims its turbine was "designed specifically for environments close to people and buildings," it is also true that there aren't a whole lot of places near buildings with consistent wind speeds that high.

As mentioned, the qr5 costs £20,000 ($32,000 at the time of this writing), not including the foundation or installation costs, which push the total to £28,000 to £32,000 ($45,000 to $51,000). This means they are a bit on the expensive side for 6 kW.

Like several new VAWTs, the qr5 design has won various awards, including the Rushlight Wind Power Award in 2007, the Business 2.0 "Bottom Line Design – Energy Award" the same year, and the Guardian Library House CleanTech 100 in 2008.

Finally, we want to briefly revisit the Warwick Urban Wind Trial Project (Chapter 5), because Crosher is critical of the study and feels the disappointing results shouldn't be extrapolated to his company's product. Crosher said the researchers should have looked at a wider variety of wind environments, including sites with stronger winds. "Warwick is one of the least windy parts of the [United Kingdom] and therefore an inappropriate place to conduct a wind trial," he explained.

"Within the study there was one rooftop location that showed reasonable wind—this was a small block of flats located on a small rise in the surrounding area. Unfortunately, the small wind turbines located on this project suffered from vibrations and were switched off, as they were the wrong type of turbine for a rooftop," said Crosher.

"Our conclusion of the trial is that the wrong technology was installed in the wrong location. There was too little understanding about how wind behaves around buildings. Rooftop wind turbines can work well when the following criteria are met: The building needs to be several stories higher than the surrounding buildings, the mast used on rooftops should be at least half as high as the depth of the building; for example, if the building in plan is 8 m × 8 m the mast should be at least 4 m tall (Quietrevolution advises a 6-meter mast, as this is appropriate for most rooftops). Underlying topography should be considered. A tall building in a dip is unlikely to work well, while a smaller building on a hilltop is likely to perform. Lastly, the choice of site should not be shadowed from the more dominant wind directions [in the United Kingdom this is all directions from south through west and the northeast]," Crosher concluded.

Crosher makes interesting points, the most important of which is the need for lots of field testing and rigorous data, at different locations and at multiple wind speeds. The overall result of the Warwick tests is to underscore how critical wind speed is, as well as durability. Several of the turbines tested broke down. That's more evidence to go with something with a good track record in the real world.

From Jellyfish to Helixes: Alternative VAWTs

A number of clever takes on the VAWT have sprung up in recent years, often to the delight of design aficionados and prize committees. We recommend a healthy dose of skepticism and caution when considering such designs. Be aware that small VAWT companies often come and go, after burning through their start-up capital, and sometimes leave a wake of disappointed owners who had unrealistic expectations of energy production.

FIGURE 8-20   Clarian Power's 48-inch-tall SmartBox VAWT has a low price of $800 and can reportedly generate 40 kWh a month with an average wind speed of 12 mph. *Clarian Power.*

Still, it's possible that someone could find the right mix of simplicity, low cost, and accessibility to make a small VAWT work. One company that's thinking outside the box (jellyfish) is Seattle-based Clarian Power, which designed a 48-inch-tall vertical axis wind turbine that resembles a robotic jellyfish. The company's SmartBox micro-turbine (Figure 8-20) is slated to go on sale by the end of 2011. It is rated at 400 watts in a brisk 28 mph wind, but can reportedly generate 40 kWh a month with an average wind speed of 12 mph. Even 12 mph is an awfully strong wind for something that appears to be designed for rooftops or other convenient places in our built environment—most of which have much lower average wind speeds.

Still, what the SmartBox does have going for it is its low price, a suggested retail of $699 to $799. If you do happen to be lucky enough to have 12+ mph winds on your roof, the micro-turbine could reportedly pay for itself in five to eight years, when you figure in the 30 percent tax credit (assuming an annual savings on your electric bill of $50 to $60, based on an electricity cost of $0.12/kW).

The SmartBox is also said to be plug-and-play, "like an appliance," and it is said to install in "one to two hours." It doesn't need a dedicated electrical panel, and everything you need is said to come in the sale package, including a built-in micro-inverter.

We tried to look into the manufacturer's claims further, but repeated attempts to solicit comment were not returned.

There's some risk that people will order a SmartBox as an impulse purchase, without checking to see if their site has reasonably good winds. As we have pointed out, most rooftops get low average wind speeds and high turbulence, and in that case the SmartBox, like any turbine, is unlikely to pay for itself in any reasonable amount of time, despite its relatively low initial price. To some, this could end up discrediting the whole wind industry. Our advice: only get a wind system if you have a good resource. Otherwise, it can take a long time to even "pay back" the embodied energy in the equipment, plus the shipping from the factory. If you want to watch something spin, get a copper lawn ornament for $100.

Advantages of the SmartBox:
- Low price ($699–799)
- Easy to install
- Package deal, with built-in micro-inverter
- Wi-Fi connection for monitoring

Disadvantages:
- Even with brisk 12 mph winds, which are unlikely on rooftops, the SmartBox only provides 40 kWh a month, 4 percent of average U.S. home energy usage, or 10 percent for an efficient home.
- Only 120 V model available at first, though 240 V is said to be next.
- No warranty information yet available.

It's also worth mentioning Poway, California–based Helix Wind, which has received a deluge of favorable press from bloggers over the past few years (Figure 8-21). Helix Wind didn't respond to multiple requests for comment for this book. According to Helix's SEC filings, as of last September the company had $41.7 million in debt and negative cash flow. A December article in *Xconomy* was titled "Helix Wind's Fate is Blowin' in the Wind."

On Helix's website, an introduction notes: "Inexpensive, reliable, simple, the hallmarks of the Helix system make it the best choice for low wind speed residential and commercial applications. The Savonius turbine–based design catches wind from all directions creating smooth powerful torque to spin the electric generator. Mounted up to 35 feet high, in winds as low as 10 mph the Helix system creates electricity to power your home or business."

We noticed a few red flags there. As we pointed out earlier in this chapter, Savonius wind turbines tend to produce relatively little energy, since they are based solely on drag, not lift. We were unable to confirm if this was the case for Helix's particular model as well, but in general you want to minimize drag and maximize lift for producing energy. Further, it's clear the company promotes the turbines for low-

Figure 8-21 California-based Helix Wind offers an attractive take on the Savonius turbine—but we're not sure how much energy is actually produced, and the company has reportedly had financial trouble. *Ahlea/Flickr.*

height applications, which rarely have sufficient winds to make much energy. The "wind from all directions" argument is overstated, we think, since HAWTs can yaw. And as we have mentioned, torque isn't particularly helpful for generating energy.

It's worth noting that Helix also makes a pair of Darrieus wind turbines, although the company's flagship is the Savonius style. It's easy to see the appeal: the gleaming white turbines resemble billowing ship sails, or perhaps playful spirals. But can they deliver good energy at typical sites?

VAWT vs. HAWT: The Final Conclusion

When it comes down to it, only you can decide what kind of turbine you want to get. For at least the near future, it seems likely that horizontal axis turbines will continue to dominate the utility-scale wind industry. When it comes to small wind, old-school players like Ian Woofenden and Paul Gipe are highly skeptical of VAWTs. "My recommendation is to steer clear of VAWTs, both for power and as an investment," Woofenden writes in *Wind Power for Dummies* (For Dummies, 2009).

FIGURE 8-22   Although many people promote VAWTs for rooftops, they still need low turbulence and strong winds to produce energy. *WePOWER*.

Dan Fink told us, "I think people should totally forget rooftops (Figure 8-22). There's not enough wind speed and there's too much turbulence. There are vertical axis machines that approach the efficiency of horizontal axis, but the problem has always been reliability, and fatigue cycle on every rotation. The commercial vertical axis machines built in the '70s and '80s have all broken."

For now, we think it's reasonable advice to be wary of VAWTs, which are largely unproven in the field and relatively expensive.

Still, if you want to experiment, or serve as a test case for someone else's invention, that's fine with us. It's possible your tinkering could even lead to new innovations. Even if you insist on getting a VAWT because you prefer the way it looks, that's not necessarily a bad thing, as long as you know what you're getting into.

One final point: casual readers frequently ask why more people don't consider putting VAWTs on top of tall towers to take better advantage of strong winds. It's an interesting question, although we'd point out that very few people have ever tried it. We suspect it's not a good value proposition. If you are going to go to the considerable expense of erecting a tower, with foundation, it is very important to recoup that investment with a machine that is going to produce solid amounts of energy. In that case, it is safer to go with a tried and true (HAWT) design, rather than risk a brand-new product from a start-up. In addition, the higher you mount something, the less of an issue it becomes for noise. And the harder it becomes to service, so you again want something with a proven track record, and then it doesn't matter much if the equipment is at the base of the turbine or in the middle; you still have to get up to it.

Finally, we suspect that anyone who is investing in a proper tower has probably done some detailed research, and has come to realize that HAWTs are better where it counts: producing energy. Though there is a strong narrative online and in show-rooms that "VAWTs are more efficient," they really aren't. A quick look at the graph from NREL (Figure 8-15) shows why. Sure, manufacturers like Quietrevolution can add lots of technology to wring better efficiency out of the design. But in the end, is it worth it? Only you can decide for your site.

Standards and Certifications

If you're like us, you want a wind system that is efficient, safe, and durable. While there are few 100 percent guarantees in life, one of the ways consumers can decrease exposure to risk is by choosing products that meet designated standards. It is no different with wind turbines, which are increasingly being produced in accordance with national and international testing safety and performance criteria (Figure 8-23).

FIGURE 8-23   A number of standards and certifications now cover small wind turbines, which can help consumers weed out dangerous and suspect products. *Kevin Shea.*

We hope our brief synopsis of small wind standards may help you move confidently through the garden of choice, without getting stuck with a big weed. We include links to various standards, guidelines, and codes that you will be looking for on manufacturers' brochures, websites, and manuals.

Most manufacturers and dealers should be able to walk you through the relevant certification systems, or point you to those who can. Please note that not all standards apply everywhere or to all projects. However, if any component appropriate for your design doesn't meet the standard and someone tells you they feel it shouldn't have to, do we have to tell you to avoid them like an approaching spinning blade?

ISO-9001 Quality Management (International)

A wind turbine is an expense that should last 20 years or more. If you are looking for a manufacturer or supplier that can reasonably support your product for that length of time, this label posted on their website or brochure may be an indication of their staying power. ISO-9001 certification demonstrates that the company has been audited and certified by the International Organization for Standardization (ISO) for having developed and implemented a quality management system for business practices. Such a system has as its goals adequate training of employees, adherence to proven procedures, and commitment to exceeding customer expectations. Certification to an ISO-9001 standard does not guarantee quality of products and services, but it does certify that formalized business processes are being applied.[1] This helps screen out fly-by-night and poorly managed companies.

CE Safety Standards (European Economic Area)

The CE (conformité européenne) marking is affixed on products certified to have met European Union consumer safety, health, or environmental requirements (e.g., noise limits). It is still not required within the member countries of the Central European Free Trade Agreement (CEFTA), although some of them are already adopting many of its standards.[2] The European Economic Area includes all members of the EU, plus a few other participating countries, such as Norway and Iceland.

RoHS Less-Toxic Electronics (International)

Don't be alarmed, but if you are reading this on any other device besides a printed page, you are very close to some hazardous materials. Paints, PVC cables, solders, circuit boards, batteries, mercury, lead—many hazardous materials are used in the manufacture of electronics. A wind turbine is no different, since it contains a number of electrical components.

In a bid to decrease the toxicity of commonly used products, the European Union adopted RoHS, the Restriction of Hazardous Substances Directive, which restricts the use of six hazardous materials—the heavy metals lead, mercury, cadmium, and hexavalent chromium, plus common fire retardants polybrominated biphenyls (PBB) and polybrominated diphenyl ether (PBDE). Outside of the EU, some countries have adopted similar guidelines, as has the state of California.[3] If you live in Europe, California, or another place with RoHS-type rules, it's a good idea to order compliant (or specifically exempt) parts so you don't give officials an excuse to shut you down.

	UL-1741 Safety Standards (United States) UL-1741 requirements govern safety standards for inverters, converters, charge controllers, and interconnection system equipment for use in both off-grid and grid-tied power systems. Other countries may have similar standards.
	IEEE 1547 Electric Standards (United States) The IEEE 1547 standard establishes criteria and requirements for interconnection of distributed resources with electric power systems. It provides requirements relevant to the performance, operation, testing, safety considerations, and maintenance of the interconnection.[4]
	National Electric Codes NEC (NFPA 70, United States); IEC 60364 (International); CEC, CE, CSA C22.1 (Canada); BS 7671 (United Kingdom); PSE (Japan); NF C 15-100 (France); RGIE (Belgium) This is a brief list of electrical code standards found around the world. Most countries, including the United States, don't have laws mandating a national electric code per se, yet there are often published national design safety standards. Many local or power authorities require that you show evidence that your system complies with the national standard. Safety is a very good reason why a certified or qualified electrician may be required to install a wind turbine system.
	IEC-61400-2-B Small Wind Turbine Standards (International) The IEC-61400-2-B is an important international standard that covers safety, quality assurance, and engineering integrity, and specifies requirements in design, installation, maintenance, and operation specific to small wind turbines.[5]
	ANSI C84.1 Grid Power Standard (United States) ANSI C84.1 refers to the American National Standard for grid power. It's important, because every country's power grid has a specific frequency and voltage. Many countries outside the United States run on 220 to 240 volts at 50 Hz. Much of the Americas run on 60 Hz, at either 220 to 240 volts or 120 volts, as in the United States. Running more voltage through a device than it is designed for can decrease performance or cause permanent damage. That is why you need those converter plugs when you travel abroad. If you are buying a system outside the country you live in, you will need to inform the turbine manufacturer or dealer of your voltage and frequency to ensure that the turbine and inverter are a good pair. It may simply require a model reflecting your service voltage or a downloadable software fix to operate within the voltage range of your utility and your appliances.[6]
	UL 6140-1 Safety Standards for Wind Turbine (United States) As of 2010, Underwriters Laboratories is investigating safety standards for entire wind turbine systems, large and small, for risk of fire and shock, control system electrical performance, and utility grid-interconnect performance. This includes blades, hubs, generators, drive trains, support structures, controls, and beyond.[7] In addition, please keep an eye out for the following: UL 6142 Small Wind Turbine Generating Systems UL 6171 Wind Turbine Converters and Interconnection Systems Equipment

Independent Certification

Certification has become a hot issue in the emerging micro- and small-scale wind turbine industry. Several countries, including the Netherlands, Canada, the United States, and the United Kingdom, have some form of certification standards. Though turbine suppliers are likely to pass the testing fees on to the consumer, it should be a good value for the end user, who now can more confidently shop for a certified system. Certification can help prevent unethical marketing and false claims, and it can make comparison of products easier.

For any certification system, testing should be based predominantly in the field. Most programs require that a turbine be tested a minimum of six months, but more typically one year, including comprehensive design analysis and a technical review. A certification scheme can include, among other things, different guidelines for wind turbines intended for rooftops, or for different climate zones and altitudes. Equally important is that output of wind turbine makes, types, and sizes be measured at a range of standard average wind speeds, such as 3 m/s (6.7 mph), 5 m/s (11.2 mph), and 7 m/s (15.6 mph).

The Small Wind Certification Council (SWCC) recently launched an effort to independently test and certify numerous small wind turbines to stringent performance and safety standards created by the American Wind Energy Association (AWEA).[8] In the United Kingdom, small turbines (less than 50 kW) can be certified by the rigorous Microgeneration Certification Scheme (MCS), which is required for users to access the country's feed-in tariffs. Turbines larger than 50 kW in the United Kingdom would fall under Ofgem (Office of the Gas and Electricity Markets) Wind Turbine Certification.

List of Certified Turbines

Small Wind Certification
bit.ly/swccturbcert

Small Wind Turbine Certification (less than 50kW)
bit.ly/UKturbcert

OfgemWind Turbine Certification (more than 50kW)
bit.ly/Ofgemcert

Summary

In this chapter, we took a closer look at arguably the most exciting part of a small wind system: the actual turbine. We took a tour through a turbine's main parts, and introduced the differences between drag- and lift-based designs, as well as upwind and downwind turbines. We took a detailed look at vertical axis wind turbines (VAWTs), which have a few advantages, but a number of significant disadvantages. Although the VAWT industry has seen a recent renaissance, the number of installations remains small. Most wind experts recommend avoiding VAWTs for most applications, especially where energy generation is the main goal.

Boosters of VAWTs sometimes argue that their products shouldn't be thought of as substitutes for HAWTs. They say their technology is intended for places a more traditional horizontal access turbine wouldn't be a good fit. Although it's certainly possible that there may be suitable sites that fit that description, we wonder how common they are. For most people, it doesn't make much sense to put a wind turbine on a roof or in an urban area, because there is usually not enough wind and too much turbulence.

Finally, this chapter introduced the most common standards and certifications that apply to small wind turbines in much of the world. These standards are constantly evolving, but in general, it is good news that consumers are getting more guidance and designs are being tested by third parties. We think this is evidence that the small wind industry is maturing. When you are ready to shop for a wind system, definitely look for relevant certifications.

CHAPTER 9

How to Choose the Right Wind Turbine for the Job

The word "energy" incidentally equates with the Greek word for "challenge." I think there is much to learn in thinking of our federal energy problem in that light. Further, it is important for us to think of energy in terms of a gift of life.

—THOMAS CARR, TESTIMONY TO U.S. SENATE COMMERCE COMMITTEE,
SEPTEMBER 1974

Overview

- What questions do I ask when considering a wind turbine?
- What are the most important differences when comparing models?
- What is the tip speed ratio, and why isn't faster always better?
- How do wind turbines mediate very high winds safely?

We're getting closer to arguably the most exciting decision of the whole process: which wind turbine will you spend your hard-earned money on, and hopefully enjoy watching spin for years to come?

But everything else being in place, it's also true that shopping for a wind generator can be a lot of fun. We're as smitten by the "cool factor" of the technology as much as anyone (Figure 9-1). But before we get too carried away, let's take a look at a brief bucket list.

When considering a wind turbine, there are five main things you need to know:

1. Swept area is the single most important factor in determining energy output.
2. It's a good idea to go with a turbine certified by an independent agency.

FIGURE 9-1   Building and owning a small wind system can be a blast. Just ask Hugh Piggott (center), who leads small wind workshops in his native Scotland, in America, and beyond. *Hugh Piggott.*

3. Hype, testimonials, and slick YouTube video demonstrations do not validate turbine performance claims; field tests do, especially third-party tests.
4. Durability is more important than performance when selecting a turbine.
5. Choose the turbine with the highest net capacity factor.

So, bringing together what we learned in previous chapters, a simplified process for deciding on a wind-electric system looks like this:

- Determine your energy need in kWh per day, month, and year (Chapter 3).
- Downsize your energy need (Chapter 4).
- Estimate your average wind speed and decide on your tower location and height (Chapter 5).

What's covered in this chapter:

- Determine rotor diameter based on predicted kWh at your site's average wind speed.
- Compare all of the products in the rotor diameter range you have selected.

One note: If you are merely looking to get a micro-turbine that only needs to produce a very small amount of energy, you will have more flexibility. However, there are still useful parts of this chapter that can help inform even that decision.

Fortunately, there have been significant improvements in small wind turbines over the past few years, so some of the things that were written just a few years ago are no longer required. It's not unlike the personal computer industry. Not long ago, owning a computer required a learning curve to install programs, ensure compatibility of a monitor with an operating system, and so on. Yet today, computers are much more intuitive to use and "plug and play."

Small wind turbines used to require more specialized, technical knowledge and skills, often borrowed from the agricultural, marine, and automotive sectors. Today, the small wind industry has become more commercialized and accessible. It is still rather immature and underdeveloped compared to large wind, but it has come a long way since the 1970s, when heightened interest first blossomed. Today, there are a number of offerings for durable small wind turbines, many of which have good track records of providing 20 to 30 years of relatively trouble-free service.

What was once primarily the domain of garage tinkerers and research engineers has developed into a consumer-facing industry that is willing to back up their products with five-year warranties and develop international standards. Established manufacturers test their equipment for many months in various climate conditions. Largely gone are the memories of controller meltdowns and rotor failures; instead, the industry focuses on boosting efficiency and reliability and lowering costs. The fact that numerous governments support small wind turbines with tax credits, grants, loans, and feed-in tariffs should provide a measure of confidence in our ability to empower ourselves.

With more than 250 manufacturers and more than 500 models of small wind turbines available worldwide, you'll discover considerable choice, on both rotors and kits. By the end of this chapter, you will be better prepared to make the tough decision—the right decision. We begin with the basics, and then we examine the types of wind turbines on the market today.

Small wind author and consultant Paul Gipe has argued that "small wind turbines should be designed for simplicity, ruggedness, and low maintenance."[1] We think that's sound advice. To help you begin to sort out which system will bring you many years of energy production, we include tables of comparison data, websites for field-tested data, and some easy calculations to evaluate some good candidates. Since cost is likely on your mind, let's begin with a rough guide.

Average turbine cost: Approximately 10 to 40 percent of total system cost

In other words, you should expect to spend roughly 10 to 40 percent of your system's budget on the star, the actual turbine. Let's take a closer look at the decision process.

Important Questions to Ask When Shopping for a Small Wind Turbine

If you aren't interested in understanding all the workings of your turbine and what makes one different from the other, that's okay, because you can have a dealer or installer do the legwork for you. We recognize that some people don't really want to know how their car works, for instance, preferring instead to trust their ride in the care of a mechanic. Others want a computer or cell phone that works smoothly, yet they don't want to ever have to open the case. If that's true for you when it comes to small wind turbines (Figure 9-2), you may choose to skip over some of this chapter. Even so, we still think it's a good idea to become an informed consumer. While the vast majority of small wind installers we've heard of seem to be honest, every industry has its share of shady characters, and knowledge is the best defense.

So, to become an informed consumer, here is a list of questions you can ask a manufacturer or dealer of a small wind turbine, as adapted from the June/July 2009 issue of *Home Power* magazine.[2] We will explain the concepts in more detail in this chapter. Please use our easy downloadable chart (see the link on the following page) to fill out the evaluation. We hope this process will at least help you rule out models that won't be a good fit for you.

FIGURE 9-2   Choose the right turbine for the job. If you have a boat or small cabin, you may only need a micro-turbine, like this Southwest Windpower Air 403. *Southwest Windpower/ DOE/NREL.*

1. Has your turbine(s) been certified for performance and safety standards by an authorized agency, such as the Small Wind Certification Council (SWCC)?
2. What is the annual energy output (AEO), measured in kilowatt-hours, in average wind speeds of 8 to 14 mph (see Chapter 3)? Is AEO calculated using real-life ("field") data (preferred) or laboratory/wind tunnel testing? (Remember, this is not the same as the rated power, which is given in kilowatts, and is not particularly useful.)
3. What is the warranty length and coverage? (The industry standard is five years, although see Chapter 6 for more details.)
4. Has the turbine and/or tower ever gone through a reliability test? By whom and for how long? What were the findings?
5. For how long was the prototype tested? By whom? In the field or in a wind tunnel?
6. How many turbines of this model have been sold, and for how long? How many are still running?
7. How frequently has the model been redesigned? What were the changes and why?
8. What problems have other customers had, and how have you dealt with them?
9. Can you provide references from other customers?

Questions to Ask About the Wind Turbine

bit.ly/MyWindquestions

What to Look For When Buying a Wind Turbine

Although manufacturer marketing materials may suggest the opposite, there are no additional silver bullets when it comes to evaluating a turbine. None of the bells and whistles touted by manufacturers on blade shapes, vibration isolators, or low wind cut-in speed are as important as the basics already covered. Still, in this section, we take a closer look at some other features, in order of decreasing priority:

1. Swept area
2. Durability
3. Annual energy output
4. Cost

5. Speed (RPM and TSR)
6. Governing mechanism
7. Shutdown mechanism
8. Sound

The following comparison matrix (Table 9-1) provides a sample of some of the more popular small wind turbines and what to look for. We'll go over this in some detail, and then cover some of the finer points of the turbine that might affect your design, or at least give you a better understanding of why designs are the way they are today.

TABLE 9-1 Sample Wind Turbine Comparison

	Bergey Excel-S	Eoltec Scirocco	Whisper 500	Skystream 3.7
Swept Area				
Swept area sq ft (m²)	415 (35.3)	265 (24.6)	176 (16.6)	113 (10.9)
Rotor diameter ft (m)	23 (7)	18.4 (5.6)	15 (4.6)	12 (3.7)
Durability				
Tower top weight lbs (kg)	1,050 (463)	450 (202)	155 (70)	170 (95)
Survival wind speed mph (km/h)	125 (200)	133 (215)	124 (200)	139 (225)
Cut-in wind speed mph (m/s)	7.6 (3.4)	5 (2.7)	6.9 (3.1)	7.8 (3.5)
AEO				
kWh/yr @ 8 mph (3 m/s)	5,000	3,496	1,474	914
kWh/yr @ 9 mph (4 m/s)	7,100	4,997	2,139	1,373
kWh/yr @ 10 mph (4.4 m/s)	9,600	6,746	2,907	1,925
kWh/yr @ 11 mph (5 m/s)	12,700	8,687	3,749	2,594
kWh/yr @ 12 mph (5.3 m/s)	15,900	10,751	4,637	3,216
kWh/yr @ 13 mph (5.8 m/s)	19,500	12,870	5,544	3,898
kWh/yr @ 13.4 mph (6 m/s)	23,300	14,983	6,445	4,578
Cost				
Grid-tied version (US$)	$29,500.00	$29,130.00	—	$10,745.00
Battery-charging version (US$)	—	—	$8,795.00	Same model
Battery voltages	—	—	24, 36, 48	24, 36, 48
Generator voltage(s) (Volt)	48, 120, 240	240	24, 32, 48, 120	grid-tied
BOS included	Yes	Yes	Yes	Yes, except controller
Tower or Installation included	No	No	No	Tower (45 ft)
Warranty (years)	10	5	5	5

	Bergey Excel-S	Eoltec Scirocco	Whisper 500	Skystream 3.7
Routine maintenance	Biannual inspection	Annual inspection and lubrication	Annual inspection	Annual inspection
RPM				
RPM	240	245	325	330
Rated wind speed mph (m/s)	27 (12)	25 (11.5)	24 (10.7)	29 (13)
Blade tip speed mph (m/s)	194.6 (87)	161 (72)	175 (78.3)	143 (64)
Tip speed ratio (TSR)	7.25	6.3	7.3	4.9
Shutdown Mechanism				
Generator type	PM	PM	PM	PM
Governing system	Side-furling	Blade pitch governor and stall regulation	Angle-furling	Dynamic brake
Governing wind speed mph (m/s)	15.6	11.5	12.1	11.0
Shutdown mechanism	Crank-out tail	Dynamic brake with blade pitch option	Dynamic brake	Dynamic brake
Turbine Noise level (dB)	52.8[3]	47.7[4]	40–60[5]	40–50[6]

Swept Area (Rotor Diameter)

When determining your rotor diameter (Figure 9-3), it's worth remembering that you may consider a larger turbine if you expect your needs to grow over time, or a smaller turbine if you expect to be downsizing your consumption later on. That said, nearly every turbine owner we've talked to has said they wished they would have gone bigger, not smaller. The only exception we can think of is Rosalie Bay Resort in Dominica, which has a sizable 225 kW turbine. The owners told us they wanted something a bit smaller, but couldn't get an installer to commit to their remote location for anything else. But for everyone else we can think of, they wish they could be producing more clean kilowatt-hours, not fewer.

Of course, in most net-metering agreements, you primarily benefit by using the electricity that you produce, not so much by selling excess. Still, in our experience, most people tend to underestimate how much electricity they use. If you produce too little with your wind generator, you will be paying retail for the difference. With a battery-based system, it is more than an inconvenience when you can't run a heater or use any of the tools in your new workshop, because the turbine takes days or even weeks to recharge the batteries after you constructed your kid's new bunk bed. Either way, it is better to be a little bigger than too small.

Note: Grid-tied systems without battery backup are often package systems sold with the turbine, controller, inverter, and frequently a tower matched and specified

Figure 9-3   Participants in a workshop held by "The Dans" Fink and Bartmann show off the wooden rotor they made to maximize swept area. *Dan Fink*.

by the manufacturer. So once you choose the turbine, the system comes along as a package deal, unless you have some special situation that can't be easily accommodated. A package arrangement is certainly convenient, and it can reduce compatibility issues, although it might limit your choice of options.

Swept area, the size of your wind collector in square feet or meters, is the single largest factor influencing turbine output—besides your average wind speed, which we have already discussed at length. It is the rotor that collects the wind energy, and depending on the efficiency of the blades and generator, a certain amount (up to 59.3 percent, the Betz limit; see Chapter 3) of the available wind energy is converted to electrical power.

If you aren't immediately told the swept area of a wind turbine, the rotor diameter permits you to calculate it, using the basic formula for the area of the circle (πr^2). Divide the diameter in half to get the radius, square this number, then multiply it by pi (roughly 3.14159265).

From that geometric formula, we see that

- As the length of the blade is *doubled*, this quadruples the swept area.
- As the swept area is doubled, it *doubles* the power output.
- A larger rotor will give you more energy, all other things being equal.

As you fine-tune your turbine selection, you should pay attention to how small change can have a big impact (Figure 9-4). The difference between a turbine with an eight-foot diameter and one with a ten-foot diameter might not seem large, but

FIGURE 9-4   Students of The Dans carefully measure a new rotor. *Dan Fink.*

the 25 percent increase in diameter represents a 56 percent increase in collector size, with a proportional increase in energy output.

We will provide an extensive list of turbines, but it is recommended that you gain the tools to assess power available for any turbine size. Not only will this help you determine what size might work for you, it will simultaneously verify suspect claims of performance.

$$\text{Power} = \frac{1/2 \times \rho}{A \times V^3}$$

For the formula provided here, first determine if you are going to use either the imperial or metric system, and then follow these instructions.

1. **Air density (rho)**   Air density is affected by the air pressure (altitude), temperature, and humidity. To simplify the calculation, plug in 1.225 kg/m³ (0.0764 lb/ft³) if your site is at sea level and with an average temperature of 14 degrees Celsius (57 degrees Fahrenheit). Otherwise, you'll have to

determine your air density based on temperature and elevation. Charts are widely available. We include a link to do an online calculation as well.

Resources

Simple Air Density Calculator
bit.ly/calcRho
(Fahrenheit)

Comprehensive Air Density Calculator (includes humidity and pressure)
bit.ly/calcairdensityCnF

2. **Swept area (A)**   Input the turbine's swept area in cubic meters.
3. **Velocity (V)**   Input the average wind speed in meters per second, multiplied to the third power. If you have yet to assess your site, no worries. Insert the global average of 5 m/s (11.2 mph), just for comparison purposes.

We have also provided a link for an online wind turbine power calculator. Later in this chapter, we will give you the know-how to accurately calculate your turbine's energy output, factoring overall turbine efficiency and the capacity factor that provides a more realistic estimate for your turbine. But first, the most important part is how big is your potential.

Resources

Simple Power Calculator bit.ly/calc_wtpower

Comprehensive Power Calculator bit.ly/calc_windturbinepower
mywindpowersystem.com/wind-power-calculator/

Durability

After swept area, the next important question you should be asking yourself is, "Will the turbine last long enough to provide a return on investment, or will it become an expensive kinetic sculpture?" Your turbine is outside 24/7, often in inclement weather. High winds don't pack a punch; they pack a sledgehammer. Salty air will wreak havoc on internal circuitry and make Swiss cheese of iron-based metals. Extreme temperatures and any form of precipitation penetrating into the nacelle can have a negative effect on mechanical equipment and electronics.

Wind writer Paul Gipe wisely notes that efficiency, price, and peak power are meaningless without durability. A machine can be "engineered," "an advance in technology," and "spectacular" in the advertisements, but when it dies on the

FIGURE 9-5   Wooden blades are relatively durable, although they do require maintenance, as students of The Dans know. *Dan Fink.*

tower or falls off the tower, all "advancements" are about as useful as a turbine made of brick.

It's important to remember that the total cost of the turbine depends on how long it lasts (Figure 9-5). Before you even take into account the savings from the annual energy output of your turbine, the price tag for a $10,000 robust system that lasts 20 years (10,000 / 20 = $500 annually) is a better deal than a $5,000 lightweight toy that fails after only five years ($5,000 / 5 = $1,000 annually). A manufacturer may tout that their product has a lifetime expectancy of 20 years, but unless they tested it on your site, you may have to give that claim some qualifications.

Turbine Weight

Do I look heavy to you? Discussing turbine weight might seem a little odd, yet weight plays a part in controlling vibrations, which can be made worse by blade imbalances, extreme weather, or even worn bearings.[7] A heavy turbine does not guarantee durability, yet more mass does help buffer vibration, which stresses internal components. The weight comes in the form of a full metal jacket, beefier alternators, and heavier bearing assemblies.

Start-Up Speed

Makers of rooftop and micro-turbines frequently boast that their products have extremely low "start-up" speeds. Some small, lightweight rotors are designed to start spinning in as low as two miles per hour. With less powerful alternators, they may even be able to start producing electricity at that speed. But you should know by now that there is very little energy available in winds of two miles per hour. Does it matter if you start producing when it's only a trickle?

In truth, it depends on your application. If you only need enough juice to charge a cell phone, that might be fine. Maybe a gentle breeze or a sneeze is enough for you. Be aware that effective turbines have a limited range of wind speeds for which they can collect energy without sacrificing efficiency. So a micro-turbine that works in very low wind speeds is not going to work well in higher winds. It is going to have to shut down to avoid spinning too fast and getting damaged. So you'll miss out on the opportunity to harvest energy when the winds are strong. At worst, the turbine can overheat and fail.

Cut-Out Speed

At very high wind speeds, typically above 45 mph, most wind turbines cease power generation and shut down. The wind speed at which shutdown occurs is called the *cut-out speed*. Having a cut-out speed is a safety feature that protects the wind turbine from damage during excessive gusts. Normal operation usually resumes when the wind drops back to a safe level. If you are considering a micro-turbine that doesn't have a cut-out speed, it better be either extremely well armored or come with a free turbine-disposal bag.

Survival (Maximum Design) Wind Speed

All wind turbines are designed for a maximum wind speed, at least "on paper." This is usually within the range of 90 mph (40 m/s) to 160 mph (72 m/s). Few manufacturers willingly sacrifice expensive equipment by scheduling field tests in hurricane zones. Sure, they can do tests in wind tunnels, but those only go so far when it comes to real-world conditions. In the field, it is essentially impossible to reliably and repeatedly test for survival wind speed by factoring out wind shear, wind chill, and wind-borne debris. It would be difficult to prove that it was the 150 mph gust that blew your turbine apart and not a wind-blown projectile. This means the maximum wind speed is a somewhat theoretical number, and leads to varied ratings.

What tends to kill wind turbines with more certainty than wind speed is turbulence, which can come even at lower wind speeds. Manufacturers issue warnings to tilt down, lock up, or remove the turbine entirely during episodes of wind approaching the survival wind speed. However, in the event that you find a turbine that has a low survival speed, or none at all, it is best to avoid it even if it is in the small-scale range.

Warranty

Most manufacturers currently provide a warranty on their turbines for at least 5 years, although good-quality systems are often expected to last up to 20 years, based on the lifetime expectancy of the composite blades, which withstand the brunt of harsh conditions. If the product warranty is less than five years, covers less than comprehensive coverage, or warns the installer to keep it out of winds approaching 30 mph—a speed most people would assume is an ideal wind condition—the turbine is a toy, and should have an Adult Safety Hazard label like "Blade debris swallowing hazard."

INTERCONNECT

Turning Household Junk into a Wind Turbine

Sometimes, handy folks ask if they can rig up a small wind turbine out of spare parts they have lying around the garage. The short answer is yes, that is certainly possible. As we have pointed out, a determined 14-year-old brought electricity to his family for the first time by scavenging parts and dreaming big in his native Malawi in 2002. In fact, mechanically minded folks have been building and servicing their own homebrew turbines for a century.

These days, one can download plans from the Internet or order a kit that can make it easier to get started. However, there are a number of limitations to this approach.

For one thing, these kinds of hand-built devices tend to produce very small amounts of power. If you live in rural Africa without access to the grid, a little juice to run a light bulb, radio, or pump could make a big difference in your life, and might justify your efforts. However, if you live in a typical home in the developed world, you're likely to be disappointed by seeing only a few watt-hours after all your hard work. Still, if you treat it like a hobby, educational exercise, or demonstration project, more power to you.

It's also worth noting that turbines built from parts that weren't specifically manufactured for that purpose often require more maintenance than commercial units. They tend to be fussy and prone to breakdowns, and they are often a bit noisy. Few offer the polished, sleek look of a professional install, and some, frankly, are downright ugly.

A functioning example of a small homebrew system is the six-foot turbine built by the team behind Windstuffnow.com (Figure 9-6). Although our e-mails seeking comment weren't returned, the website claims the downwind turbine is "a very small but quite efficient little unit." The turbine was built with a mix of custom-milled and off-the-shelf, nonwind parts, including a car alternator and chain drive. The controller unit was designed and built by Robert Nance Dee of Design Specialties. The six blades were cut from wood and painted yellow, making the finished product look like a sunflower. The leading edges of the blades were covered with durable tape in order to decrease damage from hitting raindrops.

According to the website, the chain drive was "quite noisy," due to the modified car alternator. However, the team is working on fitting it with a more powerful alternator. With everything working properly, the turbine will reportedly start spinning in about a 5 mph wind and start charging at around 7 mph.

FIGURE 9-6   This small homebrew turbine was originally made with a car alternator (not recommended) and chain drive, although it was eventually upgraded with better parts. *Windstuffnow.com.*

In a test, the small turbine started producing 19 watts in 11.6 mph winds, according to Windstuffnow.com, and up to 320 watts in rare 32 mph winds. In 12.5 mph winds, it produced 38 watts, and at 15.6 mph it produced 76.2 watts, roughly enough to light one conventional incandescent bulb. Clearly, that's not a lot of power, but it's an interesting experiment.

Windstuffnow.com also sells micro-turbine rotor kits for $375.

Predicted Annual Energy Output (AEO)

Once you have a size range in mind for your new turbine and are assured that it is durable, how much energy will it actually produce (Figure 9-7)? You'll want to ask the manufacturer for a chart of predicted annual energy output (Chapter 3) in kilowatt-hours, at a range of 8 through 14 mph (3 to 6 m/s). This won't be for your specific site, of course, but it gives you some idea of how much energy the wind turbine will produce for a given average wind speed. Remember, wind speeds are listed at hub height. Since such production numbers are taken from manufacturer's data, they should be regarded with a large grain of wind-borne sand. Not surprisingly perhaps, manufacturers do tend to overstate their performance.

Actual production will depend a great deal on the specific situation a turbine finds itself in, and remember that turbulence kills production. Some turbines perform better at certain wind speed ranges than others. In addition, AEOs are usually determined at sea level, at temperatures of 15 degrees C (59 degrees F). If your site is different, you should make adjustments.

To be conservatively cautious, if you live in an "average" location, you could simply multiply the AEO listed by about 75 percent (AEO × 0.75). It's better to underpredict AEO and be pleasantly surprised. Even better, once you have a list of turbines in your range, choose the next larger model.

FIGURE 9-7   Bergey turbines have a long track record of durability and excellent energy production. *Warren Gretz/DOE/NREL.*

Power Curves = Meh, Energy Curves = Helpful

Graphing the generator output against the varying levels of wind speed yields a *power curve*. Figure 9-8 is a graph indicating how much power in watts (not watt-hours) the Air-X small wind turbine will produce at any given wind speed.[8]

Like a roller coaster, we are drawn to look up at the peak power, believing that this is what we should expect. But these are just points on a curve, with no indication of how frequent your site will ever get this speed. At most sites, the wind

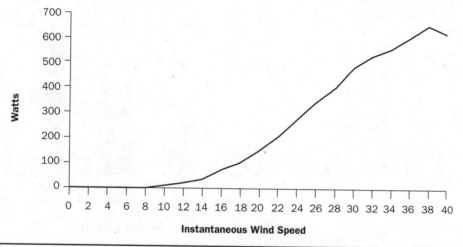

FIGURE 9-8   A typical small wind turbine power curve, in this case for an Air-X turbine. *Kevin Shea, based on data published by Brent Summerville in "Small Wind Turbine Performance in Western North Carolina."*

speed at which a turbine generates its peak power occurs only a very small percentage of the time.

Many sites, particularly home sites, have annual average wind speeds more in the 7 mph range. As you can see from the sample power curve, that means an Air-X turbine installed there is going to be generating very little energy *most of the time.* Once again, we see why it is so critical to have access to a strong wind resource.

We avoid spending much time on power curves because they are, at best, misleading, though they are a central feature in manufacturer marketing materials and on their websites. But you really can't compare turbines by the peaks, unless you happen to live on the top of Mt. Washington (Chapter 1). Instead, it is best to compare wind turbines on the area swept by their rotors (or by simply using rotor diameter).

You'll also want to compare *energy curves* of turbines (Figure 9-9), which are often available on manufacturer websites and literature, though not as prominently touted as power curves.

The first thing you should notice on an energy curve is that it will have the *monthly energy output,* which is what you are looking for. With an estimated average wind speed at your site, the energy curves can help you project the estimated energy production (in kilowatt-hours) from a particular turbine. Then you can determine how that projection matches up with your energy needs, determine which wind generator is right for your site, and get on with the job of designing and installing your wind-electric system.

If you have the average wind speed of your site and the power curve graph of the turbines, we have provided you a quick tool to quickly estimate the production of your unit.

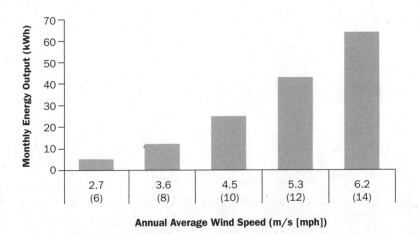

FIGURE 9-9   Better than power curves are energy curves, which show monthly energy output, not just instantaneous power. *Kevin Shea, based on data published by Brent Summerville in "Small Wind Turbine Performance in Western North Carolina."*

Resource

Energy curve calculator
bit.ly/energycurve

But a generalized formula is a generalization—it cannot take into account all factors for an individual turbine. It does, however, take into account the two most important factors—swept area and wind resource. Beyond that, we can next apply factors that tell us that a specific estimate is probable, unlikely, possible, and impossible.

Overall Turbine Efficiency

Like most things, a turbine is the sum of its parts. In Chapter 3, we introduced the concept of the Betz limit, which states that only as much as 59 percent of the energy in the wind is available to be converted to electrical energy. Actually attaining the Betz limit would require perfect conditions, including ideal temperature, pressure and air density, and perfectly smooth airflow. However, in the real world, we have turbulence, wind shear, elevation, changing weather, and other factors that interplay to reduce the usable energy.

Further, as we move along the chain from mechanical to electrical components—generator, shaft, bearings, sensor circuitry—we shall see that some of that energy is being used by the system components or is converted into waste heat. The overall turbine efficiency can range as wide as 10 percent up to just shy of the Betz limit. For most small wind turbines on the market, it is commonly 15 to 25 percent of the potential or rated power output.[9] If any turbine reports less than 10 percent efficiency, it is advisable to look for another turbine.

Large utility-scale turbines tend to have higher efficiency, and many are approaching the Betz limit. But they have the advantage of sophisticated, computer-managed active controls, which make many small adjustments to blade pitch in real time. Some are even using lasers to gauge incoming wind speed and direction.

Capacity Factor

Besides the efficiency of the turbine system, you would like to get a realistic idea of the actual output capacity of your turbine (Figure 9-10). Naturally, no two wind sites are exactly the same. Some are dominated by occasional powerful gusts, while others have longer periods of more steady winds, plus everything in between. Each turbine design performs differently at different wind speeds and wind behavior.

The *capacity factor*, or the turbine performance at the rated power, compares the turbine's actual production over a given period with the amount of power the turbine would have produced if it had run at full capacity (100 percent). The *rated*

FIGURE 9-10   Participants in a workshop led by The Dans test a hand-built turbine rotor. Note the alternator magnets embedded in the rotor. *Dan Fink*.

capacity is the full-load output in kilowatt-hours if the wind was pushing steadily at the manufacturer's rated power.

The net capacity takes into account most of the unused capacity due to the nature of wind's intermittency, although some portion could be the result of shutdowns, equipment failures, and routine maintenance. It also includes the offset energy needed for control, monitoring, display, or maintaining operation.

To estimate the *actual* output, you can match up the data for annual wind history for the site with the performance (power) curves of the proposed turbine.

Researchers at Colorado State University did just that for a site in Stonington, Colorado.[10] The Governor's Energy Office installed and monitored a 50-meter anemometer tower there in 2007 and 2008. Scientists took the observation data on the wind and matched it to performance curves of some common turbines of various sizes and various heights. For the larger turbines, the researchers extrapolated the wind data to higher tower heights in order to allow enough clearance for the big blades.

Let's consider the example of Northern Power's NW 100/20 turbine, with a rotor diameter of 20 meters and rated power of 100 kW. Research suggested a hub height of 50 meters, which is also the top of the anemometer tower, so it wouldn't require extrapolation. At that location, the wind speed at this site was a brisk 17.28 mph. Even with such a strong resource, the turbine would have zero output 16.7 percent of the time, and rated output none of the time. The average net energy output for this turbine was estimated at 234,900 kWh per year.

To find the annual capacity factor for this turbine, take the energy generated during the year (kWh) and divide by wind generator–rated power multiplied by the number of hours in the year.

$$\frac{234{,}900 \text{ kWh (actual)}}{(365) \times (24 \text{ hours/day}) \times (100 \text{ kW rated capacity})} = 0.268 \sim 26.8\%$$

You can use the capacity factor data to help you choose a turbine that would be best suited to your wind resource. A higher percentage number is better—it means the turbine is running closer to its full capacity. A capacity factor of 25 to 40 percent is common, although higher capacity factors may be achieved during windy weeks or months.[11,12]

 Power Up! *Figure 9-11 suggests that larger turbines tend to have higher capacity factors, which isn't surprising, since we have already mentioned that large turbines tend to be "more efficient" than smaller ones. However, even among the smaller turbines listed, there are some that show higher capacity factors than others for the test site. That's because all turbines do not perform equally at different speeds and wind distributions, based on their design tolerances. This shows that it can pay to optimize your turbine*

Turbine	Rotor Diameter meters	Rotor Power kW	Hub Height meters	Hub Height Wind Speed mph	Time At Zero Output percent	Time At Rated Output percent	Average Net Power Output kW	Average Net Energy Output kWh/yr	Average Net Capacity Factor %
Bergey Excel-R	6.7	7.5	30	15.63	12.19	7.27	2.34	20,500	31.1
Bergey Excel-S	6.7	10	30	15.63	5.92	3.96	2.53	22,100	25.3
Bergey XL.1	2.5	1	30	15.63	2.19	10.19	0.35	3,100	34.8
Southwest Skystream 3.7	3.7	1.8	30	15.63	11.24	0.00	0.62	5,500	34.6
Southwest Whisper 500	4.5	3	30	15.63	12.15	8.72	1.09	9,600	36.5
Northern Power NW 100/20	20	100	50	17.28	16.71	0.00	26.8	234,900	26.8
Vestas V47 - 660 kW	47	660	65	18.25	9.86	0.98	232	2,030,200	35.1
GE 1.5s	70.5	1,500	80.5	19.10	12.89	10.90	525	4,596,100	35.0
Vestas V80 - 2.0 MW	80	2,000	100	20.03	12.32	6.44	813	7,126,000	40.7
GE 2.5xl	100	2,500	110	20.46	9.04	16.66	1,127	9,869,000	45.1

FIGURE 9-11 Comparison of popular wind turbines, showing energy outputs and capacity factor. *Mike Kostrzewa/Colorado State University 2008 Mechanical Engineering.*

choice for your particular site to maximize your ROI. If there are no turbines with at least a 10 percent capacity factor, you may want to reconsider whether wind is a good choice for your site.

If you haven't assessed your site yet but want a survey of turbine performance, there are test fields around the world you can use to see the average performance at the rated power of various turbines. Access to those sites is provided at the end of this chapter.

How to Calculate Net Energy Output for Any Wind Turbine

 Tech Stuff *We have been showing you how to estimate actual energy output, but each time we seem to share one more item that will reduce the ideal rated output. Now that you are developing an understanding of the parts of your wind collector and how each part factors in to how much energy will be produced, let's bring it down to a more complete picture. How does the collection of selected parts—environment, blades, alternator, electronics, and so on—work together to get your juice blender the wattage it needs to make your morning margarita?*

Each of the elements of the performance formula has its own distinct contribution to total wind turbine power output and resulting yearly energy yield. If there is any single equation that the beginning wind enthusiast should memorize, this is it.

Wind Turbine Power:

$$P = [1/2 \times \rho \times A \times V^3] \times Cp \times Ng \times Nb$$

You can hopefully recognize the formula in the brackets as the power availability formula, discussed earlier in this chapter and in Chapters 3 and 5. To briefly review, ρ (rho), is the air density (about 1.225 kg/m³ at sea level, less higher up). A is the swept area of the rotor, typically calculated in meters squared. V is the wind speed, let's say in meters per second.

The rest of the formula factors in the losses that affect the system, which are worth considering so you can better size your unit.

- *Cp = Coefficient of performance*, the aerodynamic efficiency of conversion of wind power into mechanical power, often called the power coefficient. This value has a theoretical maximum of 59 percent (0.59), the Betz limit (Chapter 3). Utility-scale turbines are approaching the Betz, while a "good" design for small wind turbines is around 35 percent (0.35).
- *Ng = Generator efficiency*, the conversion efficiency of the turbine generator (alternator). Typically, this will be 80 percent or more for a permanent magnet generator. With a car alternator it would be around 50 percent.

- *Nb = Bearings/gearbox efficiency,* the conversion efficiency of the bearings and gearbox, if one is present. In a well-made system, this could be as high as 95 percent.

We weren't able to find any premade online tools for these calculations, so we decided to create one. Check out the second worksheet of the spreadsheet titled "online calculator."

Resource

Turbine Power and AEO Online Worksheet
bit.ly/AEO_xl

Cost

Prices in the chart are the manufacturer's suggested retail price and generally include the turbine, controls, adapter for the towers, and the inverter. Battery-based systems might also include the batteries, charge controller, and divert load.

The versions field in the comparison chart is to let you know what system types are available and their price differences. As we have mentioned, batteryless grid-tied is normally the most cost-effective choice. Most such systems are configured to be connected to a standard 120 or 240 VAC single-phase utility service connection. Some turbine kits, such as Skystream 3.7 and the newer model Skystream 600 (tentative name), are also available for connection to a three-phase utility service. You might need a dedicated transformer if one is not nearby, but the installation and cost are generally the responsibility of the power authority with which you have a net-metering or feed-in tariff agreement. In addition, if you're determined to have backup for utility outages, some battery-charging turbines can be grid-tied via a battery-based inverter that also synchronizes its output with the utility grid. Check with the manufacturer.

Battery voltages are listed for battery-charging turbines so you can choose the right turbine voltage for your battery bank. Most modern, whole-house, battery-based renewable energy systems use a 48 V battery bank with an inverter to supply the house with 120 or 240 VAC.

BOS included lists what components of the balance of system you receive when you buy an off-grid turbine—components such as a controller, batteries, dump load, and metering. If they are not included, don't forget to add them into your cost estimates—these components can be expensive. Most grid-tied wind systems are usually sold as a package, with the wind turbine, controls, and inverter.

Tower or installation included. For smaller systems, the price of the tower and its foundation can easily eclipse that of the turbine. Therefore, it's not surprising that

FIGURE 9-12 Byers Auto recently installed a 100 kW Northwind wind turbine at its car dealership in Columbus, Ohio, with the help of a $200,000 grant from the state. *Byers and Renier Construction/DOE/NREL.*

more and more manufacturers are offering full-package deals that include the tower (Figure 9-12).

The all-in-one package option has various advantages for the consumer. It takes out the process of subcontracting the construction or locating the appropriate tower. The added cost of transporting a tower from the manufacturer or distant distributor is compensated by the fact that the warranty now includes the tower and foundation, assuming it is installed by a crew certified by the manufacturer. In fact, several manufacturers are now insisting on selling the whole package and using certified installers in order to avoid tower-related problems. Otherwise, less experienced installers frequently underestimate the lateral thrust of the wind. Some people put turbines on the darndest things (telephone posts, trees, plastic plumbing pipes).

Warranty is not only an indication of the manufacturer's confidence in the machine; it tells you what is covered—usually it's equipment only and not the costs of replacement shipping or labor, which can be significant. If the manufacturer provides less than five years of warranty coverage, it might be possible to extend the warranty for an additional cost. See Chapter 6 for more on warranties.

Speed (RPM and TSR)

RPM (revolutions per minute) gives the turbine's blade speed at rated power output.[13] Since most manufactured small turbines are direct-drive, it is also the alternator/generator rpm. Although we stated early in this chapter that you want a turbine with blade speeds (driven by lift) faster than the oncoming wind speed, it's also not necessarily true that faster is always better. For one thing, the RPM rate relates to two other important characteristics of wind generators: durability and noise.

Just as you wouldn't drive your car down the interstate in first gear because of the increased engine wear and poor fuel economy, you don't want a wind turbine that spins at an excessively high RPM. The faster that rotor turns, the quicker the parts will wear out. That's because higher RPMs mean more vibration, friction, centrifugal forces, and noise. As if wind shear weren't enough, as a machine wears and the blades get nicked and abraded, imbalances will become more pronounced at higher speeds. This will naturally put more stress on the turbine and tower, which can lead to early failure in any of these components.

A faster turning engine will produce more friction, which means more of your precious wind energy will be lost to waste heat. And heat affects other parts. Bearings are one of the few moving parts that can fail on a direct-drive wind turbine. Several factors affect the life of a bearing, with heat, dirt contamination, and inadequate lubrication as the major culprits. Another side effect is the centrifugal force that constantly tries to tear the rotor apart. Finally, a slower rotor speed in a given class of turbines will generally mean a quieter turbine.[14]

Lower RPM does not necessarily mean lower energy production, just as higher RPM doesn't always mean higher production. What you want is a good design that matches the alternator RPM and rotor speed to get as much energy out of the wind as possible, without sacrificing integrity of its components (Figure 9-13).

Tip Speed Ratio

A concept closely related to RPM is *tip speed ratio* (TSR). Browse through a glossy wind turbine catalog, and you'll likely encounter numerous references to it, because it is important to the efficiency and longevity of your machine.[15] The tip speed ratio describes how fast a turbine's blade tips are moving compared to the oncoming wind speed. For example, if a 20 mph (9 m/s) wind is blowing on a wind turbine and the tips of its blades are rotating at 80 mph (36 m/s), then the tip speed ratio is $80/20 = 4$. This means that the tip of the blade is moving four times faster than the wind driving the airfoil.

For a drag-based design like an anemometer, the TSR will always be 1 or less, because drag machines can never go faster than the wind. With the added push of lift, however, a turbine could see TSRs of 11 or higher.

Figure 9-13 National Wind Technology Center researcher Dave Corbus outfits a 10 kW Bergey turbine for testing. *Warren Gretz/DOE/NREL.*

Some engineers say a TSR of 8 would be the optimal balance of lift over drag. As tip speed ratios get smaller, drag tends to increase, since each blade encounters more of the wake of the previous blade. As TSR gets higher, blades get louder, and more stresses are placed on the equipment. As it turns out, most engineers decide that TSRs of 5 or 6 are optimal in the wind turbine industry, since that range is viewed to give the highest efficiency of energy extraction, balanced with longevity and avoidance of excessive noise.

It's also worth pointing out that the fewer the blades, the faster the rotor has to rotate to cover its swept area in order to extract the maximum amount of energy (Figure 9-14). That means the fewer blades a turbine has, the higher its TSR tends to be.

 Tech Stuff *If you are considering a turbine with a number of blades not shown in Figure 9-14 and would like to calculate what the optimal TSR should be to achieve maximum power extraction, use this formula. The variable* n *is the number of blades:*

$$\lambda max\ power = \frac{4\pi}{n}.$$

In Figure 9-15, we can see how the number of blades affects the coefficient of performance, also called the power coefficient.

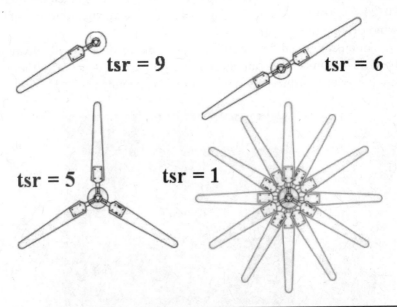

Figure 9-14   The fewer the blades, the faster the rotor spins, and the higher the TSR tends to be. Higher TSR means more noise and stresses, while lower TSR means more drag. *Kevin Shea.*

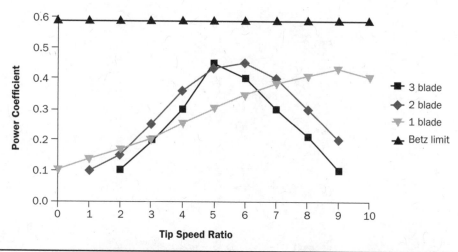

FIGURE 9-15   A comparison of power coefficient (coefficient of performance) and tip speed ratio for optimal wind turbines with one, two, or three blades. Most people choose three blades as a compromise of performance, stability, and durability. *Kevin Shea.*

Imagine that the lines in Figure 9-15 are like cargo nets, and anything underneath each net has been captured, while anything above has been lost. The maximum theoretical power extraction is 59.26 percent of the windstream—the Betz limit. Small turbines at their optimal tip speed ratio can achieve a range of 35 to 45 percent conversion efficiency due to losses and inefficiencies. Furthermore, in general, a one blade-turbine performs 10 percent less than a two-blade turbine (Figure 9-16), which in turn performs 5 percent less than a three-blade turbine for the same swept area.[16]

If a turbine frequently exceeds its optimum TSR, it can wear out faster and cause noise and excess vibrations, which can make for angry neighbors.[17] If a turbine is below its optimum TSR, much of the wind will pass undisturbed through the gap between the rotor blades. In addition, the blades will tend to *stall*, a term borrowed from aviation that means the drag force is overpowering the lift force. This happens if the angle of attack of the airfoil is too big. It cuts into the power extraction and can disable grid-connection with the power authority.

So, your takeaway with regard to tip speed ratio is to ensure that your turbine's optimal value approximates the numbers in Figure 9-15.

Power Up!   *Manufacturers sometimes tout higher TSRs as an attractive "feature." We think this is because as rotor RPM increases due to increased TSR, alternator size can decrease and still get the same output at a given wind speed. Faster spinning machines allow for smaller, cheaper alternators for a given output power level. For the manufacturer, this means less cost to fabricate their particular product, which translates to a more competitive spot in the marketplace.*

FIGURE 9-16 Two-bladed turbines like this model from Gaia Wind at the National Wind Technology Center tend to spin fast, though they introduce powerful cyclic forces on the hardware. *Lee Jay Fingersh/DOE/NREL.*

Calculating Tip Speed Ratio

Tech Stuff *In the event that the turbine manufacturer doesn't provide the tip speed ratio, or you want to know how this number has been calculated, there is a simple process to extract this information.*

The tip speed as blades make a revolution is expressed in "meters per second," and is calculated by multiplying the diameter (in meters) by π, and then multiplying this product by the RPMs divided by 60 (seconds), to get meters per second. For example, if we had a four-meter-diameter rotor that had 300 RPM at the rated output speed,

$$\text{Tip Speed} = (D\pi)\,(\text{RPM} / 60)$$
$$= 4\pi \times (300/60)$$
$$= 63$$

With the tip speed, you can now get the TSR if you divide it by its rated wind speed:

$$\text{TSR} = \text{Tip Speed of Blade} / \text{Rated Wind Speed}$$
$$= 63/\,12\ (\text{m/s}) = 5.25$$

The tip speed ratio is chosen carefully so that the blades do not pass through too much turbulent air left by the previous blade, yet they also do not allow too much air to pass between the blades without being "harvested."

Resources

TSR discovery activity
bit.ly/tsrdiscover

Technical Article
bit.ly/tsrRagheb

Governing Mechanism

As you know, high wind speeds carry a lot of kinetic energy. However, turbines are built to operate only up to a specific speed (Figure 9-17). Anything from mechanical damage to catastrophic failure can occur when this limit is exceeded. Therefore, all turbines have some manual and/or electronic controls that allow the rotor to be slowed down or stopped (stalled). In fact, there are several reasons for having such a *governing mechanism*:

- Optimizing aerodynamic efficiency
- Enabling maintenance
- Reducing noise
- Keeping the generator within its speed and torque limits
- Keeping turbine and tower within their strength limits

FIGURE 9-17   Students and staff from Appalachian State University install a new 2.5 kW ARE 110 turbine with Robert Preus from the manufacturer, Abundant Renewable Energy. *Appalachian State University/DOE/NREL.*

The *governing wind speed* is the average point at which the turbine starts the mechanism(s) that deflects some of the wind. A lower governing speed suggests that the turbine designers were conservative, preferring a turbine to last through the rare times of excessive winds. A higher governing wind speed indicates that it was designed to spin a bit fast and furious, eking out a bit more yield, possibly at risk of reducing longevity.

In general, there are three main types of governing systems: stalling the blades, reducing the area of the rotor facing the wind, or changing the pitch angle of the blades. Let's take a closer look.

Furling	
	When the wind exceeds a determined force on the turbine, a furling system directly or indirectly turns the rotor at an angle out of the wind, either horizontally or vertically. The desired effect is to turn the blades so the edge is facing the wind. This provides less surface area so that wind passes over the blades and the rotor slows down. As the wind speed slows, the rotor turns back into the wind. Springs or a shock-absorbing piston may be utilized to provide some dampening, increasing its longevity and maximizing its face-time with the wind.

Variable Blade Pitch	
	The principle of variable blade pitch is similar to the vertical slat blinds that pivot open and closed to allow sun to pass through. During low and intermediate winds, the blades have default pitch for best start-up and energy harvesting. However, when the rotor starts running too fast, the electronics, weights, or springs that act on centrifugal force rotate the blades to a less efficient aerodynamic angle, which slows them down. In extreme winds, the controls turn the blades nearly parallel to the wind current, called feathering, to protect the unit from damage. The blade basically stalls, thus decreasing lift and limiting the rotor speed.
	A blade tip pitch is another variation, in which the outer tip of the blade responds to centrifugal force instead of the whole blade.
	Models: Most large commercial turbines have pitched blades, but so do select small turbines, including all Proven Energy models, several Endurance Wind Power models, Eoltec's Scirocco E 5.6-6 (6 kW), Kestrel's SWT, Ventura Energy's VT10-240, and the original Wincharger Jake (see chart at end of chapter).

Governing Mechanisms

Furling	Pitch Control	Stalling*
Principle: Decrease the rotor's angle of attack by moving the rotor axis out of the direction of the wind	**Principle:** Decrease the blade's angle of attack by rotating each individual blade	**Principle:** Aerodynamic stall occurs when the blade's angle facing the oncoming wind becomes so steep that it starts to eliminate the force of lift, decreasing the speed of the blades
Green Rhino Energy	*Green Rhino Energy*	*Green Rhino Energy*
Types:	**Types:**	**Types:**
• Active: Vertical furling driven by hydraulic, spring-loaded, or electric motors	• Active control (only): Blades rotate out of the wind when wind speeds are too high	• Passive: Fixed-angle blades designed with a twist so that the winds above a certain speed will cause turbulence on the upwind side of the blade, inducing stall
• Passive: Horizontal furling with yaw		• Active: Based on power output, electronically pitches the blades into the wind, inducing stall
Advantages:	**Advantages:**	• Hybrid: Manually adjustable pitching to reflect site's particular wind conditions
• Tower and guy wires receive less of the wind force	• Keeps optimal power output at the concurrent wind speed	**Advantages:**
Disadvantages:	• Less noise	• Optimizes energy capture
• Produces very little output power while furling	• Reported as more reliable than furling in protecting the machine[18]	• Passive system has fewer parts to maintain
• Wind hitting the blades at an angle can cause increased noise	**Disadvantages:**	• Hybrid system permits you to optimize aerodynamics for your location
• More components to maintain	• Complexity	**Disadvantages:**
	• Increased up-front cost	• High winds hitting the blades at this angle cause increased noise and vibration
	• More parts to maintain	• Active system has more components to maintain
		• Usually for larger commercial systems only

Furling System Designs

There are several furling system designs, and not all are equal in their performance.

Tilt-Back The wind generator is hinged just behind the nacelle. When wind speed gets too high, the entire nacelle, hub, and blade assembly tilts back out of the wind, to nearly vertical, while the tail and vane follow the wind direction (Figure 9-18). As the wind slows down, it returns to normal horizontal operating position by springs, wind action on a tilted tail, or a counterweight. If you want to see what it looks like, follow the link to the YouTube video.

bit.ly/tiltback

- Advantages:
  - It is time-tested, works effectively, and deflects large wind forces.
- Disadvantages:
  - It cannot be used for very large turbines.
- Models: Many micro-, residential, and commercial wind generators use this method.

Figure 9-18 This Whisper H40 demonstrates tilt-back furling to mitigate high winds. *Dean Davis/DOE/NREL.*

Flexible/Coning Blades The blades flex or pivot on a hinge back from the wind like an umbrella with a big wind gust (Figure 9-19).

bit.ly/coning

- Advantages:
  - They can be inexpensive to make.
  - They can relieve stress on the tower during high or gusty winds.
- Disadvantage:
  - The flexing may fatigue the blades.

FIGURE 9-19   Proven Energy wind turbines are among those with flexible blades, which are inexpensive and relieve stress in high winds. *Proven Energy.*

       – With downwind machines (see later), tower shadow is a problem, since the rotor blade actually passes behind the tower. This can cause turbulence and increased fatigue on the unit.

       – Increased noise as blade shears wind.

    • Models: Proven Energy's 2.5 kW, Southwest Windpower's Skystream 3.7 and Skystream 600

Side Furling or Furling Tail   The generator is mounted off-center horizontally from the yaw. The tail and/or vane is hinged and set angled back and to one side. The weight of the tail will want to turn it down. Figure 9-20 shows a small wind turbine that uses both side furling and tilt back.

bit.ly/av_
furlingtail

    A common misconception is that the tail folds. Actually, when the wind on the rotor is strong enough to overcome the weight of the generator, making it want to yaw, the angled tail stays aligned with the wind and the nacelle yaws sideways, turning the rotor away from the wind direction. When wind speeds drop, the nacelle is returned to normal operating position by gravity or additional springs.

    • Advantages:

       – It requires less stress on the system than tilt-back control.

FIGURE 9-20 Invented in 1927 by two brothers in Iowa, the Wincharger was popular on farms before grid electricity. Note the air brakes perpendicular to the blades. *The Tin Box, www.thetinbox.com.*

- Disadvantages:
  - The vane gets the brunt of the gale winds and can eventually lose integrity.
- Models: Residential and commercial-size turbines utilize this system, including many homemade designs. Examples include Bergey's Excel and XL.1 and Fortis Energy's Montana.

Air Brakes   Early small wind turbines included metal "cups" that extended from the hub from centripetal force during high winds to slow the machine down; they retract back into the hub when the wind slows.

bit.ly/
wincharger

- Advantages:
  - Simple and accessible parts for DIY kits.
- Disadvantages:
  - Many components to maintain.
  - Noisy and full of vibration.
- Models: Air brakes are sometimes found in homemade designs, and were common in older models, like the classic Wincharger from the 1930s (Figure 9-20). Invented by brothers John and Gerhard Albers in Iowa in 1927, the Wincharger could charge a six-volt battery. The Albers started a company, and their design caught on across rural America. In 1935, they sold the firm to Zenith, which packaged the small turbine with a radio for $10. They also added 12, 32, and 110 volt versions. Vintage Winchargers can still be seen on farms across the Americas.

Shutdown Mechanism

Shutdown mechanism refers to the method used to completely stop the turbine, such as for service, in an emergency, or when you just don't need or want it to run—like before a trip or an approaching storm. After shutdown, it is not allowed to spin at all, and should be able to survive extremely violent winds in this condition. Generally, larger, more expensive wind turbines have more reliability and redundancy built into their shutdown mechanisms, and may use more than one braking method.

Shutdown Mechanisms	
Mechanical Shutdown	**Electrical Shutdown**
Principle: These systems physically brake the wind generator, or force it out of the wind by turning the tail parallel to the blades. The rotor stops, spins slowly, or stalls. Generally, a cable is attached to a hinged tail, with a small hand winch or crank located at the bottom of the tower. Even commercial models can have such emergency shutdown systems.	**Principle:** An electrical system shorts out the main AC power-out leads (three phases) together, creating a very strong braking force on the rotor.
	Types:
	Manual: A simple manual switch at the tower base or even inside the house. In addition, a shorting plug can be installed.
Some wind turbines use a mechanical disk brake to stop the rotor, though this is very rare on small wind turbines.	**Automatic:** Many small wind turbines rely on dynamic braking, the electrical shutdown mechanisms included with permanent magnet alternators. Once the RPMs exceed the limit for a certain period, an electrical shorting of the windings is activated. To prevent overheating, the system requires a delay between repeated braking.
Advantages:	
Mechanical brakes are usually more reliable than dynamic braking (electric).	
Disadvantages:	**Advantages:**
None; however, many smaller turbines simply have no mechanical means to shut them down.	• Electric shutdown systems are generally reliable.
	• Manual system switches are easily accessible.
	• If designed well, works well in high winds.
	Disadvantages:
	If poorly designed, extremely high winds during storms can overpower the magnetic braking, and the system can be damaged or burn out.

Sound

We discussed the issue of noise and wind turbines at length in Chapter 2. As we pointed out, to try to get a handle on the often-subjective nature of noise annoyance, many ordinances specify the level of allowable sound from a wind turbine at the adjacent property line, the nearest dwelling, or at a certain distance. Typically, 50 to 60 decibels is considered the maximum allowable, with some exceptions for short-term events.

One variable you might see advertised for a turbine is *sound power level,* which is defined as the sound level at a distance of 1 m (3.3 ft) from the source, which for turbines we take as the center of the rotor, or in other words, hub height.[19] Remember, as a person gets farther and farther away from the source, the intensity of the sound they hear reduces as the square of the distance.

The Small Wind Certification Council defines AWEA's Rated Sound Level as such:

The sound level that will not exceed 95% of the time, assuming an average wind speed of 5 m/s (11.2 mph), a Rayleigh wind speed distribution, 100% availability, and an observer location 60 m (~200 ft.) from the rotor center. The final decision as to what is appropriate sound levels is between you and your local ordinances.

So, what used to be purely dependent on marketing pitches or anecdotes is becoming more transparent and quantifiable. (It's also true that turbines have been designed to be quieter.) Now, turbine manufacturers can shout about how quiet their machines are, and you in turn can show noise level certifications to your angry neighbors.

That said, it's worth pointing out that if noise is critical to you, there are small wind turbines that are quieter than others. Ask manufacturers for their noise ratings, and visit some examples of their turbines installed in the field during different wind conditions.

If you don't need a lot of juice, you might consider the small Swift turbine, which is said to operate at less than 35 decibels, compared to a more typical range of 40 to 55 decibels for small turbines. According to the inventor, Scotland-based Renewable Devices, the lower noise is partly due to what they call a *ring diffuser*, the outer ring that goes around the tips of the five blades (Figure 9-21). This reportedly decreases turbulence, which is often the source of blade noise.

FIGURE 9-21   Swift turbines are said to be especially quiet because of the ring around the blades, called a "ring diffuser," which is supposed to decrease turbulence. However, rooftop installations usually don't produce much energy. *Judith/Wikimedia Commons.*

Swift turbines are made and sold for the North American market by Michigan-based Cascade Engineering, and they are priced aggressively low (around $8,500). The company says the grid-connected Swift has a peak production of "over 1.5 kW," or 1.0 kW at 24.6 mph (11 m/s) winds. However, you should know by now that wind speeds that high are quite rare on almost all sites, especially in urban and suburban areas, where the turbine is marketed toward. Cascade says the Swift can produce 1,200 kWh a year on a site with 11.2 mph (5 m/s) annual average winds, which is still quite high. It isn't that common to find such good annual average wind speeds even with a 200-foot tower, although the Swift is marketed primarily for rooftops.

According to Cascade's website: "A structure-mounted SWIFT turbine is mounted on an aluminum mast with a minimum blade-roof clearance of approximately two feet. It is optimally mounted at the highest point of a roof, in a position which benefits from maximum prevailing wind, but it will work effectively in almost any location … Now you can have clean, quiet energy generated right from your rooftop."

This marketing makes the Swift controversial in the eyes of some, who fear that it won't be able to produce enough energy to be worthwhile. But if you have enough wind and a small load, it might be worth a look.

Time to Select Your Wind Turbine

To help you get started on your comparison of small wind turbines on the market, we put together the matrix chart in Table 9-2. We tried to include a helpful list of variables to compare, as well as the most popular models from major manufacturers. Please note that all machines that are registered for certification are highlighted. Happy browsing!

To make selection easier, consider some guidelines:

1. Start by selecting a table based on your (estimated) energy requirements now and in the future.
2. Select your better matches based upon the monthly energy output (MEO) and your ultimate goal (savings and/or earnings).
3. Grid or Off-Grid is your next choice. There might be turbines that are not designed for grid connection.
4. Utilize the rotor diameter or swept area for calculations at your site and the hub height chosen.
5. Blade numbers and noise emission levels should be considered if at least one of your stakeholders is concerned.
6. Cost is always important. Note if the price tag includes the wind-in-a-box kit or just the turbine.

TABLE 9-2   Representative Small Wind Turbines

The List

Small wind turbine standards, such as the Microgenerations Certification Scheme (MCS) in the UK and the Small Wind Certification Council (SWCC) in the U.S., are still in their early days and few turbines have completed their testing at this time. We have included a separate table of turbines pending approval from the latest SWCC standards or achieved MCS status. However, there is variety out there, and you are in the market right now. So, in addition we created a list of turbine candidates that meet a series of common sense criteria:

· The manufacturer has sold and installed more than twenty production versions of their turbine
· The manufacturer can demonstrate a turbine that has been running in the field for three years
· There is a network of dealers in the relevant geographic locations who are trained/experienced in servicing the machine
· The turbine is recognized by industry experts as safe and has an adequate level of performance standards
· At minimum there is warranty information, third-party verified energy data and a power curve
· The wind turbine does not appear to break any fundamental laws of physics, such as the Betz limit

The Tables

We have generated a series of 5 tables based on size class and general purpose.

Wind Turbine Size Classes			
	Rotor Diameter m (ft)	Swept Area m² (ft²)	Standard Power Rating* kW
Micro	0.5-1.25 (2.0-4)	0.2-1.2 (2.0-13)	0.04-0.25
Mini	1.25-3.0 (4.0-10)	1.2-7.1 (13-76)	0.26-1.40
Household	3.0-10 (10.0-33.0)	7.0-79 (76-845)	1.40-24.9
Small Commercial	10.0-20 (33-66)	79-314 (840-3,400)	25-100
Micro	Inland boats, motor cruisers, caravans, motor homes, small devices, small battery-charging		
Mini	Pumping water, small home utility, cabins, electric heaters, grid via battery banks		
Household	On-grid, stand alone, hybrid, water pumping		
Small Commercial	Larger homes, nurseries, factories, power equipment, wind farms, all with either feed-in tariff or net-metering agreements.		

*Standard Power Rating: 200W/m²

(continued)

TABLE 9-2 Representative Turbine Candidates (continued)

Certified Small Wind Turbines (Pending)

Manufacturer www.smallwindcertification.org/	QR	MEO kWh	Rated Power kW	Rotor Diam m (ft)	Swept Area m² (ft²)	Hub Height m (ft)	Grid	Voltage VDC	Speed control type	Shutdown type	Blade #	Noise db	BOS incl.	Cost (USD $)
Airdolphin GTO Zephyr Corporation americanpyr.com		100	1.1	2.54 (8.3)	5.1 (54.9)	10 (33)	•	24, 48, 250	Regenerative electro-magnetic Braking System	Regenerative electro-magnetic braking system	3	44.83	•	$5,995
Skystream 3.7 Southwest Windpower bit.ly/skystream		426	2.4	3.72 (12)	10.9 (115.7)	13.7 (45)		12, 24, 36, 48	Electronic stall regulation	Dynamic braking	3	50	•	$12,250* $6,875
Kestrel e400i 250V Eveready Diversified Products (Pty) Ltd. bit.ly/kestrelE40oi		465	3	4 (13.1)	12.6 (445)	12.2 (40)		48, 110, 250	Passive Blade Pitching	N/A	3	45		$3,500
Bergey 5kW Bergey Windpower bergey.com		710	5	5.2-8 (17-9)	TBD	12.2 (40)	•	TBD	Side furling	Manual	3	TBD	TBD	TBD
Endurance S-343 Endurance Wind Power bit.ly/s343		808	5.2	6.37 (20.9)	31.9 (343)	18.2 (60)	•	120/240 VAC 60hz	induction generator	Dual fail-safe disc brakes; manual brake	3	46.7	•	$50,000***
Ventera VT10 Ventera Energy Corporation bit.ly/VT10-2		893	10	6.7 (22)	35 (380)	15 (50)	•	240 VAC 50/60hz	Centrifugally controlled blade pitching	Electro/dynamic brake	3	—	•	$12,800; $21,200* $31,800-$45,300**
Evance R9000 Evance Wind Turbines bit.ly/evancer9000		915	5	5.5 (18)	23.75 (839)	12.2 (40)	•	220/240 50/60 hz	Reactive Pitch control	Electric brake	3	40-45	•	$12,500; $22,000-25,000**; $40,000****
Bergey Excel-S Bergey Windpower bit.ly/ExcelS		1133	10	7 (23)	38.5 (1,360)	15.2 (50)	•	208/240vac	Side furling	Manual furling cable	3	52.1[1]	•	$25,770 - $31,770*

Certified Small Wind Turbines (Pending)

www.smallwindcertification.org/	MEO	Rated Power	Rotor Diam	Swept Area	Hub Height	Grid	Voltage	Speed control	Shutdown	Blade	Noise	BOS incl.	Cost
Xzeres- Xzeres Wind Corporation bit.ly/xzeres	1,290	10	7.2 (23.6)	41 (442)	18.2 (60)	•	Single Phase; Three Phase	Stall regulation	Manual regenerative braking switch	3	TBD	•	$35,000*
10kW Hummingbird Potencia Industrial bit.ly/hummingbird10k	1,750	10	7.5 (24.6)	44.23 (476	21 (68.9)	•	120/240VAC	Active yaw gravity-activated furling mechanism dynamic braking	Dynamic braking	3	TBD	•	$27,800*
Renewegy VP-20 Renewegy, LLC renewegy.com	2,026	20	9.5 (31.2)	70.9 (763)	30 (98.5)	•	480VAC-3P 208VAC-3P 240V-1P	Active pitch active yaw	Active pitch; disc brake	3	48	•	$85-95,000***
Evoco 10kW Evoco Energy bit.ly/evoco	1,808	9.55	9.7 (31.8)	74 (797)	12.1 (40)	•	0-500	Passive blade pitching	Manual electronic brake; manual pitch,	3	52.9	•	$49,502
GW 133 – 11kW Gaia Wind Ltd. bit.ly/Gaia133	2,092	11	13.3 (42.6)	133.3 (1428)	18 (59)	•	480VAC-3P 208VAC-3P 240V-1P	Passive stall of blades	amechanical brake, passive pitch tips	2	45	•	$54,500**
P10-20 Polaris America LLC bit.ly/polaris20	2,333	20	10 (33)	78.5 (855.3)	21.3 (70)	•	N/a	Regenerative/ dynamic braking	Dual caliper disc brake	3	50	•	—
AOC 15/50 Seaforth Energy bit.ly/aoc15-50	5,000	50	15 (49.2)	177 (1905)	24 (80)	•	12, 24, 48	Blade stall	Tip brakes; dynamic braking	3	64	•	$175,000
Enertech E13 Enertech, Inc. bit.ly/enertechE13	6805	40	13 (44)	141.3 (1521)	24.3 (80)	•	N/A	Stall	Electro-disc brake, variable pitch tip	3	TBD	•	—
P15-50 Polaris America LLC bit.ly/polarisP15-50	7,917	50	15 (49.2)	177 (1905)	30 (100)	•	N/a	Regenerative/ dynamic braking	Dual caliper disc brake	3	50	•	—

(continued)

TABLE 9-2 Representative Turbine Candidates (continued)

Certified Small Wind Turbines (Pending)

www.smallwindcertification.org/	MEO	Rated Power	Rotor Diam	Swept Area	Hub Height	Grid	Voltage	Speed control	Shutdown	Blade	Noise	BOS incl.	Cost
Suelflow 100 Talk, Inc. bit.ly/windtalkinc	—	—	—	—	—	•	—	—	—		—	—	—
EddyGT Urban Green Energy bit.ly/eddy1kw	104	1	1.8 (5.9)	4.62 (50)	7-20 (23-66)	•	0-600V	PVI-3000 Inverter	Electrical brake	3	38	•	7060
Windspire – 800040 Mariah Power bit.ly/windspire	166	1.2	1.219 (4.0)	7.43 (80)	6.1 (20)	•	120/220	Redundant electronic braking	Redundant electronic braking	3	42	•	14,000***
UrWind O2 UrWind Inc. bit.ly/urwind	208	1.7	2.7 (8.9)	7.3 (78.5)	8 (26)	•	—		Electric/manual braking	3	—	•	—
UGE-4K Urban Green Energy bit.ly/uge-4k	416	4	3 (9.85)	13.8 (149)	7-20 (23-66)	•	0-600V	PVI-4200 inverter	Electrical brake	3	36	•	19150
TTK-10kW- Taisei Techno Co. bit.ly/ttk10kw		10	6.0 (29.5)	—	12.8 (42)	•	280DCV	—	—	3	—		$100,000
Vbine 10-05 VBINE energy vbine.com	—	—	—	—	—		—	—	—		—	—	—

Micro Wind Turbines

For inland boats • Motor cruisers • Caravans • Motor homes and other battery-charging

Manufacturer	QR	MEO kWh	Rated power kW	Rotor diam. m(ft)	Swept area m² (ft²)	Tower clear. m(ft)	Grid	DC voltage VDC	Speed control	Shutdown	Blade #	Noise db	BOS incl.	Cost ($)
Rutland 503 Marlec bit.ly/marlec503		6.8	0.03	0.51 (1.7)	0.2 (7)	10 (33)	•	12	n/a	Manual safety brakes	6	—	•	$382
Rutland 910-3F Marlec bit.ly/rutland513		21.75	0.09	0.91 (3)	0.65 (23)	10 (33)		12,24	n/a	Manual safety brakes	6	73		$625
Ampair100 bit.ly/ampair100		22.6	0.1	0.93 (3)	0.68 (24)	10 (33)		12, 24, 48	Blade pitch control	—	3	—		$1,810
AirBreeze Southwest Windpower airbreeze.com		35.9	0.2	1.17 (3.8)	1.1 (38.85)	10 (33)		12 ,24, 48	Electronic stall control	Manual switch	3	55		$699
Ampair 300 bit.ly/ampair300		37.8	0.3	1.2 (3.9)	1.1 (38.85)	10 (33)		12, 24, 48	Blade pitch control	—	3	—		$2,933
Superwind 350 superwind.com		37.8	0.35	1.2 (3.9)	1.1 (38.85)	10 (33)		12, 24	Blade pitch control	Blade pitch control; stop switch	3	—		$1,599

(continued)

TABLE 9-2 Representative Turbine Candidates (continued)

Mini Wind Turbines

| Manufacturer | QR | MEO kWh | Rated Power kW | Rotor Diam. m | Swept Area m² | Tower Clear. ft | Battery-charging, grid, water pumping, communications | | | | Blade # | Noise db | BOS incl. | Cost ($ in Th) |
							Grid	DC Voltage	Speed control	Shutdown				
Kestrel E150 bit.ly/KestralE150		59.1	0.6	1.5 (4.9)	1.8 (63.6)	10 (33)	•	12, 24, 48, 110, 200	Rotor turbulence	—	6	30-54	•	—
Ampair 600 bit.ly/ampair600		76	0.6	1.7 (5.6)	2.27 (80.2)	10 (33)	•	24, 48, 230	Blade pitch control	—	3	—	—	$2,394
Airdolphin GTO Zephyr Corporation zephyreco.co.jp		85	1.1	1.8 (5.9)	2.8 (98.9)	10 (33)	•	24V, 48V, 250V	Regenerative energy braking system	Regenerative energy braking system	3	44.83	•	$5,995
Rutland 1803 Marlec bit.ly/rutland1803		85	0.39	1.87 (5.9)	2.6 (91.8)	10 (33)	•	12, 24, 230	Furling tail	N/a	3	—	•	$2,772
AVX-1000 AeroVironment bit.ly/AVX1000		88	1	1.83	2.6	10 (33)	•	250	—	—	5	—	•	—
Bornay 600 J. Bornay bit.ly/bornay600		104	0.6	2 (6.56)	3.14 (109.47)	10 (33)		12, 24, 48	Tilt-up furl, electronic regulator	Auto-braking	3	—	•	$3,245
Whisper 100 Southwest Windpower bit.ly/whisper100		115	0.9	2.1 (6.89)	3.6 (38.75)	10 (33)		12,24,36,48	Side-furling	Electric brake	3	—	•	$2,567
Swift Renewable Devices Ltd. renewabledevices.com		140	1.5	2.1 (6.89)	3.46 (123.6)	4.9 (16)	•	—	Angle furling; dynamic braking	Dynamic and mechanical braking	5	32	•	$8,925

Mini Wind Turbines

	QR	MEO	Rated Power	Rotor Diam.	Swept Area	Tower Clear.	Grid	DC Voltage	Speed control	Shutdown	Blade	Noise	BOS incl.	Cost
								Battery-charging, grid, water pumping, communications						
Whisper 200 Southwest Windpower bit.ly/whisper200		175	1	2.7 (8.86)	5.73 (61.7)	10 (33)	•	12,24,36,48 high voltage available	Tilt-furl	Electric brake	3	57?		$3,405
Bergey XL1 Bergey Windpower bit.ly/bergeyXL1		219	1	2.5 (8.2)	4.9 (52.7)	18 (60)		24	Side-furling	Electric brake	3	—	•	$3,170*
Falcon 1.2 WePOWER bit.ly/wepower1		87.58	1.2	1.78 x 2 (5.8 x 6)	2.4 (25.83)	5.5 (18)	•	48, 150-400	Automatic aerodynamic variable pitch	Automatic electromagnetic	4	432	•	—

Household Wind Turbines

Manufacturer	QR	MEO	Rated Power	Rotor Diam.	Swept Area	Tower Clear.	Grid	Voltage	Speed control	Shutdown	Blade	Noise	BOS incl.	Cost
								Grid, battery-charging, hybrid, Agriculture, Landowners, Commercial premises, Small wind farms						
		kWh	kW	m ft	m²	ft	•	V			#	db	•	($ in th)
XZERES 110 Xzeres Wind bit.ly/xzeres110spec		376	2.5	3.6 (11.8)	10.2 (110)	12.2 (40)	•	48	Furl-	—	3	—	•	$29,000*
Proven 7 bit.ly/proventurbine		392	2.5	3.5 (11.48)	9.6 (339)	18.25 (59.9)	•	24,48,120, 300	Passive yaw, pitch and coning control		3	-48	•	$0
Westwind 3 bit.ly/westwind3kW		398	3	3.7 (12)	10.8 (381.4)	12.2 (40)		48, 96, 110, 120	Auto tail furl	—	3	—	•	—

(continued)

TABLE 9-2 Representative Turbine Candidates (continued)

Household Wind Turbines

	MEO	Rated Power	Rotor Diam.	Swept Area	Tower Clear.	Grid	Voltage	Speed control	Shutdown	Blade	Noise	BOS incl.	Cost
							Grid, battery-charging, hybrid, Agriculture, Landowners, Commercial premises, Small wind farms						
Skystream 3.7 Southwest Windpower www.windenergy.com	426	2.4	3.72 (12)	13.7 (45)	13.7 (45)	•	12, 24, 36, 48	Electronic stall regulation	Dynamic Braking	3	50	•	$12,250* $6,875
Kestrel e400i Eveready Diversified Products bit.ly/kestrelE40oi	465	3	4 (13.1)	12.6 (445)	12.2 (40)		48, 200, 300	Passive blade pitching	N/A	3	45	—	—
TSW4000 TechnoSpin Inc. bit.ly/TSW4000	508	4	4.2 (13.8)	13.8 (487.3)	13 (42.7)	•	12, 48, 120, 240	Aerodynamic tail; furling	Short circuit system	3	40	•	
Whisper 500 Southwest Windpower bit.ly/whisper500	546	3	4.56 (15)	16.4 (579.1)	13 (42.7)		24, 36, 48, 120	Side-furling	Short circuit system	2	—	•	$6,695 $14,321*
S250 Endurance bit.ly/EnduranceS250	550	5	5.5 (18)	23.8 (256)	19.2 (63)	•	120-240 VAC	Stall control (constant speed)	Mechanical and Electronic Braking	3	—	•	
Hummer 5kW Hummer Wind bit.ly/hummer5kW	576	5	6.4 (20.9)	20.9 (225)	12.2 (40)	•	240V-50/60hz	Active yaw	Electro-magnetic braking	3	34	•	$25,000*
API 5kW Free Advance Power Inc bit.ly/API5kW	724	6	5 (16.4)	19.6 (211)	12.1 (40)	•	240V	Tail-furling electronic controller	Manual electronic/ mechanical brake; auto electric brake	3	—	•	$11,200*
Montana Fortis Wind bit.ly/montana5kW	726	5	5 (16)	19.6 (211)	12.2 (40)		48, 120	tail-furl	Short circuit system	3	—	—	—

Household Wind Turbines

	MEO	Rated Power	Rotor Diam.	Swept Area	Tower Clear.	Grid	Voltage	Speed control	Shutdown	Blade	Noise	BOS Incl.	Cost
Proven 11 Proven Wind Turbines bit.ly/Proven11	745	5.2	5.5 (18)	23.8 (256)	12.2 (40)	•	24,48,120V; 300VAC	Passive yaw, pitch and coning control	—	3	44.8	•	$25,000*
Scirocco Eoltec bit.ly/Scirocco6kW	895	6	5.6 (18.3)	24.7 (265)	12.2 (40)	•	230V-50/60hz	Aerodynamic blade stalling	Remote electric brake	3	35.4	•	$32,482*
Nova-Wind 6 bit.ly/Nova6K	919	6	6 (19.7)	28.27 (304.3)	12.2 (40)	•	120/240/ 230/ 400V-50/ 60hz	Passive pitch side-furling	Passive pitch electromech. disc brake	4	35	•	$46,985*
Excel-S Bergey Windpower bit.ly/excel_info	1,133	10	7 (23)	35.5 (382)	18 (60)	•	220/240 VAC; 60/50 Hz	Side-furling	Mechanical/ electrical brake	3	52.1	•	$31,770*
Excel-R Bergey Windpower bit.ly/excel_info	1,762	7.5	7 (23)	35.5 (382)	18 (60)		24/48/120/ 240 VDC	Side-furling	Mechanical/ electrical brake	3	52.1[2]	•	$25,770 - 26,870*
Aircon 10S- Aircon International bit.ly/aircon10s	1,769	9.8	7.13 (23.3)	39.6 (426)	18 (60)		240 VAC, 60 Hz, 220 VAC, 50 Hz	Active yaw; generator azimuth adjustment	Hydraulic disc brake	3	< 60	•	—
442SR Xzeres Wind bit.ly/442SR	1,819	10	7.2 (24)	40.7 (438)	18 (60)		48vdc, single/ three phase	Side-furling	Dynamic braking	3	—	•	$39,950
AirForce10 Future Energy bit.ly/airforce10	1,853	10	8 (26.2)	50.3 (541.4)	12.1 (40)	•	12, 24, 36 or 48vdc	Active yaw	Electronic/ mechanical disk-brake	3		•	$73,084

(continued)

TABLE 9-2 Representative Turbine Candidates (continued)

Household Wind Turbines

	QR	MEO	Rated Power	Rotor Diam.	Swept Area	Tower Clear.	Grid, battery-charging, hybrid, Agriculture, Landowners, Commercial premises, Small wind farms						
							Grid	Voltage	Speed control	Shutdown	Blade	Noise	BOS incl.	Cost
Proven 35-2 Proven Energy bit.ly/proventurbine		1,933	12.1	8.5 (27.88)	56.7 (610)	15 (49)	•	300VAC	Passive yaw, pitch and coning control			52.3		
GW 133 – 11kW Gaia Wind Ltd. bit.ly/Gaia133		2,042	11	13 (42.6)	133 (4697)	18 (59)		400	Passive stall of blades	Mechanical brake, passive pitch tips	2	45	•	$49,500*
Jacobs 31-20 Wind Turbine Ind. bit.ly/Jacobs31-20		3,110	20	9.4 (31)	61.3 (660)	18 (60)		40 – 180 VAC	Centrifugal variable pitch governor auto-furl	Manual disc	3	—	•	$63,513*
API-5KW-VAWT Advanced Power Inc bit.ly/api5_VAWT		296	5	4.0 x 4.6 (13 x 15)	12.3 (132.4)	5.5 (18)		110vdc		Automatic braking	5		•	$17,700*
Big Star Vertical Ropatec bit.ly/ropatec20		844	20	8 x 4.3 (26 x 14)	22.9 (75.1)	12.1 (40)	•	360V 3-phase; split phase		Brake control	4	42	•	—

Small Commercial Wind Turbines

Manufacturer	QR	MEO	Rated Power	Rotor Diam.	Swept Area	Tower Clear.	Nursery, Wind Farms, Feed-in Tariffs, Power Equipment, Factory						
		kWh	kW	m(ft)	m²(f)²	m(ft)	Grid	Voltage	Speed control	Shutdown	Blade #	Noise db	BOS incl.	Cost ($ in th)
Jimp 25 BMP bit.ly/jimp25		4,670	25	10.8 (35.4)	91.6 (300.5)	24 (80)	•	400	Passive pitch and stall control	—	3	—	•	$86,118
HY30-AD11 Huaying Wind Power bit.ly/huaying30kW		4,845	30	11 (36)	95.3 (312.7)	24 (80)	•	380VAC	Passive pitch	Postive blade pitch	3	—	•	—

Small Commercial Wind Turbines

	MEO	Rated Power	Rotor Diam.	Swept Area	Tower Clear.	Grid	Voltage	Speed control	Shutdown	Blade	Noise	BOS incl.	Cost
Aeolos 5kW Aeolos Wind bit.ly/aeolos30kW	6,256	30	12.5 (41)	122.7 (1320)	24 (80)	•	320-480 V	Electric active yaw and brake system	Hydraulic brake system	3	41.5		$73,790
44A Enertech bit.ly/Enertech44A	6,767	44	13 (42.7)	133 (1431)	24 (80)	•	480 vac, three phas 240 vac, single phase	Variable pitch tip brakes	Electro-mechanical disc self-adjusting	3	—	•	—
Endurance E-3120 Endurance Wind Power bit.ly/G-3120	7500	35	19.2 (63)	289.53 (3120)	24 (80)	•	Single phase 240VAC, 3 phase, 480VAC	Passive pitch control	Mechanical disc brake	3	Inaudible at 200 metres		—
WES 18 Wind Energy Solutions Canada bit.ly/WES18turbine	8,500	80	18 (59)	254 (2734)	30 (100)	•	230/400 V	Passive blade pitch	Active yaw	2		•	$97,446
USW KCS56 PowerWorks Wind bit.ly/pworks100kw	8,500	100	18 (59)	229 (2,463)	30 (100)	•	3 phase, 480 VAC, 60/50 Hz	Variable pitch blade design	Remote and automated control braking system	3	—		—
AOC 15/50 Seaforth Energy bit.ly/aoc15-50	9,009	50	15 (49.2)	177 (1902)	24 (80)	•	12, 24, 48 volts	Blade stall	Tip brakes; dynamic braking	3	64	•	$140,000
NPS 100 Northern Power bit.ly/NPS100	12,803	100	21 (68.9)	346.4 (1136)	37 (121)	•	3 phase, 480 VAC, 60/50 Hz	Variable speed, stall control	Dynamic braking; motor-controlled calipers	3	55	•	$450,000
ReDriven 50kW ReDriven Power, Inc. bit.ly/redriven50kW	14,317	50	18 (58)	254.5 (2739)	30 (100)	•	600/480/240	Electric active yaw	Normally closed braking system	3	59	•	$161,070

1 http://production-images.webapeel.com/bergey/assets/2010/11/4/75837/Excel-S_Acoustics_Report.7.14.2010.pdf

2 http://production-images.webapeel.com/bergey/assets/2010/11/4/75837/Excel-S_Acoustics_Report.7.14.2010.pdf

The Variables

- **MEO** (monthly energy output) is how much energy your turbine might produce in a month in kilowatt-hours. The data was tabulated from the reported AWEA Rated Monthly (or Annual) Energy at the mean wind speed of 5 m/s. Otherwise, it was tabulated data from the manufacturer's annual energy production (AEP) curve at sea level air density at a standard tower height of 10 meters (33 feet) with the Raleigh Wind Distribution factor of 2, and with the site roughness set at a grassy field. In both cases we divided the results by 12 months to get the turbine's monthly output.
- **Rated Power**: When the manufacturer didn't provide an official rated power, we provide the AWEA Rated Power at 7.5 m/s. This is to your advantage, because many manufacturers tend to overrate their turbines.
- **Tower Clearance**: We take the standard tower height to be at least 30 feet above obstacles. This would make the minimum tower hub height around 10 meters (33 feet) (knowing that, for those who are doing a boat, cell tower, or roof mount, this doesn't necessarily apply).
- **AWEA Rated Sound Level**: The sound level at 60 m (197 ft) that the wind turbine will not exceed 95 percent of the time at a 5 m/s (11 mph) average wind speed site.
- **BOS**: Balance of system, in this case meaning the manufacturer sells the turbine in a package that also includes the inverter, charge controller, stop switch, wiring, and perhaps disconnects and meters.
- **Cost**: Generally the starting cost includes the turbine only. If the balance of the system is included (excluding the tower, battery bank, and installation) the price will be followed by an asterisk. If the tower *is* included, it will have 2 asterisks. If it includes installation, 3 asterisks.

Notes:

1. Please note that due to time constraints and constant subsidy changes, we cannot include a feature for financial incentives for each turbine. It is advisable for you to locate the turbine of your choice and check for it in your local and national government programs.
2. If the manufacturer failed to provide the information for any variable for whatever reason, we simply place a dash in that cell. This is not to suggest that it is not important or that it doesn't apply to that turbine. If it doesn't apply, it would have a "n/a" in the cell.
3. For those that want that extra assurance, we have highlighted the rows of wind turbines that have achieved either or both the MCS or SWCC (temporary) certification.

A *regenerative brake* slows a machine by converting its kinetic energy into another form, which can be either used immediately or stored until needed.

Resource

For more information on how to select your wind turbine, check out this *Home Power* magazine article by Ian Woofenden and Mick Sagrillo:

2010 Homepower's Wind Generator Buyer's Guide
bit.ly/2010windbuyersguide

Summary

In this chapter, we tried to give you the tools to critically compare one wind turbine with another. The turbine, of course, is a central part of a wind system, and that choice often dictates much of the balance of system. As we pointed out, some of the key specifications to consider are swept area, durability, annual energy output, cost, speed, governing mechanism, shutdown mechanism, and sound.

We introduced the concepts of tip speed ratio and RPM, and explained the ways in which turbines dissipate excess energy during high winds. Going forward, you should be well armed to critically examine manufacturer claims, to make sure you get all your questions answered, and to focus on what's really important. Otherwise, it's easy to get distracted by nifty videos and slick brochures.

CHAPTER 10

Balance of System: Batteries, Inverters, Controllers, and More

The stupidest thing we can do is put a turbine, large or small, in a place that has no wind. It's also stupid to drill in Saudi Arabia when we have such great wind here.

—Morten Albaek, Senior Vice President for Global Marketing of Vestas

Overview

- Do I need an inverter, and what does it do?
- Should I get batteries for my system?
- How do I wire everything together?
- How do I protect the system from surges and shorts?

You could have the sleekest turbine in the land on the highest tower, but if that's all you've got, you aren't getting any energy. You've also got to have what is often called the *balance of system*: bits of wire and various electronic do-dads that connect the turbine to your load and/or the grid. The balance of system (BOS) also has components that help ensure the whole system functions safely and smoothly.

This is also the part where we take a close look at batteries. You may have wondered why we haven't spent a lot of time discussing batteries through most of this book. That's because whether you ultimately decide to get batteries or not doesn't have a big impact on many of the other topics we covered. You still need a good wind resource, a sturdy tower, permits, and a good generator, whether you store your energy on-site or not.

There are a few things to be aware of when it comes to batteries, however, and we'll cover them in this chapter. We'll also examine the other devices you need to

FIGURE 10-1   A National Renewable Energy Laboratory technician inspects the Trace inverter (left) and charge controller (upper right) of an off-grid solar/wind hybrid system. *Warren Gretz/DOE/NREL.*

condition and distribute your hard-won energy, including charge controllers, inverters, disconnects, breakers, wire, and more (Figure 10-1).

Inverters

The first question you may have when it comes to inverters is whether you even need one. If you are reading this as an e-book on a handheld device or notebook, you are technically using direct current (DC) power. If you are reading this book under house or office lighting, most likely you are using alternating current (AC). Most of our stuff runs on AC. True, some applications do operate solely on DC power, such as many recreational vehicles (RVs), most boats, and some isolated cabins and water-pumping stations.

There has actually been a fierce debate, spanning more than a century, about which type of power is better, DC or AC. We won't get into the details here, although it's worth pointing out that many people, from electrical engineers to do-it-yourselfers, have strong opinions about their relative merits and best applications. Thomas Edison sacrificed cats, dogs, and even a circus elephant to the (literally) fiery debate. For the most part, most of the appliances and electronics we want to run take AC power out of the box, even if many of them have an adapter that converts incoming line power into DC to actually run the device (computers are like that).

Most small-scale power generators produce DC (although with wind, it's even a bit more complicated than that). So if you want to run something like an AC-powered gadget from a DC car battery in a mobile home, you need a device that will convert DC to AC—an *inverter*. If you want to go the other way, AC to DC, you don't use an inverter. You use another electrical component called a *rectifier*. Rectifiers are usually based on diodes, and most wind turbines have them.

Because wind is a variable phenomenon, it doesn't drive the rotor at a constant RPM. This means the system's alternator doesn't spin at a regular rate. So instead of producing an even source of current, as the grid does, it tends to produce AC that constantly varies in voltage and frequency. This type of current is often called *wild AC*, or three-phase AC, and it isn't very usable. So most wind systems send this current into a rectifier, which then converts it to DC.

In batteryless systems, the voltage coming from the turbine rectifier could be anywhere from 100 to 600 volts DC. In a battery-based system, it is most likely at 48 volts DC, or 12 volts DC for a small system. (Older systems sometimes produced 24 volts DC, but that is getting rare.)

We mentioned in Chapter 8 that it is possible to make a wind turbine with a DC generator, although that is quite rare, since they tend to be more expensive and larger than alternators. Either way, you end up with DC power that you can use directly if you are driving a water pump or wiring up a wilderness cabin. But most folks take the next step to AC.

If you wanted to make a simple inverter, all you need is a mechanical switch that reverses the flow of current. Reversing a battery on a simple loop circuit would do the trick, for example (all batteries work on DC). If you were to flip that battery or that switch 60 times a second, you would approximate the frequency of the AC current in the North American grid. In fact, the simplest types of inverters are like that, and the output of current they produce would show up on a graph as square waves, with each peak or trough indicating the direction of current.

Square wave inverters are still on the market, and they are the least expensive, though a lot of devices won't run on them, they waste power, and they can be hard on equipment. Better is a *sine wave inverter*, which produces a more gradual switching of current. Sine wave inverters cost more, but they more closely replicate the grid, so they don't wear out your toys.

In between a square wave inverter and a sine wave inverter is something called a *modified sine wave inverter*. In these devices, the output goes to zero in between switching from positive to negative, instead of flipping abruptly from one direction to the other. The energy is still a bit "rough," but it is usable in many devices, though sometimes not in some sensitive equipment like laser printers. Prices are often (~$0.10USD/Watt), compared to (~$0.50 to $1.00USD/Watt) for sine wave inverters (also called *pure sine wave inverters*).

Ian Woofenden cautions consumers to beware of marketing claims that boast "true sine wave," since that isn't a regulated term. Sine wave inverters are solid-state devices, so they can't ever produce perfectly smooth patterns of electricity, the way power companies can with their moving magnets. Instead, sine wave inverters use a device called an H-bridge, which is really a series of switches, much like our thought experiment earlier with the battery switching directions (Figure 10-2). The more finely tuned the switching structure in the inverter, the closer the resulting waveform will be to an actual sine wave. And the more expensive the inverter will be.

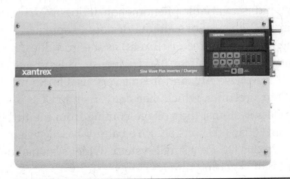

FIGURE 10-2   A Xantrex SW Plus Inverter/Charger, a sine wave inverter and battery charger combination unit. *Xantrex Technology/DOE/NREL.*

Woofenden suggests that if you are really "nitpicky," look for the lowest *total harmonic distortion,* preferably less than 3 percent, which indicates how far off the wave is from the ideal.

If you only need to power up some basic stuff, like lights, power tools, and kitchen appliances, you may be fine with a basic modified sine-wave inverter, though be aware that most loads will draw more power than when connected to the grid. However, you may not be able to run dimmer switches, variable-speed drills, sewing machines, battery chargers, or anything else where the current varies. That's because the switches on such devices are used to looking for the sine wave pattern of grid power to control operation, and they have trouble "reading" the abrupt zero point from a modified sine inverter.

With a modified sine inverter, you also may feel a bit like you entered the Twilight Zone, because fluorescent lights and stereos may buzz and plug-in clocks may have trouble keeping time. The lower the total harmonic distortion, the more everything will behave like it is on grid power, although you'll pay more for the inverter.

Of course, it's also worth remembering that all electrical components "lose" a little of your precious energy as waste heat. But good equipment loses less, retaining 90 percent or more.

For many people shopping for wind energy systems these days, there isn't a lot of choice when it comes to inverters. More and more manufacturers and dealers are selling their products as package deals, with inverters that are prematched to the system's other components.

Tech Stuff   *Even so, it's important to know a few things about inverters. One thing you may notice if you start researching them is references to different types of internal technologies. Inverters may use newer high-frequency transformers, conventional low-frequency transformers, or no transformers at all. A* transformer *isn't just a robot toy; in electrical*

engineering it is a device that steps up or down voltage, usually through coils of wire that induce a field in each other. In our experience, when it comes to inverters, these technical differences are not make-or-break features, but if you want to geek out on inverters, more power to you.

Although an inverter's primary job is to convert DC to AC, it may also serve some other functions in your system. For instance, they can aid in charging, start generators, and provide data logging.

In general, there are three main areas to consider when selecting an inverter:

- **Input voltage**   You will likely be considering ranges of 100 to 600 volts DC for batteryless systems or 12, 24, or 48 volts DC for battery-based systems. Clearly, you need to check with your turbine's manufacturer to see what type of voltage it will be producing. You don't want to fry your brand-new inverter the first time the wind blows!
- **Output voltage**   This is typically 120 or 240 volts AC, but may also be 208, 277, or 480 VAC.
- **Wattage**   The peak capacity the inverter can handle, or the maximum amount of energy you can use or sell back to the grid at any moment. In a grid-tied system, the peak wattage of the inverter needs to be at or above the absolute peak power of the turbine.

While the terminology and content will vary by manufacturer, inverter data-sheets also often include the following information:

- **Peak efficiency**   This value indicates the highest efficiency the inverter can achieve. Most grid-tied inverters on the market have peak efficiencies of 94 percent or greater. No system is going to have 100 percent, because some energy is always going to get lost in the shuffle, usually as waste heat. There usually isn't much difference between inverters when it comes to peak efficiencies, so this isn't so much a comparison tool as it is an awareness tool. If you are counting watts, the efficiency tells you how much you need to subtract from the input to the output. For example, if your inverter has an efficiency of 95 percent, and you need 5,000 watts out of it, you'll have to put in 5,263 watts.
- **CEC weighted efficiency**   You may have noticed the word "peak" in the preceding entry, and that might have left you feeling there is still guesswork in the system. The CEC weighted efficiency attempts to close that gap, and is published by the California Energy Commission (CEC) on its GoSolar website. It's an average efficiency that better represents actual operation.
- **Peak power tracking voltage**   This represents the DC voltage range in which the inverter's maximum-point power tracker will operate.

- **Start voltage** This value indicates the minimum DC voltage that is required for the inverter to start working.
- **NEMA rating** The National Electrical Manufacturers Association rating indicates the level of protection the device has against water intrusion. Most inverters are NEMA 3R, which means outdoor rated for most situations.
- **P56 rating** (rest of the world) Similar to the NEMA, indicating suitability for use outdoors.

Now let's take a look at the two main types of inverters:

- Grid-tied inverters
- Off-grid or stand-alone inverters

Note that you have to make the decision about which type of inverter you want at the onset. Inverters come as either-or deals, and you can't simply switch over or "upgrade" an inverter to the other type.

Grid-Tied Inverters

A grid-tied inverter (or grid-interactive or grid synchronous inverter) is used to convert direct current into alternating current and feed it into the grid (Figure 10-3). This is done to either offset electricity costs or supplement residential or commercial power usage.

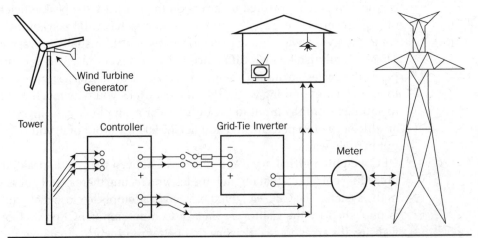

FIGURE 10-3   In a grid-tied system without batteries, current flows from the turbine through a charge controller, into an inverter, and then through a meter and out to the grid. *WindEnergy7.com.*

Grid-interactive inverters are the most efficient way to produce renewable energy, because the least amount of energy is wasted. They send energy to the grid, and you are credited for each watt, per your net-metering agreement. You don't have to worry about any energy that is dissipated over time in batteries. However, grid-tie inverters must have an active connection to the grid to function. If the grid is down, they cannot be used at all. They have an auto-shutoff feature that disables them, so that way any line workers won't get shocked by your system.

Before a grid-tied inverter can send power to the grid, it must synchronize its frequency to 60 hertz in North America. It does this through its built-in components, including a local oscillator. The inverter must also synchronize the voltage to the grid, which is usually at 240 or 120 VAC (volts AC). Typical modern grid-tied inverters have a *fixed unity power factor*, which means output voltage and current are closely lined up, with a phase angle within one degree of the grid. The inverter typically has a built-in computer that senses the grid's waveform and matches it. Therefore, a grid-tied inverter is always a sine wave inverter.

In the United States, grid-tied inverters must also meet specific technical provisions in the National Electric Code. Most other countries have similar rules.

Off-Grid Inverters

Off-grid inverters, of course, work without the presence of a grid. They take the energy you produce with a wind turbine and convert it into a form that is useable in your home, business, cabin, or boat. In that case, you have a choice of square, modified sine, or sine wave models, and we discussed the advantages and disadvantages of each earlier.

We also mentioned that battery-based wind generators typically produce 12, 24, or 48 VDC. That means off-grid inverters have to step up the voltage to something more commonly useful, usually either 120 or 240 VAC.

In most cases, when you have an off-grid situation, you are going to want energy storage, for when the wind isn't blowing strong or the sun isn't shining. To get storage, you have to have batteries, as well as a battery-based inverter.

With a battery-based system, the battery voltage you choose will be the inverter's input, while the output will be the standard load you want to run. You have to choose the inverter's wattage, although note that it should always be equal to or larger than the turbine's peak output.

One thing to keep in mind is that the larger an inverter is, the more power it will eat up during operation. That could be anywhere from 2 to 20 watts continuously. So when it comes to sizing your inverter, we hope you followed our advice in Chapter 4 and assessed how much energy your place uses.

You probably don't need to get an inverter that can run absolutely everything you own at once, but you do want something that can handle what you use during the course of a typical day. Otherwise, you may end up with angry family members, constantly bickering over who gets to use the hair dryer or the Xbox. Note that many devices use more energy at startup than during operation, although the good news is that most off-grid inverters are designed to be able to handle brief surges.

Rex A. Ewing wrote in *Countryside* magazine that he runs his house, and a "voracious 1.5 horsepower well pump," with a Xantrex SW 4024 sine-wave inverter, offering 4,000 watts of continuous output and a 120/240-volt transformer. Ewing explained that his guest cabin "does nicely with a Xantrex DR 1524 modified sine-wave inverter, capable of 1,500 continuous watts, while the tools in my small work-shop work handily with an inexpensive Aims 2,500-watt modified sine-wave unit." He concluded, "Three different inverters, for three different jobs. It's just a matter of knowing what each one will, and won't, do."

You'll also have to determine if you want an inverter that can charge the batteries through the use of a gas-powered generator. Almost everyone living off the grid has a supplemental fuel generator, usually diesel or gasoline fired. Being able to recharge your batteries with that generator can boost flexibility.

In fact, many high-end inverters have multistage battery charging capabilities built right in, assuming the job of a charge controller (see later). However, such a combination unit does add cost and weight. During normal operation, the inverter will send power to the loads that are running at the time and then charge the batteries. High-end inverters also usually come with sophisticated programs for monitoring and controlling the system from your computer.

Grid-Tied with Battery Backup

This type of system tends to confuse people, who often assume that it combines the best of both worlds, without realizing that it also has drawbacks. If you have batteries at all, you need to have an inverter that can work with them. That means a battery-based inverter, even if you are also hooked up to the grid (Figure 10-4).

This type of system will keep working even if the grid goes down, which is good, but it also means you have the extra expense of batteries, plus the effort to maintain them, and you lose power over time that is sent to your batteries but not used.

We discuss the relative merits of this choice in greater detail in the section on batteries that follows.

Batteries

If you live off-grid, you are going to want batteries, unless you have very restricted applications (Figure 10-5). Otherwise, your lights are going to flicker and your computer is going to shut off every time there's a lull in the wind. That does not sound like fun. If you live on the grid, you may *think* you want batteries for backup during a blackout. But read this first.

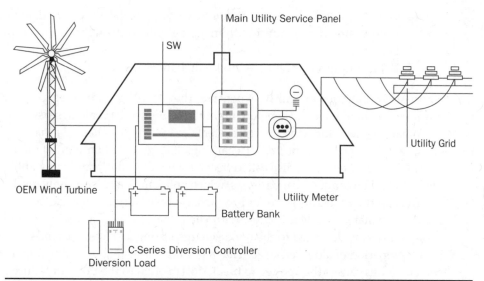

FIGURE 10-4   In a grid-tied system with batteries, current flows from the turbine through a charge controller, then into the batteries, a dump load, or an inverter.

FIGURE 10-5   This bank of 20 Trojan L-16 deep-cycle batteries is mounted in a vented box inside a Colorado home, which is powered by a 2.88 kW solar system and two AIR 403 wind turbines on tilt-up towers. *Warren Gretz/DOE/NREL.*

Five Things You Need to Know About Batteries

1. Batteries need more maintenance than most other components in a wind system.
2. The bigger the battery bank, the higher the cost and the lower the efficiency.
3. Battery banks are primarily recommended only for off-grid living.

4. Grid-connected users are required to use special inverters that prevent energy stored in batteries from going into the public utility during a blackout.
5. Public utilities traditionally frown on battery banks.

In short, batteries are fairly expensive, they take up a lot of space, and they require more maintenance then you might think. But we'll break these issues down.

What you need for wind energy systems are what are called *deep-cycle batteries*, which are designed for repeated discharging of a significant amount of their stored energy. These batteries are also called *traction batteries*. What you don't want is *starting batteries*, like common car batteries, which aren't meant to be heavily discharged.

To keep their weight down and to increase their surface area, car batteries are made with many thin lead plates, with each one having a porous texture like a sponge. They are designed for heavy discharge lasting just a few seconds, followed by a long period of slow recharge. That's perfect for a car starter. But deep cycling such batteries causes the sponge to break down, often to the point of failure in just 30 to 150 deep cycles.

That is not what you want for a wind system. Better are deep-cycle batteries, which have thick plates made of solid lead. This means they have less surface area, so they can't release as much energy as rapidly. Instead, they are designed for hours of heavy discharge each day, followed by relatively fast recharge in only a few hours.

In your travels through battery stores, online or in person, you will likely encounter *marine batteries*. As the name implies, these products are intended for boats, although they have a wide range of quality and design. In general, marine batteries tend to be intermediate between starting and deep-cycle batteries, though a few brands like Rolls-Surrette and Concorde are built like "true" deep cycle. In an intermediate marine battery, the plates are midway between thin and thick, and may or may not have a sponge texture.

Unfortunately, there aren't really batteries designed specifically for small wind systems, because the industry is too small and niche to support robust manufacturing of them. Instead, most batteries used with turbines were originally designed for electric golf carts, forklifts, or floor scrubbers.

People new to wind energy often ask if the batteries are sophisticated *lithium-ion* types, like the ones found in laptops and cell phones. As of this writing, we aren't aware of any folks who are using those powerful batteries, because they are quite expensive. For comparison's sake, the lithium-ion batteries on board hybrid cars cost thousands of dollars. The price for lithium-ion has been decreasing steadily, so that may change in the near future.

It's true that some people use *NiCad* batteries with solar or wind systems, but they tend to be expensive as well (less so than lithium). NiCad batteries can be difficult to dispose of because of the toxic cadmium inside. In general, most experts suggest you avoid them for wind energy, unless you have to work in extremely cold temperatures, where they perform better than alternatives.

Types of Batteries

Today, most wind systems rely on tried-and-true *lead-acid batteries,* like the big one in your car, except formulated for deep cycling instead of starting. Lead-acid batteries come in two general types, described in the following sections.

Flooded Batteries

In a *flooded battery,* the cells are open, and you have to add distilled water occasionally (Figure 10-6). The liquid inside a battery is called electrolyte. It's not Gatorade; it's a mix of sulfuric acid and water. The level tends to go down over time in flooded batteries, due to slow evaporation and during a process called *battery gassing.* You never want the liquid level to fall below the plates, because they will be permanently damaged through contact with air.

You have to check the levels of these batteries every few months, and if they are low, add distilled water. If you are the kind of person who can't keep a houseplant alive, even a cactus, that might be a problem for you.

The pros keep a temperature-adjusted *hydrometer* handy, and they routinely test the specific gravity of the electrolyte. You should do this when the battery is fully charged. Don't use it for several hours, and then discharge it for a few minutes to remove the surface charge. Then you can insert the hydrometer and take a reading. If the specific gravity is below 1.225, you could have *sulfation* happening.

Sulfation is a chemical reaction that deposits sulfates on the lead plates. That's a problem, because the sulfates act like insulators and diminish the battery's ability to store and release charge. At an advanced stage, sulfation will ruin the battery, and the only remedy is replacement. Sulfation can occur when batteries are less than 100 percent charged, so for this reason, experts usually suggest keeping lead-acid batteries "full" at all times.

FIGURE 10-6   These flooded lead-acid batteries have hydrocaps, which are supposed to recombine the oxygen and hydrogen released while charging, reducing water loss up to 90 percent. *Byron Stafford/DOE/NREL.*

Hydrometer readings shouldn't vary by more than 0.5 between cells. A fully charged battery should generally be at 1.265, while a battery at 75 percent will be at 1.225, 50 percent will be at 1.190, 25 percent will be 1.155, and discharged will be 1.120.

Advantages of flooded batteries:
- Cheaper
- Last longer

Disadvantages of flooded batteries:
- Must be periodically topped off with distilled water
- Require occasional cleaning
- Vent flammable hydrogen and oxygen gas when charging (and potentially toxic hydrogen sulfide gas)

Hazard   *Given the last point, it's critical to make sure the space that holds your batteries is well ventilated. If flammable or toxic battery gases are allowed to build up, it can spell trouble or even death.*

Sealed Batteries

Sealed batteries have their cells completely closed off, so you can't add more liquid, even if you wanted to. The good news is that sealed batteries don't require as much maintenance, but the bad news is they cost more and don't last as long. They also require more careful charging, so you have to make sure your charge controller is properly set up. If you overcharge a sealed battery, it will force out some moisture, and there's no way to replace that. (As you know, they are sealed.)

Sealed batteries come in a few different stripes. Some are *gel* (also called gelled), which are made with silica gel (Figure 10-7). In that case, the battery acid is bound up inside the gel, sort of like a jelly doughnut. An advantage of gel batteries is that it is impossible to spill acid, even if the case were to break in half. However, gel batteries must be charged more gradually, at a slower rate and lower voltage, to avoid damage.

A newer technology, *AGM (absorbed glass mat) batteries* use fiber separators to absorb the electrolyte, and are generally more durable. These batteries also won't spill acid, since it is bound up in the mat. They are more rugged than gel batteries and can be charged faster.

All gel batteries and most AGM ones are "valve regulated," which means that a tiny valve keeps a slight positive pressure. However, AGM batteries vent very little, since they are *recombinant*, meaning oxygen and hydrogen recombine inside the battery. AGMs also have a low self-discharge rate, typically from 1 to 3 percent per month. This means they will hold their charge for a long time. AGMs are also quite resistant to freezing, since they have very little liquid inside.

FIGURE 10-7   A typical valve-regulated gel lead acid battery. *Koyosonic.*

The main disadvantage of AGMs is their price: they tend to cost two to three times as much as standard flooded batteries of the same capacity.

Advantages of sealed batteries:
- Less maintenance (don't have to put water in them)
- Vent less flammable and toxic gas

Disadvantages of sealed batteries:
- Cost more
- Don't last as long

Basic Battery Chemistry

 Tech Stuff *At this point, we thought it would be helpful to provide a brief primer on how batteries actually work. Batteries are electrochemical devices that store and release electrical energy through chemical reactions.*

When a lead-acid battery is discharged, the electrolyte (sulfuric acid) and some of the lead on the plate react, producing water, lead sulfate, and current. Thus, the chemical equation for discharge is: PbO_2 Pb $2H_2SO_4$ -->$PbSO_4$ $2H_2O$ + Electrical energy.

During charging, the equation would run in reverse, producing lead, sulfuric acid, and heat, plus hydrogen gas. In flooded batteries, the hydrogen

FIGURE 10-8   Dave Corbus from the National Wind Technology Center inspects a bank of Trojan L-16 deep-cycle batteries that are connected to a small turbine. *Warren Gretz/ DOE/NREL*.

gas vents to the outside (unless it is partially caught and recirculated via a hydrocap). In sealed ones, it recombines with oxygen.

Sometimes, if the plates are damaged or the electrolyte levels get too low, hydrogen sulfide gas can form. That's the gas that smells like rotten eggs, and it is poisonous at high concentrations. In sealed batteries, the gas shouldn't leak to the outside, but it is still possible through small cracks. So having well-ventilated batteries is critical (Figure 10-8).

Battery Ratings: Amp-Hours

We introduced the concept of *amp-hour* (Ah or AH) in Chapter 3: one amp-hour is one amp of charge flowing through a wire for one hour. So amp-hours are amps times hours. This is how all deep-cycle batteries are rated. So if you have something that pulls 20 amps and you use it for 20 minutes, then the amp-hours used would be 20 (amps) × 0.333 (hours), or 6.67 Ah.

Most batteries are rated at 20 hours. This means the amp-hours put out by a battery are measured over a period of 20 hours, down to the point of total discharge. By the way, batteries have 10.5 volts at full discharge, due to their internal chemistries. Many manufacturers also list amp-hour ratings for other periods, such as 6 hours or 100 hours.

In case you are wondering, a battery *cycle* is usually defined as one complete discharge and recharge cycle. If you need a technical standard, it is often considered discharging from 100 percent to 20 percent, and then back to 100 percent. However, there are often other ratings.

Cost of Batteries

As we mentioned, batteries aren't cheap. One general estimation is to add $2 or more per watt to the cost of a system when you add batteries. That means that batteries for a 10 kW system can cost as much as $20,000 if you want to store all the energy. This would include the cells, the basic extra equipment needed to charge them, and the storage location.

Individual batteries cost $150 to $300 each for a heavy-duty, 12-volt, 220 amp-hour, deep-cycle product. Larger-capacity batteries, those with higher amp-hour ratings, cost more. A typical 110-volt, 220 amp-hour battery bank, with a charge controller, costs at least $2,000.

Power Up!   *And remember, the larger your battery bank, the less efficient the system, because the more energy you "lose" over time through self-discharge and internal resistance.*

Energy Loss from Internal Resistance

Everything in life has its price, and using batteries is no exception. A rough observation is that if you use 1,000 watts from a battery, it may take 1,050 or 1,250 watts to fully recharge it. Much of the energy loss during charging or discharging is due to what's called *internal resistance*, which usually manifests itself as heat. The lower the internal resistance, the less energy gets wasted as heat, so the more efficient the battery.

Typical lead-acid batteries are 85 to 95 percent efficient, while deep-cycle AGMs can approach 98 percent. NiCads are around 65 percent. However, note that the slower the charge or discharge, the less energy is wasted as heat, so efficiency is higher. A battery rated at 180 amp-hours for 6 hours might do 220 Ah at 20 hours or 260 Ah at 48 hours.

Battery Longevity

Now that you have some idea how much batteries cost, chances are good that you want to make the ones you have last as long as possible. It turns out that there are many factors that affect battery longevity, including charging patterns, temperature, maintenance, and so on. Less deep and less aggressive charging and discharging cycles tend to make for longer-lasting batteries.

Although deep-cycle batteries are designed for heavy discharging, running them all the way down repeatedly does decrease their lifespan. This is why many battery manufacturers suggest trying not to discharge below 50 percent.

According to Northern Arizona Wind & Sun, if a battery is discharged to 50 percent every day, it will last about twice as long as if it is cycled to 80 percent *depth of discharge* (DOD). If it is cycled only to 10 percent DOD, it will last about five times as long as if it were cycled to 50 percent. Although that last number sounds impressive, you are also buying batteries to use them, so 50 percent is often considered a reasonable compromise. (Note that DOD is the opposite of how much energy is left in the cells. A DOD of 80 percent means you have a battery that is 20 percent "full." See more on proper battery charging later.)

By the way, don't think that you can just set up your batteries and forget about them. Like plants, they need regular feeding, because inactivity can be harmful to their internal chemistries. Ian Woofenden suggests making sure your batteries are fully charged up at least a couple of times a week (Figure 10-9). Never buy a battery to store for later use. Either use it or, at a minimum, put it on a slow and steady trickle charge.

As far as temperature, even though battery capacity increases with higher temperatures, battery life decreases. This explains why your grandma kept her batteries in the refrigerator, next to the meatloaf. So if you can keep your batteries in a cool space, they will last longer.

You can also increase the longevity of your batteries with regular maintenance. If you have flooded batteries that need refilling, stay on top of it. Never let the plates get exposed to the air.

Speaking of plates, it turns out that the thickness of the positive plate is one of the more important factors when it comes to battery longevity. The positive plate

FIGURE 10-9   This Colorado home is powered by a hybrid solar and wind system, plus the battery bank in Figure 10-5. *Warren Gretz/DOE/NREL.*

gets eaten away over time, so the thicker the plate, the longer it tends to last. Incidentally, the negative plate expands a bit during discharging. This is actually why most batteries have separators that can take some compression, such as paper or glass mat.

According to Northern Arizona Wind & Sun, car-starting batteries have plates around 0.040 inch (4/100 inch) thick, forklift batteries have plates more than a quarter-inch" thick, and golf cart batteries have plates around 0.07 to 0.11 inch thick. Concorde AGM's are 0.115 inch, the Rolls-Surrette L-16 type (CH460) is 0.150 inch, the U.S. Battery and Trojan L-16 types are 0.090 inch, and the Crown L-16HC size is 0.22 inch thick.

Northern Arizona Wind & Sun has provided a rough guide of minimum and maximum lifespans for the following battery types, *when used in deep cycle service, such as would be found with a wind electric system:*

- Starting batteries: 3 to 12 months
- Marine batteries: 1 to 6 years
- Golf cart batteries: 2 to 7 years
- AGM deep cycle: 4 to 7 years
- Gel deep cycle: 2 to 5 years
- Deep cycle (L-16 type, etc.): 4 to 8 years
- Rolls-Surrette premium deep cycle: 7 to 15 years
- Industrial deep cycle (Crown and Rolls 4KS series): 10 to 20+ years
- Telephone (float): 2 to 20 years. These are usually special-purpose "float service," but often appear on the surplus market as "deep cycle." They can vary considerably, depending on age, usage, care, and type.
- NiFe (alkaline): 5 to 35 years
- NiCad: 1 to 20 years

Sizing Your Battery Bank

Regardless of what type of batteries you use, it's critical to size your *battery bank*—your collection of batteries—properly. You're not going to want a massive bank, because that is going to get quite pricey when you have to replace them every few years. You also don't want to waste too much energy over time through seep and inefficiency.

It turns out that sizing a battery bank can be pretty tricky (Figure 10-10). Your first impulse may be that you'd like to have enough batteries to run everything in your house for at least a week without generating any power. That's a nice idea, and we hate to smash your dreams, but that's not very realistic. It would just take too many batteries to be practical. You will also run into problems if it takes you too long to replenish your battery bank from normal operation of your wind generator. Since you lose some of the energy you put in batteries, you don't want to

Figure 10-10   Sizing a battery bank can be tricky; batteries are expensive and require maintenance, and the more you have, the more energy you lose through self-discharge. *Warren Gretz/DOE/NREL.*

end up on a downward spiral of always trying to fill the things up and never quite catching up.

More typically, you'll probably want to size your system to run critical loads for a couple of days. After that, if the winds don't pick up or the sun doesn't shine on your solar panels, you're going to have to crank up the old diesel generator.

Batteries come in many different sizes, and may be given a wide range of codes that refer to their physical heft or intended use. For example, we often see GC for golf cart batteries or FS for floor scrubbing batteries. As we mentioned, there aren't any WS batteries for wind systems. Even so, the actual battery size and designation isn't what's usually critical. It's more important to map out the specifications and make sure the battery matches your system and your objectives.

So, there are a few things to consider when thinking about battery sizing. All lead-acid batteries supply around 2.14 volts per cell when fully charged. Since standard 12-volt batteries are made with six cells, that means the total would be 12.6 to 12.8 V. In addition, a typical 6-volt golf cart battery can store about 1 kilowatt-hour of useful energy:

$$6 \text{ volt} \times 220 \text{ amp-hr} \times 80\% \text{ discharge} = 1,056 \text{ watt-hours}$$

To give you an idea of what that means, that's roughly enough juice to power two 50-watt light bulbs for ten hours.

$$2 \times 50 \times 10 = 1,000 \text{ watt-hours}$$

As you can see, unless you live like a monk, you will probably need more than one battery. Remember that batteries are heavy (around 65 pounds for that golf cart number), so you need some space. You also need good ventilation, and preferably a floor that could withstand the occasional splash of acid.

Although 220 amp-hour golf cart batteries are often used, some others prefer larger "L-16" deep-cycle batteries with ratings of 350 amp-hours. The good news is that you probably have considerable choice, although note that a lot of discount batteries don't hold up very well.

Another thing to keep in mind is that battery capacity (how many amp-hours it can hold) goes down as temperature goes down. Compared to the standard rating at room temperature, going down to –22 degrees F would tend to drop the capacity in amp-hours by 50 percent. At freezing, capacity is typically reduced by around 20 percent. At 122 degrees F, battery capacity would tend to be 12 percent higher.

We bring this up because it can affect the sizing of your system, particularly if you live in places with extreme temperatures. You may also need to think about insulating your battery bank. The good news is the bank itself is a considerable thermal mass, and usually takes some time to adjust to air temperature.

Writing in *Backwoods Home Magazine*, Jeffrey Yago, a licensed professional engineer, gives more insight into precise sizing of a battery bank. Yago suggests the example of needing to power a 12-volt DC lighting system, consisting of four fluorescent fixtures, in an off-grid cabin. Each light is rated at 36 watts and draws 3 amps (36 watt/12 volt = 3 amp). If everything is lit up at once, that gives a lighting load of 12 amps (4 lights × 3 amps).

Yago reasonably suggests that you might use these lights from 5:00 P.M. until 11:00 P.M. each night, or six hours per day. That means your battery system must supply 72 amp-hours (12 amps × 6 hours) of energy each day, ignoring efficiency losses.

If we have a 350 amp-hour battery, it will deliver 175 amp-hours (350 × 0.50) before needing recharging if discharged to 50 percent, the standard recommended level. That means the lights can be used for two days (175 amp-hr/72 amp-hr) before needing recharging.

Yago also points out that there's a problem with his example, namely that the 6-volt battery isn't producing enough voltage to run the 12-volt lights. He explains that if we simply wired two of the batteries in series, we would end up with 12 volts, but only the same capacity of battery, because multiple batteries wired in series have the same amp-hour capacity as each individual battery. Wiring batteries in parallel gives multiples of the amp-hour capacity, but does not increase the voltage.

The solutions to this electrical mindbender are to increase the system voltage to 12, 24, or 48 volts, and/or to wire the batteries in *series-parallel*. Outlining exactly how to do the wiring is beyond the scope of this book, but the good news is there are many plans online. Better yet, your installer can show you the ropes, or hire a licensed electrician.

Yago points out that there are a few things to watch out for with battery wiring. He notes that failure of individual cells can drag the system down, so it's a good idea to be diligent about maintenance and monitoring. Yago also says every wire that passes through the positive terminal of a battery should go through a protective fuse—and one rated for DC. He notes that DC fuses have to be more robust than AC fuses.

To bring it all together, Table 10-1 is a handy chart to help you size your battery bank, written by Chris Brown for www.btekenergy.com.

TABLE 10-1   How to Size a Battery Bank

Step	Process	Example
1	Identify total daily use in watt-hours (Wh)	6,000 Wh/day
2	Identify days of autonomy (backup days); multiply Wh/day by this factor.	3 Days of Autonomy: 6,000 × 3 = 18,000 Wh
3	Identify depth of discharge (DoD) and convert to a decimal value. Divide result of Step 2 by this value.	40% DoD: 18,000 / 0.4 = 45,000 Wh
4	Derate battery bank for ambient temperature effect. Select the multiplier corresponding to the lowest average temperature your batteries will be exposed to. Multiply result from Step 3 by this factor. Result is minimum Wh capacity of battery bank: Temp. in [degrees] F. Factor 80+ 1.00 70 1.04 60 1.11 50 1.19 40 1.30 30 1.40 20 1.59	60° F. = 1.11 45,000 × 1.11 = 49,950 Wh
5	Divide result from Step 4 by system voltage. Result is the minimum amp-hour (Ah) capacity of your battery bank.	49,950 / 48 = 1,040 Ah

Organization of Battery Banks

Most people put their battery bank in a basement, garage, or storage building. Don't put batteries directly on a cold floor, because that will decrease their capacity (although it would increase their lifespan). Try to keep clutter from encroaching on their space, and if you need to wipe off a battery, use only water (Figure 10-11). Never use any kind of solvent around them.

People often build enclosures for their batteries to keep things tidy. It's a good idea to control the ventilation, too, which should exit to outdoor air. One popular solution is to build a wood frame, line it with plywood, and then add PVC piping for ventilation. It's a good idea to cover the wood with several coats of fire- and acid-resistant paint.

FIGURE 10-11   Inverters and battery bank at Manzanita Indian Reservation near Boulevard, California. *Warren Gretz/DOE/NREL.*

    Hazard   *Keep open flames away from battery banks. During certain stages of their operation, they may release flammable hydrogen and oxygen.*

INTERCONNECT

Building a Useful Small Wind Turbine for $140

Michael Davis is a small wind enthusiast who shows how a hobby can have some tangible benefits (Figure 10-12). A few years ago, the amateur astronomer bought a piece of property in the wilderness of Arizona that was perfect for stargazing, but far removed from any grid service. On the plus side for Davis, this meant freedom from the light pollution that obscures the heavens. But he still wanted a little power to run a few devices, like his laptop and camera chargers.

Davis answered our questions via email, and he has posted extensive information (and videos) about his project on his website, www.mdpub.com/. From scrap and off-the-shelf parts, Davis has built his own telescope, solar panels, and even a jet engine. On his remote Arizona property, Davis installed some of his solar panels, built a wood gasifier, and in 2006 he erected a homebrew wind system, which he wrote, "cost hardly anything."

Davis told us that his little system has held up very well, and said he was surprised that "even such a crude design could work very well." He wrote, "If you have some fabricat-

FIGURE 10-12   An amateur astronomer and tinkerer, Michael Davis built a small wind system for his remote Arizona property for just $140 in parts. It powers his computer, cell phone, air mattress pump and other small devices. *Michael Davis www.mdpub .com/Wind_Turbine.*

ing skills and some electronic know-how, you can build one too." Interestingly, Davis said his website has become "insanely popular," taxing its server with more than a million visits a year. He said he gets "dozens of requests for help every day," mostly from folks who want advice in building their own wind turbines.

Davis began building his turbine by doing research online, and seeing what has worked for other handy folks. He noticed that some people use permanent magnet DC motors as generators, instead of the more customary AC alternators. He suggests that what can work well is a motor that is rated for high DC voltage (more than 12 V), low RPMs and high current. Such motors were historically used in computer tape drives. In his view, car alternators aren't a good choice because "they have to spin at very high speed to produce useful amounts of power." For his project, Davis scored an old Ametek tape drive motor on Ebay for $26, which he says works great.

Davis cut his blades out of ABS plastic drainage pipe, which he sanded down to a smooth shape to give an airfoil. He bolted the blades to a metal hub, and fastened the motor to a piece of 2x4, adding a bit of sheet metal for a tail (to keep the turbine facing into the wind). For the tower, Davis selected 1¼-inch diameter steel EMT electrical conduit, which he got from a big box store. He used pipe fittings at each end, and a slightly thinner, 10-inch pipe serves as the connector and yaw bearing for the top. For the base, Davis used a piece of plywood, with holes for staking into the ground.

Davis ran wires from the generator through the center of the conduit, to his controller, which he built himself based on schematics he found online. He then connected lead acid batteries (Davis suggests always connecting to batteries first, then to the wind turbine, to avoid damaging the electronics; similarly, unhook the turbine first if you are taking it down). For a dump load, Davis wired together several high-wattage, 2-Ohm resistors. He also added a 120V inverter.

On site, Davis is able to easily raise and lower the 10-foot tower, which is supported by four guy lines (at first nylon rope, later steel cable). On most stays on his remote property, Davis' small wind turbine produces all the juice he can use for his laptop, air mattress

Figure 10-13 Davis made his small wind system from off-the-shelf parts, and it meets his needs on his off-grid property. *Michael Davis www.mdpub.com/Wind_Turbine.*

pump and electric razor, plus charging of his camera and cell phone batteries. When he brings his pop-up camper along, he has power for a light bulb and vacuum cleaner (though not quite enough for his girlfriend's hair dryer).

Davis gets all this convenience for a total cost of $140 in materials for his wind system, plus some of his time and a little ingenuity. Recently, he added a small home-built solar panel. His systems are cheap and easily portable, and they show that if you have modest electrical needs, a very small turbine may be a good solution (Figure 10-13).

Davis added, "If I had it to do over again, knowing what I know now, I would have made the tail larger, and used steel guy wires. Maybe I would have looked for a generator with higher output."

He told us, "Many of my neighbors have small wind turbines. They are more popular than urban people may realize." However, he warned, "It won't power your whole house. Small wind turbines are a great supplement to a solar power system. If you are going to live off-grid, you are going to have to seriously change your lifestyle and drastically reduce your power consumption."

Charge Controllers

If you have batteries, you need to have a *charge controller,* or you will quickly ruin the batteries by under- or overcharging from your wind generator. Even systems that are grid-tied but also have batteries need a charge controller, primarily for when the grid is down. Otherwise, the inverter effectively acts as a charge controller, sending only as much power to the batteries as needed to keep them full.

Fortunately, charge controllers aren't particularly expensive, and are often in the $100-to-$200 range. The primary job of a wind turbine charge controller is to constantly monitor the state of the batteries. It does this by watching the voltage coming out of the battery bank and looking for what are called *set points.*

When battery voltage approaches a maximum set point (called the *float voltage*), the controller turns on the dump load (also called a shunt or a diversion load). Most charge controllers have adjustable set points, which are often chosen in the 13.2 V to 15.2 V range for 12 V charging systems, or 26.4 V to 30.4 V in 24 V charging schemes.

When the battery voltage falls back below another threshold (typically 0.2 V to 0.5 V below the *float voltage*), the dump load is switched off, and juice starts flowing into the battery bank again.

As we stated in Chapter 1, the purpose of the shunt is to dissipate excess energy without overcharging the batteries or leaving the turbine unloaded. If you overcharge the batteries, their lifespan will be reduced, and if you unload the turbine, it could dangerously freewheel. As mentioned previously, the most common dump load is an electrical heating element placed in water or open to the air. Some small systems use high-wattage car headlight bulbs, and some clever folks rig up electric sawmills or other devices that can put the excess energy to use doing work.

In a grid-tied, batteryless system you don't need a dump load, because you can always send juice to the grid. If the grid goes down, your system will stop. If you have a grid-tied system with batteries, you do need a dump load for when your batteries are full but the grid is down.

A charge controller also usually has a blocking diode in it that stops reverse current from seeping out of your batteries toward your wind generator. Many controllers also provide automatic battery equalization (see later) and built-in data logging. Most controllers have a built-in heat sink to dissipate heat that builds up during operation, particularly from relays (Figure 10-14). Sometimes the heat sink is a noticeable series of metal fins, although in some models it looks more subtle.

FIGURE 10-14   Part of the balance of system inside the Colorado home from Figure 10-8, showing two stacked inverters (left) and charge controllers (upper right). Note the fin heat sinks across the tops of the latter. *Warren Gretz/DOE/NREL.*

Will Solar Charge Controllers Work for Wind Systems?

In general, you do not want to use a charge controller that is designed for solar panels with a wind system. That's because a wind turbine must always be connected to a load, unlike solar panels, which are more forgiving. If solar panels start to produce too much juice, the controller can just short-circuit them, and nothing will get damaged. If you just cut off a wind turbine, it won't have a load, and it could start to freewheel out of control.

It is true that many new charge controllers are marketed as working well with both small solar and wind systems. Such units must be able to transfer current to a shunt.

Many of these controllers come with two switchable modes: *shunt mode* or *diversion mode*. In shunt mode, the dump load is powered directly from the battery. This will draw down the energy stored in the battery. You might choose to do this if the batteries got overcharged, for example. In diversion mode, the battery retains its charge, while the power coming out of the generator goes to the dump load.

Note that although they can be designed without them, some solar systems are set up with dump loads. One reason to do this is to avoid going through a switching transistor in a regular solar charge controller, since those devices rob a bit of precious energy.

AC and DC Controllers

If you are like Michael Davis (see the Interconnect section on page 351) and your wind turbine has a DC generator, you can use a DC charge controller. However, as stated previously, most wind turbines don't have DC motors. Most have alternators that produce wild AC, also called three-phase AC. Before this can be used it is usually rectified to DC, then used to charge batteries or sent to an inverter.

Many charge controllers have built-in three-phase bridge rectifiers, which convert the wild AC into DC. Some wind turbines have built-in rectifiers, so in that case you may not need this feature, and may be able to get along with a DC controller. In some setups, separate rectifiers are put in place before the charge controller. The important thing is to know how all the parts of your system are going to interconnect and how they will work together. The good news is there are options out there.

The Effects of Temperature Changes

We mentioned previously that temperature affects batteries, so if your system is subject to wide changes in temperature, you may need to make some adjustments. Battery-charging voltage varies from about 2.74 volts per cell at –40°C to 2.3 volts per cell at 50°C. Since there are six cells in each battery, this means correct charging voltage drops by about 0.15 V for each 5°C rise in ambient temperature.

FIGURE 10-15   A technician tests the 230-volt VRLA batteries of a 50 kW system at the National Wind Technology Center. *Warren Gretz/DOE/NREL.*

One strategy is to protect your batteries from temperature extremes by placing them in a conditioned space (Figure 10-15). You can also add insulation around batteries.

Also note that some charge controllers come with temperature compensation (such as Morningstar). One thing to remember is that this will only be effective if the controller is subject to the same temperatures as the batteries. If one is inside while the other is outside, that might be a problem.

Protecting Your Batteries: Proper Charging and Equalization

Fortunately, lead-acid batteries aren't subject to the *memory effect*, unlike nickel-based technology. Over time, nickel batteries will lose capacity unless they are fully discharged at least once a month.

However, as we mentioned, lead-acid batteries are subject to *sulfation*, a chemical reaction that deposits sulfates on the lead plates. This diminishes the battery's capabilities, and could ruin it. Since sulfation can occur when batteries are less than 100 percent charged, experts usually suggest keeping lead-acid batteries "full" at all times.

This is another job for your charge controller. When set up properly, it will continuously *float charge* the batteries to compensate for gradual *self-discharge* and keep them at 100 percent. Of course, if you have a grid-connected system with batteries, this means you are also continuously using a bit of your energy just to keep your batteries topped off, even if you rarely use them.

In addition to float charge, there are two other general stages of most battery charging, plus a few related concepts.

- **Bulk Charging** Starting from an empty battery, *bulk charging* is usually the first step. In this process, current is sent into the battery at the fastest rate it can safely accept, typically at 10.5 to 15 volts (assuming a 12 V system). Bulk charging normally ends when the battery voltage hits 80 to 90 percent of the full charge level. Note that with gel batteries, all charging is usually done at two-tenths of a volt less than with flooded batteries in order to reduce water loss and damage.
- **Absorption Charging** During *absorption charging*, voltage remains constant and current gradually tapers off, while the battery's internal resistance rises. During this stage voltages are typically around 14.2 to 15.5 volts.
- **Float Charging** When the battery reaches a full charge, the process should stop, because overcharging a battery will damage it. However, batteries lose charge over time through self-discharge, and that's where float charging comes in. Float charging "tops off" the battery as needed, typically at a lower level of 12.8 to 13.2 volts. This is also called a *maintenance* or *trickle charge*.

     The controller may use *pulse width modulation* (PWM), in which it senses tiny voltage drops in the battery, and then compensates by sending short pulses of charge. It is called "pulse width" because the width of the pulses varies, from a few microseconds to several seconds. For long-term float charging of batteries that are rarely used, some experts recommend keeping the float voltage around 13.02 to 13.20 volts.
- **Float Charging vs. Trickle Charging** Strictly speaking, there is a difference between trickle charging and float charging. A trickle charger sends a low level of current continuously, regardless of the state of the battery. So leaving a trickle charger connected to your battery indefinitely will overcharge it. A float charger, however, includes circuitry that monitors the battery voltage. When it senses that the battery is full, it stops charging. This means float charging can be left on indefinitely, as long as it is set up properly. As mentioned previously, you may need to make an adjustment for temperature changes (many controllers have programs that make this easy), because the correct float voltage drops by about 0.15 V for a 5°C rise in ambient temperature for a 12-volt system.
- **Equalization** To extend the life of your flooded batteries, you should apply an *equalizing charge* every 10 to 40 days. An equalization charge is about 10 percent higher than normal full charge voltage, and it is applied for 2 to 16 hours. The goal is to ensure that all the cells are equally charged and that the gas bubbles mix the electrolyte. Otherwise, what happens over time is that the liquid in the battery can become "stratified," often with stronger solution on top and more dilute solution on the bottom. This can degrade your battery over time. AGM and gelled batteries require less frequent equalizations, though some experts say doing it two to four times a year is a good idea. Do check your manufacturer's guidelines.

- **C/8, C/20, and C/4**   If you start researching batteries, you may see references to maximum recommended charging rates. For flooded batteries, experts recommend that you avoid charging at more than the C/8 rate, which is the battery capacity at the 20-hour rate divided by 8. For example, a 220 Ah battery would have a C/8 rate of 26 amps. Gel batteries should be charged at no more than the C/20 rate, or 5 percent of their amp-hour capacity. Otherwise, the gel can form cavities that never heal. AGM batteries can be charged quicker, to C/4. Of course, check with your battery manufacturer for specific specs.
- **Maximum Power Point Tracking (MPPT)**   This popular controller feature adjusts the operating voltage of a wind generator at different speeds to maximize energy output.

Most of the time your system is operating, the controller will be in float charging mode (and MPPT if you have it). Of course, there are going to be times when it runs in bulk charge and absorption mode, such as when it's time to replenish batteries that have been spent. It's also a good idea to get into the habit of regular equalization sessions.

Building a Charge Controller from Scratch

Amateur astronomer and electronics tinkerer Michael Davis built his own charge controller from spare parts to work with the 12-volt small wind system he built for just $140 (see the Interconnect section on page 351). He mounted the parts of his controller on a wood board, along with a terminal block for connections and his dump load, which is a series of high-wattage resistors (Figure 10-16).

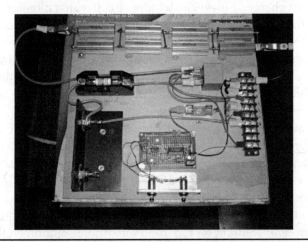

FIGURE 10-16   Michael Davis built a charge controller for his turbine from spare parts (center bottom). The rectangles on the top are high-wattage resistors, his dump load. Below and to the left of that is the main fuse, and the cube to the right is an automotive relay for switching the circuit. *Michael Davis www.mdpub.com/ Wind_Turbine.*

Davis built the main circuit of the controller based on a free schematic he found online, assembling it on a standard perfboard about the size of a paperback novel. Davis wired it to a 40 amp SPDT automotive relay, which routes the turbine power either to the batteries or to his shunt.

Davis set the circuitry up so that if the battery voltage drops below 11.9 volts, the controller switches the turbine power to charge the battery. If the battery voltage rises to 14.8 volts, the controller switches to his dump load resistors.

Davis writes that he chose 11.9V and 14.8V "based on advice from lots of different websites on the subject of properly charging lead-acid batteries." He explained that many people recommended slightly different voltages, so he averaged them. When the battery voltage is between 11.9V and 14.8V, the system can be switched between either charging or dumping, thanks to a couple of buttons he added. He also added LED lights to indicate when the system is charging or dumping.

How to Connect or Unhook a Controller

It's critical to remember that a wind turbine must remain loaded at all times so it doesn't spin too rapidly and get damaged. So it always needs to be connected to the grid or your battery bank. When you are working with a charge controller, always connect it to your battery bank first, then to the wind turbine. If you do it the other way around, the wild electrical swings coming from the wind generator won't be smoothed out by the load. The controller could start to behave erratically or even get damaged.

Similarly, always disconnect the wind turbine first and the battery last when you are taking things apart.

Voltage Clamp

Particularly with large turbines, some systems are designed with an additional regulator before the charge controller. This *voltage clamp* can be inside the turbine itself or a separate piece of equipment, though it is often provided with the turbine. The job of the clamp is to smooth out the voltage to a level that's safe for the equipment in the system.

Many charge controllers also effectively function as a voltage clamp, so there may be no need for an additional component.

Disconnects

You are going to need *disconnects*—aka switches—in your electrical system to protect it from fires and so you can safely isolate individual components for servicing. Disconnects are standard electrical equipment not unique to wind turbines, though it's important to make sure you have the right voltage and amperage ratings for the loads.

If you have a battery, there must be a disconnect between the bank and the inverter. Most often, this will be in the form of a DC-rated breaker mounted in a metal box. This protects the circuit from electrical fires and allows easy servicing.

In many countries, an additional AC disconnect must be located outside the house or building, near the panel and the utility meter. This is also usually a breaker mounted in a metal box, only it is rated for AC. The main purpose of this disconnect is extra insurance to protect line workers from your electricity during repair of the grid. Strictly speaking, it shouldn't have to be used, because once a grid-synchronous inverter senses that the grid is down, it shuts off automatically. But utilities like to have the disconnect in place.

Speaking of utilities, we mentioned earlier in the chapter that utilities often frown upon battery banks. They like to be in control of "their" systems throughout the whole lines, and they can be wary of anything added to the grid by third parties. That's why it's critical to follow the guidelines from your power authority, and to make sure you are meeting relevant code and certifications. You don't want to set up your dream system, only to receive a takedown order a few weeks later.

Breaker Panels

Breaker, breaker. If you've ever experienced an electrical surge from lightning or from your visiting in-laws trying to blow dry and curl their hair at the same time they're doing a load of laundry, you are probably familiar with your AC breaker panel. In modern construction, it will be full of circuit breakers, black plastic switches that "break" the flow of electrons if they sense a potentially dangerous surge. The breakers can also be flipped to safely work on a home's wiring (Figure 10-17). There are usually several switches for different lines in the building, and

FIGURE 10-17   From left to right, a wind system's inverter, breaker box, and powersync inverter. Under the latter is a disconnect and electric meter. *Trudy Forsyth/DOE/NREL.*

hopefully they are marked as to what they govern. There is often a central switch that will disable the entire system as well.

Older systems may still use fuses, which must be replaced if the thin wire they contain melts during a protection event. Professional engineer Jeffrey Yago suggests routing every wire connected to the positive terminal of a battery through a DC fuse.

The AC breaker panel is also where a building's wiring meets with the grid and/or the wiring for the wind energy system. The panel is usually a wall-mounted metal box that's placed in a basement, garage, or utility room (or in a hall in crowded urban apartments). In most cases, the electrical output from the inverter will be routed into the building's main AC breaker panel, with its own breaker switch.

The Wire

It may seem like the simplest part of your wind electric system, but wire isn't the cheapest, the least important, or the easiest to choose (Figure 10-18). You need to have wire to send your precious energy from the turbine through the charge controller, inverter, any batteries, and out to your home and/or the grid. Properly sizing wiring is important, and not always easy.

Except for very long runs, the majority of wiring used today in electrical circuits is made of copper, which is an excellent conductor. For wiring that will be protected in walls or in conduit, stranded copper is often preferred because it is highly flexible. For wiring that has to go underground, solid copper is usually better, because stranded wires can wick moisture that leads to corrosion or other problems. The biggest drawback to copper is its price, which currently runs $20,000 to 30,000 per quarter mile of wire.

Starting around the 1940s, people also began using aluminum to make wire, and it is fairly common, even in some wind energy systems. Aluminum wire is cheaper than copper but not as good a conductor. Copper is a 60 percent better conductor, so wire made from that metal can be thinner than aluminum wire to

FIGURE 10-18 Choosing the right wire for your system is critical, but it isn't always easy. *Patrick Corkery/DOE/NREL.*

carry the same current. One advantage of aluminum is that it is lighter, which is why it is often preferred by utilities for long-distance runs, because they can use fewer towers to support it.

Aluminum wire has a bad reputation, because for decades it suffered from a number of problems. Pure aluminum will quickly oxidize in contact with the air, forming a high-resistance film that interferes with conductance. Most aluminum wire produced today is made with alloys that make it resistant to oxidation, so this is less of an issue.

Aluminum is also more sensitive to cold, and tends to "creep." This was often a problem, because aluminum wires would routinely come loose. But today, better joining methods are used that reduce this problem. Aluminum isn't as strong as copper, so that might be one consideration.

According to some sources, modern aluminum alloy wire is 25 to 40 percent cheaper than copper wire. In terms of the raw commodities, as of this writing, copper costs $4 a pound and aluminum costs $1 a pound.

Note that a recent survey of U.S. electrical contractors found that they preferred copper 20 to 1 over aluminum for wiring. The only reasons stated for choosing aluminum were lower cost and lightweight. Copper was considered superior in all other respects.

Whichever type of wire you choose, for most applications the metal core will be surrounded by an insulating sheath. (Except in the case of ground wires, which are naked.) For most runs, wires will also be snaked through electrical conduit (Figure 10-19). This protective tubing is most commonly made of metal or plastic, though it can also be fiber or clay.

Hazard   *Wiring between batteries in a bank is no place to cut corners, though you do want to make the runs as short as possible to minimize loss. Avoid automotive, marine, or welding cables, which are not UL-listed for residential wiring. Instead, get UL-listed battery interconnects made from #2/0 multistrand flexible copper cable. They are widely available online and from renewable energy dealers, and often come in strands of 9 and 12 inches, with machine-crimped copper terminals.*

Remember that current passing through wire creates heat, which can be dangerous if not properly planned for, and also is a wasteful use of your energy. If the wire is too thin for the rate of charge flow, the high temperature can damage the insulation and increase the risk of fire. Also, if the wire run is too long for the size of the wire, a lot of energy will be wasted as heat.

Hugh Piggott points out that the DC current of a small wind turbine can be estimated by taking the machine's rated power and dividing by the system voltage. However, he points out that many systems can exceed their rated power in high winds, so it's critical to get the maximum current possible from the manufacturer.

Figure 10-19   Unless you are running ground wire, underground runs should be protected inside conduit, like this project in Indonesia. *Susan Thornton/DOE/NREL.*

Further, most electrical codes mandate an additional safety margin for wiring, so you'll need to make sure you follow those guidelines. Piggott also suggests remembering that 12 V systems must have four times the charge flow of 48 V systems to deliver the same amount of power. In short, sizing wiring is not an easy task.

We mentioned that most small wind turbines produce wild three-phase AC. In that case, three wires actually move this down to the rectifier/charge controller, where it is converted to DC. At any moment, only two of these wires conduct current. Piggott says the AC current in each wire will be 82 percent of the DC, while energy loss in the three wires will be similar to loss in two wires of the same size carrying the full DC current.

One important specification to keep in mind is the *ampacity*, the maximum current a wire can carry without getting damaged. Ampacity is based on the wire's insulation quality, electrical resistance, and ability to dissipate heat, and it is affected by ambient temperature and frequency of current (in the case of AC). Always ensure that the ampacity of the wires you use is higher than the maximum possible current through the circuit.

Piggott also suggests calculating the voltage drop in the system's wires to gauge your energy efficiency. This is possible because a turbine must produce extra voltage to push charge through the wire and into the load. However, there is always loss in every system.

So if you measure that your 48 V circuit drops 2 V across its run, then that means the wind turbine would actually need to produce 50 V, and the energy lost in the wires would be 2 / 50, or 4 percent of the total. Piggott gives this formula for the loss in a given situation:

$$\text{Percentage loss} = (100 \times \text{feet of one-way wire run} \times \text{amps}) / (\text{feet per ohm} \times \text{system voltage})$$

As an example, Piggott suggests a 1.2 kW, 24 V turbine, with rated current 50 ADC (1,200 / 24) and a maximum of 75 A. AC current coming out of the generator would be a maximum of 62 A (75 × 0.82, for the 82 percent max in the three wires). Piggott checked a table of wire properties, and saw that for 62 A, the minimum safe size would be #6 American Wire Gauge (AWG). If the one-way wire run is 200 feet, we apply our formula as follows:

$$\text{Percentage loss} = (100 \times 200 \text{ ft.} \times 50 \text{ A}) / (1{,}127 \text{ ft. per ohm} \times 24 \text{ V}) = 37\%$$

That's a high loss. Fortunately, the turbine will actually generate much less current most of the time—say, 10 to 15 A, meaning it will operate at a 7 to 11 percent voltage drop. But you still have to size for the maximum.

Ian Woofenden suggests trying to keep your voltage drop within 1 to 5 percent. Wire is typically sized in one of four ways:

1. AWG (American Wire Gauge)
2. Diameter in inches
3. Diameter in millimeters
4. Cross-sectional area in square millimeters

The American Wire Gauge naming convention is widely used in the United States and assigns a whole number to each wire size, based on diameter. The higher the gauge number, the thinner the wire. For example, a wire with AWG gauge 000 has a diameter of 10.4 mm, while a wire with AWG gauge 10 has a diameter of 2.59 mm.

Typical household copper wiring is AWG #12 or 14, while telephone wire is usually #22, 24, or 26.

Fortunately, it is easy to convert from one type of wire size to another, with a table like Table 10-2 or handy online calculator.

Resource

www.reuk.co.uk/AWG-to-Square-mm-Wire-Size-Converter.htm

As mentioned earlier, a wind turbine must be loaded at all times, so it always must be connected to the grid or your battery bank. To avoid damaging your charge controller, connect it to your battery bank first, then to the wind turbine. Similarly, always disconnect the wind turbine first and the battery last when you are taking

TABLE 10-2   Wire Size Conversion Table. *Data from reuk.co.uk.*

American Wire Gauge (AWG)	Diameter (inches)	Diameter (mm)	Cross Sectional Area (mm²)
0000	0.46	11.68	107.16
000	0.4096	10.40	84.97
00	0.3648	9.27	67.40
0	0.3249	8.25	53.46
1	0.2893	7.35	42.39
2	0.2576	6.54	33.61
3	0.2294	5.83	26.65
4	0.2043	5.19	21.14
5	0.1819	4.62	16.76
6	0.162	4.11	13.29
7	0.1443	3.67	10.55
8	0.1285	3.26	8.36
9	0.1144	2.91	6.63
10	0.1019	2.59	5.26
11	0.0907	2.30	4.17
12	0.0808	2.05	3.31
13	0.072	1.83	2.63
14	0.0641	1.63	2.08
15	0.0571	1.45	1.65
16	0.0508	1.29	1.31
17	0.0453	1.15	1.04
18	0.0403	1.02	0.82
19	0.0359	0.91	0.65
20	0.032	0.81	0.52
21	0.0285	0.72	0.41
22	0.0254	0.65	0.33
23	0.0226	0.57	0.26
24	0.0201	0.51	0.20
25	0.0179	0.45	0.16
26	0.0159	0.40	0.13

things apart. (Before you disconnect the turbine, make sure it is stopped with a mechanical brake.)

Power Up!   *It's also important to remember that the resistance of a wire decreases as the square of the diameter. So thick wire has less resistance than thin wire, which means less of your electricity will be lost as waste heat. However, thick wire costs more, so you are going to have to weigh your options.*

Sizing wire isn't easy, and you may do well to consult the equipment manufacturers, dealers, other wind system owners, and electricians.

Grounding

A wind turbine sits on a high tower, above everything else around it, so it is at significant risk of lightning strikes. Fortunately, it can be reasonably well protected with proper grounding, although the process often isn't easy. The main goal of grounding is to move as much energy as possible down into the ground and away from your equipment as quickly as possible. To that end, you don't want to use just a single ground point. Even a few ground points may not be enough if they are too close together to quickly disperse the charge.

Instead, you should install in the ground a radial network of flat copper strap around the tower, with ground rods protruding from the ribbon at twice the length of the rods. That means 10-foot ground rods would be spaced 20 feet apart. The reason you want to use copper strap instead of regular wire is because lightning is a fairly complex burst of electricity that includes high-frequency energy. Some of this energy tends to stay at the surface of the conductor, in what's called the skin effect, and straps are better at dispersing this than tubular wire is. The strap should be at least one and a half inches diameter and at least #26 AWG (0.0159 inch) thickness.

The copper loop should be buried at least 8 inches, and preferably 18 inches, underground. If the copper runs are too long or too short they may not effectively disperse charge, so they should be between 50 and 75 feet long. There shouldn't be any sharp bends, either, because the lightning will look for an easier path. If any run passes within four feet of a metal object, the object should be wired in. Experts suggest putting in several radials, but usually not more than four.

Note that ground rods generally disperse charge into the ground in a cylindrical pattern with a diameter that extends roughly twice their length (hence, the spacing rule). Ideally, ground rods should be covered in copper, which makes them last longer than regular or galvanized steel. Stainless steel rods will last even longer, but they are more expensive. Rods should be welded to the strap with exothermic bonds or brazing. If that's not available, you can use high-pressure clamps and joint compound, although that will need to be periodically inspected.

It is usually easier to put in a shallower, more spread out system of grounding strap and rods, but if that isn't an option, you may be able to "go vertical" by making it deeper, using 50-foot long rods. If your site is over solid rock or very dry soil, which doesn't conduct well, you may need a more advanced solution and heavy-duty equipment.

Studies show there is no need to run a ground wire from the top of the tower to the earth, because there is sufficient metal in a tower to act as a conductor. Putting a copper wire in contact with the steel of the tower will cause it to rust faster, since the copper will react with the zinc in the steel. You do need to properly ground every one of the tower's connection points with the soil, however. This means each tower "leg" and every guy wire.

Each individual guy wire needs to have a direct connection to the ground, preferably above the turnbuckles, even if you have several wires that converge into a single anchor. Make sure you don't have any copper touching regular or galvanized steel, as that will lead to corrosion. Instead, use an intermediate, like stainless steel. Cover any joints with corrosion-resistant joint compound. Also run ground wire from all guy anchors to the center foundation block, 18 inches underground, with a ground rod every 20 feet. It's a good idea to run an outer perimeter of ground wire around the guy points too. Use copper strap or, at a minimum, solid copper wire #4 or larger AWG.

You can also take advantage of *Ufer grounds,* which are named after an engineer who discovered them in World War II. Herbert Ufer proved that reinforced concrete is good at dissipating a lightning strike, since it is quite porous and has relatively high surface area.

To set up an Ufer ground, the rebar in the concrete needs to be connected together and to the rest of the ground system, ideally with solid copper wire. Don't leave any jumps, because lightning can flash across them and crack the concrete in the process. As long as the solid copper is encased in concrete, it won't react with the steel. Don't use stranded copper, as that can allow moisture to seep in. It's also better to case the concrete directly into the ground, without using forms, which act as insulators, or piling on too much backfill, which often isn't a very good conductor.

If your house or other building is anywhere near the turbine, you will have to install a perimeter ground around that structure, preferably of copper strap, and with regularly spaced ground rods. This perimeter will then be connected to the ground perimeter of the turbine tower, as well as the standard (code-mandated) house ground and any grounds for the inverter. The reason you do this is because you don't want all that lightning energy you just channeled into the ground from your tower to make its way into your home's electrical system, where it could fry all your fancy electronic toys. By setting up a grounding perimeter there, you will disperse any charges.

Surge Arrestors

Even the best-laid ground wires won't necessarily be able to stop 100 percent of a direct lightning strike. Plus, there are other potential sources of high current bursts that could affect your system. For these reasons, you still need *surge arrestors* (commonly called surge protectors by the general public).

No surge arrestor is perfect, and there are several different designs, each with pluses and minuses. They include the following:

- **Gas discharge tubes (GDTs)**   In these devices, a spark gap is enclosed in a chamber filled with gas. GDTs can handle very large currents, but they are slow, so they allow a relatively large amount of current to flow through before they trip.

FIGURE 10-20   These metal oxide varisters (MOVs) were destroyed by a power surge, but the arrestors successfully protected the inverter of the Xiao Qing Dao Island wind project in China. *Jerry Bianchi/DOE/NREL.*

- **Metal oxide varistors (MOVs)**   These devices have a metal oxide that acts as a resistor, and gets more and more conductive as current rises. They let less current through than a GDT, but they can handle less of it (Figure 10-20).
- **Semiconductors**   Semiconductor-based arrestors let the lowest amount of electricity through, but they can also handle the least amount.

Devices rated specifically to deal with a direct lightning strike are sometimes called *lightning arrestors*, or are rated *10/350*, which is a technical number describing how fast the current ramps up then down, in µs.

When it comes to a wind energy system, Solacity recommends installing a lightning arrestor before the alternator (your manufacturer may have done this; check with them for details). If the turbine's wiring then goes underground for 50 feet or more, a less powerful second surge arrestor can be used before the rectifier, charge controller, batteries, and inverter. This arrestor could be rated at *8/20*, meaning it is designed to handle a secondary strike, such as lightning hitting nearby.

According to Solacity, an arrestor should be used on the grid side of the system as well. If it is unlikely to get a direct strike, an 8/20 device may suffice. But if there is a large overhead line nearby that could attract lightning, it's a good idea to use a full lightning arrestor (10/350).

Utility Interconnection Equipment

As part of your net-metering agreement, your utility will typically provide the equipment you need to connect to the grid. The most important piece is the electric meter. You may need a new one, because it must be bidirectional so it can track electricity coming from the grid and sent to the grid. Most utilities provide these free of charge as part of the interconnect agreement.

You may need transformers and relay switches to ensure that the proper voltage and frequency is sent to the grid. These are also usually provided by the utility.

System Meters

You will definitely need monitoring and tracking capabilities for your system, so you can see what is going on at all times, make adjustments as needed, and ensure your system is performing well (Figure 10-21). System meters will track how full your battery bank is, how much electricity your wind generator is producing, and

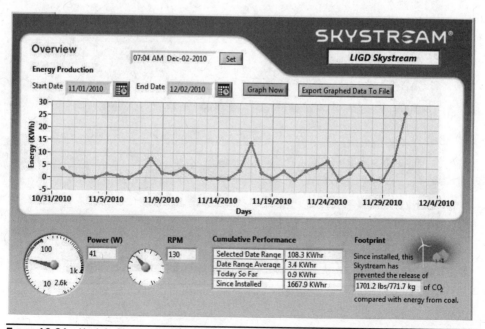

Figure 10-21 Kevin's Skystream came with easy monitoring and tracking software. *Kevin Shea.*

how much electricity is in use. Operating your system without metering is like running your car without any gauges—although possible to do, it's always better to know how much fuel is in the tank.

At a minimum, you will need to have a voltmeter and an ammeter. Many wind system components also have monitoring and data logging built in, as is the case with many charge controllers and inverters. This can give you a quick idea of what is going on, although it is also a good idea to have a centralized system meter that tracks everything. Many system meters now interface with home computers or even mobile devices, allowing instant data analysis and control.

The good news is that most small wind turbines now come with system meters as part of the package. Even if they don't, a meter isn't very expensive ($100). But it's definitely important.

Many smaller wind systems now come with integrated system analysis tools (Figure 10-22), which monitor performance in real time and stream the data via Wi-Fi. Skystream products have this capability, for instance.

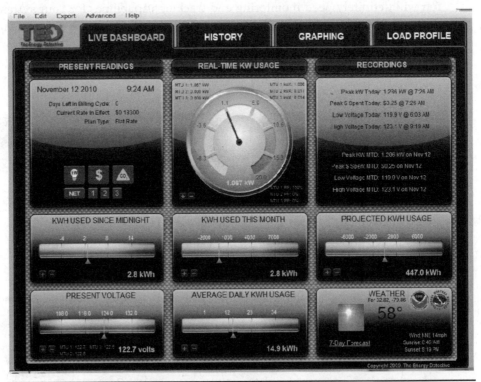

FIGURE 10-22   It's important to track your energy use and production. You wouldn't drive a car without gauges, would you? *Kevin Shea.*

Backup Generator

We don't spend a lot of time covering backup generators here, but suffice it to say that almost all off-grid systems have them. The generator is required for when you can't produce enough energy with your turbine (and/or solar panels) to run all your necessary loads. Most generators burn diesel (Figure 10-23) or petroleum fuel, although it's also possible to use natural gas, propane, kerosene, coal, or even wood.

Most generators produce AC current, so you need a separate battery charger or a charge controller and/or inverter that can accept this line and then convert it to DC to charge your batteries.

Note that the primary goal for most folks with generators is to run them as infrequently as possible. Generators tend to be noisy, smelly, and polluting. Fuel can be expensive, and you have to transport and store it.

However, system design is a balance when it comes to the generator versus size of the wind turbine. Your first impression may be to try to size the turbine large enough to never need a generator, but this is rarely practical. Such a system would likely be prohibitively expensive and overly large, and in most areas you don't make any money on surplus energy beyond what you use (or a small percentage over that). So you'd just be giving away some of your power to the man. Even then, the biggest turbines aren't any good if there's no wind.

So if you are off-grid, you'll need to have a generator. We'd suggest looking for one that is quiet and produces cleaner emissions. Even some folks with grid-connected wind systems with battery backups opt to buy a generator to have as an alternate power source if the grid is down and the batteries get depleted. It's another expense for something you may rarely use, unless you live in an area with spotty grid service. Much of the developing world is, in fact, like that, and we are certainly aware of places where the grid goes down an average of a few times a

Figure 10-23 Diesel generators are common backup power supplies for small wind systems, which can't produce energy when the breezes aren't blowing. *Ken Olson/DOE/NREL.*

week, often for days at a time. If that applies to you, a generator can be an important piece of a functioning system.

Summary

In this chapter, we learned that there are a number of other components required for a fully functioning wind system, beyond the turbine and the tower (Figure 10-24). These include a charge controller and batteries (if you want them), an inverter, rectifiers, disconnects, breaker panels, surge arrestors, grid-interconnection equipment (if there is a grid), wire, and system meters. Depending on your setup, many of these devices can be very important.

In some cases, you won't have much choice, because you will be getting the components packaged with the turbine. If you are building your entire system from scratch, you will have to make, improvise, or buy these parts. In most cases, these are not areas to skimp on. You don't want to start an electrical fire because you saved a few bucks on some cheap wire.

If this chapter seems a little overwhelming or you are feeling intimidated by electrical devices, call in professional help. Hire a reputable installer with experience in the field. Even certified electricians sometimes shy away from wind energy systems if they haven't worked on them before, because they may not be familiar with all the moving parts.

Getting everything wired up properly takes some careful attention to detail, but it also isn't as difficult as you might think. Literally millions of wind turbines have been making electricity for more than 100 years, and the safety record is extremely good.

FIGURE 10-24   Inverters, charge controllers, and a battery bank for a system that powers a portable reverse-osmosis water filtration plant. The unit was sent to the Middle East for use in water-stressed parts of Jordan, Israel, and the Palestinian Territories. *Warren Gretz/DOE/NREL.*

CHAPTER 11

Installation, Maintenance, and Troubleshooting

He caused the east wind to blow in the heavens, and by His power He led out the south wind.

–Psalms 78:26

Overview

- How much work is involved in installing a wind system?
- How are foundations and guy wires installed?
- What kind of maintenance is required?
- What are solutions to common problems?

Installing a wind turbine is a big job that can't be accomplished alone. At a minimum, you need some friends to pitch in, and better yet, mentors who have experience with small wind systems. You may also need a crane and earth-moving equipment, depending on the size of your system and the details of your site.

Once your system is properly installed and working, you can't just forget about it, like you might a driveway or a new roof. Like a swimming pool or a garden, a wind energy system requires regular maintenance and monitoring. Wind energy requires more hands-on involvement than solar energy, and in general there are more things that can go wrong.

Still, with regular maintenance and checkups, a properly designed, sited, and installed wind system should provide years of relatively reliable service (Figure 11-1). Of course, life is complicated, and "things" always happen. But in this chapter, we provide some solutions for some of the most common issues, and show you how to begin troubleshooting so you can solve your own problems, and head off others.

FIGURE 11-1   Climbing a wind turbine tower is serious business, though it can also be fun, as guests learn at Rosalie Bay Resort in Dominica. *Rosalie Bay Resort.*

Installing a Tower

So you got a great deal on a turbine package. Congratulations! But as you know, it's not much good until you raise the rotor up into the strong winds, where it can prove its worth. But that "stairway to heaven" is going to cost you. A tower typically ranges from one-third to one-half the total cost of the entire system. That is one of the reasons why people often make the mistake of installing too short a tower, but if you skimp on that, you only cheat yourself.

So let's get started in getting your turbine off the ground!

First, there are five things you need to keep in mind about towers:

1. The turbine's size and capacity should dictate the minimum tower height to perform well.
2. The higher the turbine, the more wind velocity, and, therefore, potential revenue.
3. The higher the turbine, the higher the up-front cost, although the payback period will often be shorter.
4. *Never* underestimate the power of the wind when installing a tower.
5. Towers do present a barrier to maintaining and repairing a turbine due to height, so you must have a plan in place.

We covered the first three of those points in detail in the first five chapters of this book and in Chapter 8, and we'll cover the remaining two here.

Safety First

 Hazard *The fun stuff is really getting started. But before you begin working on or around wind turbines and their towers, we need to remind you of several general hazards. No list can cover every potential problem or contingency, but major risks from wind systems include:*

- *A fall from a tower*
- *Being hit by something or someone falling from a tower*
- *Being struck from a whipping guy wire during a mast collapse*
- *Being electrocuted while touching the tower or guy wires from faulty wiring or during an electrical storm*

It is essential that you take safety seriously, and follow all guidelines from your manufacturer, dealer, and local code. We also put together this general safety protocol in Table 11-1.

TABLE 11-1 Safety Guidelines

Always	Never
· Install towers at least two lengths away from utility lines or occupied buildings*	· Work on the tower alone
· Install engineered towers only	· Drop or throw anything from the tower
· Work with trained installer(s)	· Stand or work directly below someone who is working on the tower
· Wear a safety harness, tool belt, and antifalling climbing cables and safety device	· Work on the tower during bad weather, impending bad weather, high winds, or after ice storms
· Wear OSHA-approved hardhats if you are in the vicinity of installation or maintenance	
· Use a hoistable tool bucket with a closure	
· Clear work area of unnecessary vehicles, equipment, and persons	
· Shut off wind turbine and secure alternator before hoisting into the wind or performing maintenance	

*During a mast collapse, it is possible for guy wires to whip around, so guyed towers in particular should be located at least two tower-lengths away from an occupied building and power lines.[1]

Installing a Small Wind System

We discussed the three main types of towers in some detail in Chapter 5. However, we wanted to briefly summarize the differences here (Table 11-2), because the type of tower you choose will affect your installation procedure. We put together the following quick guide, modified from the article "Wind Generator Tower Basics" by Ian Woofenden, from the February/March 2005 issue of *Home Power* magazine.

TABLE 11-2 Choose Your Tower

Tower Type	Advantages	Disadvantages
Tilt-up (lattice and tubular)	No climbing ever (can't climb)	Large base footprint
	Easy on-ground maintenance	Four sets of guy wires
	Medium cost	Requires relatively level site
	Pipe available locally	May not look as tidy or sleek
		Takes longer to assemble
		Tends to be on shorter side (below 100 feet)
Fixed guyed (lattice and tubular)	Modest base footprint	Three sets of guy wires
	Lowest cost	Must climb
	Uneven sites okay	Crane cost (if required)
		May tempt kids or others to climb (rare)
Freestanding (Lattice and monopole)	Smallest base footprint	Highest cost
	No guy wires	Must climb
	Uneven sites okay	Cost of crane installation
	Safest installation (crane)	
	Sleekest look	May tempt kids or others to climb (rare)
		Most embodied energy (concrete, steel)

To map out what type of tower you want, you will need to have a good understanding of your site. For one thing, you are going to need to know how level it is. Tilt-up towers cannot be installed on sharp inclines, though they can be placed to take advantage of gentle slopes. You also can't use a tilt-up tower if you don't have enough room to lower the turbine down or to accommodate the folding structure. If your site is quite space constrained, you may not have room for guy wires, either, and a freestanding tower may be your only option.

In this book, you may have noticed that we spend most of our ink on tall towers, for the reasons we have outlined. However, for folks who have more modest energy needs, or who are approaching small wind as a hobby or demonstration project, it's true that they may have some smaller-scale options. It is possible to make a tower for a micro-turbine out of a felled tree, for example, or a flagpole. As we mentioned in Chapter 10, Michael Davis built his out of steel electrical conduit. The main drawbacks with these types of towers are limited height and limited strength, so it restricts you to a small system that is only going to be able to produce a small amount of energy.

Regardless of what you decide, you are also going to need to know a bit about the ground. If you have very rocky soil, you may need bigger excavation equipment. So let's examine tower foundations.

The Foundation and Anchors

If you have ever tried to install a picket fence, a pole for a birdfeeder, or even a volleyball net, you probably have a feeling for how critical a good foundation is. For something as large and complex as a wind system, you can't just stick one end in the ground and hope for the best. Your foundation needs to be well engineered for the specific tower and weight of your turbine. Excavating and pouring a foundation is not easy, and we strongly recommend that you get professional help (Figure 11-2). At a minimum, you should follow the guidelines from your manufacturer.

Most foundations are poured concrete, which costs money and is very energy- and greenhouse gas–intensive, so you want to avoid mistakes and do-overs. If your soil is very wet, you may need more robust anchors than in dry soil so things don't slip out.

For a tilt-up tower, you are going to need five holes, one for the center tower and four anchors around the sides, at 90-degree angles. Seen from above, the footprint of a tilt-up tower looks like a kite, because the tower extends farther out over one of the guy anchors when it is tilted down.

Figure 11-2   High school science teacher Jim Jones digs the foundation for a Skystream 3.7 wind turbine by hand. Burlington High School is one of six in Colorado that received a grant for a small turbine in 2009. *Michael Kostrzewa/DOE/NREL.*

Tilt-up towers are carefully designed for their size and load they can carry, and you want to be sure to follow manufacturer guidelines, as well as any engineers that may be on site. One tip from the pros is to position the two side anchors a few inches closer to the lowering side of the tower so the guy wires get looser upon tilting down and tighter upon tilting up.

In both tilt-up and fixed guyed towers, the *guy radius*—the distance between the tower and the guy anchor—must be worked out through engineering, though it is generally between 50 and 75 percent of the tower height. In towers with guy wires, the foundation of the central tower itself is relatively small, and usually smaller than the anchor foundations. In a fixed guyed tower, the anchors are placed at 120 degrees from each other, at an equal distance away from the center.

A freestanding tower is easiest to site, because the only support it needs is right below the tower. As a result, this type of system is going to have the deepest, biggest foundation. The foundation may be a single block of concrete, or it may be an underground slab, on top of which rests a pedestal. Figure out where you want it, then mark it out with stakes and strings. Double-check that it is a good distance away from obstacles.

When it's time to dig your hole(s), unless you have a small system, you are most likely going to have to hire a backhoe. Try to find an operator who will make as clean an excavation as possible. As we mentioned in the previous chapter, if you can avoid using forms for pouring concrete, you will get a snugger fit that dissipates any lightning better. You should also make the bottom of the hole wider than the top, so the surrounding soil helps press the foundation in place. Set the concrete foundation below the frost line.

It's also a good idea to get all your digging done at once, so you should go ahead and dig trenches for wire, both transmission and grounding. Try to minimize disturbances to the natural environment by keeping the excavation zones as tidy as possible.

Once your hole is ready, you'll want to lay the rebar in a grid, as specified by the engineers (either in your employ or from the manufacturer and/or installer). Guy anchors, which are usually made of galvanized steel, also have to be positioned. Orient them so they are 90 degrees from their attachment points on the tower. In all cases, make sure no metal will touch any soil, as that can promote corrosion. It's also a good idea to use turnbuckles that are complete loops, not hooks, when connecting the guy wires.

When everything is firmly in place, carefully pour the concrete around the reinforcing bars, taking care not to shift them. You will likely want to use a concrete vibrator to make sure the thick mixture settles in every cranny. It's important to use the proper grade of concrete, so follow your instructions closely. When that's done, smooth out the top of the pours to make a nice even look. After the concrete sets, pile the soil back over the holes (Figure 11-3).

FIGURE 11-3   In 2003, the Navajo Tribal Utility Authority installed 44 small wind/PV hybrid power systems in Arizona, New Mexico, and Utah. *Sandia National Laboratories/DOE/ NREL.*

Remember that wind systems can experience strong lateral forces, and it's critical to get the foundation right, because improperly secured towers do fall down from time to time. It has happened to several small wind owners. Large utility-sized wind turbines have also fallen over in Oklahoma, North Dakota, and upstate New York. On December 27, 2009, a 329-foot wind turbine, from base to blade tip, collapsed at the Fenner wind farm in Madison County, New York.

Lifting the Tower (and Turbine)

When you are ready to hoist the tower, make sure you are completely familiar with all the steps, parts, and tools needed before you begin. This is not a job for one person, either, so make sure you have a good team assembled. Lay out all the pieces, and join anything that you can together while everything is on the ground. This includes any guy wires.

Remember that guy wires are essential structural elements that hold the tower up. If they fail, there's a good chance the tower will come crashing down. So pay attention to them and make sure you get them right.

Many tower kits come with precut guy wires. If that doesn't work for your site, because you have uneven ground, for instance, or if you have to size your own from the onset, you can use the trusty old Pythagorean theorem: tower height squared plus the guy radius squared equals the guy wire length squared. (Who said you would never use high school math again?)

Guy wires are most commonly made out of aircraft cable or something called *guy strand*, which is what utilities usually use for holding up their poles. Aircraft cable is more flexible and is widely available, since it has many uses, from lighting to towing. Both these metal products are tough, so you'll need good cable cutters, a grinder, or a torch for the job. Please don't try to use your mom's scissors.

Hazard   *Do always use thick gloves and eye protection when working with guy wire, because it can whip around or send off sharp shavings.*

In order to connect the guy wires to the anchors, you have a few options. You may order guy wires with ends that are pre-pressed into eyes, called *swaged ends*. Or, you can get *guy grips*, preformed eyes that grab onto cables. Also common are *cable clamps*, which have a U-bolt, a saddle that goes over the bolt, and two nuts for the bolt. If you turn a cable back on itself, the clamp holds the two together. You should use a minimum of three clamps per turn, and set them so they are oriented the same way.

Always put the U-bolt side on the dead end of the cable (a free tail) and the saddle on the live end (the part that goes to a load). Never do it the other way around. Remember the simple memory device: "Never saddle a dead horse."

Whenever you turn a cable on itself, always do it over a *thimble*, a teardrop-shaped metal loop that keeps it from twisting. Also, Ian Woofenden suggests that you always unroll guy wire off the spool, instead of pulling it, so you don't introduce a twist.

Once you have everything put together, with guy wires connected, you have a few options to actually raise your wind machine (Figure 11-4). Of course, if you decided to go with a tilt-up tower, you will only need to tilt it up once you have the generator secured. That process typically requires a power winch, and it can be a bit tricky to master at first. Your manufacturer can help you.

Otherwise, your options include the following:

- **Hire a crane**   This option is going to cost some money up-front, but it is the easiest and safest choice. A good crane operator is essential. You also need to have enough open space around the tower for the crane to maneuver and good enough road access. Theoretically, you could use a helicopter, if you had tens of thousands of dollars to spend and a willing pilot.
- **Use a gin pole**   If you and your team are left to your own devices, without heavy equipment, and your tower doesn't tilt, your best option is probably going to be a *gin pole*, which is a pole you slide up the tower to push things

FIGURE 11-4   Volunteers from H&H Utility Contractors, Inc., hoist a Skystream 3.7 into place at an Idaho middle school as part of NREL's Wind for Schools program. *Stephanie Lively/Boise State University/DOE/NREL.*

up. You can use a gin pole to lift up higher sections of the tower, which you can then fasten, or use it to lift the rotor and other pieces.

Although the metal of a lattice tower looks heavy, each ten-foot section typically weighs less than 80 pounds. That's still difficult to handle when you are high up off the ground, but a dedicated, careful team with proper climbing gear can manage it. A good wind generator can easily weigh 75 to 250 pounds or more, so you usually can't just carry it up the tower on your back. For that, you need pulleys and a crew who can hoist it up.

You are better off putting as much of the generator together as you can on the ground, where you have much more maneuverability. One reason why one- or two-bladed turbines are convenient is because they can be totally assembled on the ground. If you can't do that, you might want to try a dry run of assembling the turbine on the ground, then detach what you absolutely have to in order to get it up on the tower. In some cases, you may be able to get the whole turbine together, even if you have to do it on top of a ladder. You may be able to lift the completed machine up, using tag lines that secure each blade and are manned by crew on the ground. That way, the blades won't get damaged by smacking into the tower.

Once the generator is on top, you'll have to bolt it down and make the electrical connections. If you have a lattice tower, it will probably actually rest on what's called a *stub tower*, a short pole that emerges from the top of the lattice. Of course, you'll need to follow your manufacturer's precise instructions.

You can see why installation normally takes several days.

Safe Climbing

Of course, to bolt a turbine to a tower, you have to get up there yourself. Climbing is an inherently dangerous activity, although many people do it every day, including arborists, antenna technicians, and firemen (a former job of Kevin's). As you probably know, many people also enjoy recreational climbing, which employs similar skills and gear.

We strongly recommend that you get hands-on training and practice before you attempt to climb a wind energy tower (Figure 11-5). There are many books on climbing equipment and technique, so we will only give a very brief overview here.

It is essential to have good climbing gear, which must be well maintained and inspected before every use. You will need a harness that fits well, preferably one that gives full body support, even if you are knocked unconscious. That should be paired with a fall arrest device over your chest, which connects to a secure line at all times. That way, if you slip or get knocked over, you will be held in place. Make sure to always be connected to the tower at all times. No exceptions.

Of course, you'll also need good climbing-grade rope. And you'll want extra D-rings, carabiners, and lines to hold or hoist up tools and parts. Closeable bags are handy, as are pulleys.

You need a *rescue line* that is at least twice as long as the tower is tall, in case you need to quickly rappel down. You may also opt for lines with steel cores, to protect against accidental cutting by sharp tools.

Good climbers know to use their legs to push them up where they need to go. They rely on their arms as little as possible, primarily for balance, not for pulling. Otherwise, you will tire out quickly.

FIGURE 11-5   When climbing a turbine tower, you must be connected at all times. You should wear a full-body harness and have a fall arrest device. *Rosalie Bay Resort.*

Even so, take it slow and easy. And don't go it alone. Have help on the ground watching your lines and lifting stuff up to you.

Making the Connections

You want to run your transmission wires down the tower through plastic or metal conduit, often one to two inches thick, in order to protect them. If you have to install the tower in sections, you can do the same with conduit, making watertight connections between the sections with slip-joints.

Conduit should be secured in many places. You'll also need to have some *kellums grips*, loops that lessen the strain on the wires so they don't snap.

Once you have the transmission wire all set, you'll have to hook it up to the rest of your system, including the inverter, meter, breakers, main panel, and so on (Chapter 10). If you have batteries, you'll need to wire them in. Do that part last, although make sure that your turbine is safely braked at all times.

 Hazard *Take extra care with the batteries because they are "always on," and carry the added risk of chemical burn. Always use eye and hand protection around them, and don't mix up the terminals.*

Balance of system components are fairly complicated electrical devices, though it's also true that more and more turbines are being sold as packages. Many components are also being designed to be more "plug and play," so you may only have a few wires to hook up.

Once your components are wired in, they will have to be set up. Charge controllers and inverters usually require some amount of programming, similar to a programmable thermostat.

If you get stuck, call the manufacturer or an electrician who has experience with renewable energy. Those who normally just work on standard home wiring may not be that helpful.

Once your system is all set up and you double-checked all the connections, remove the brakes, and let her spin!

Especially in the first days and weeks, keep a close eye on the system (Figure 11-6). You may have to make adjustments in float charge levels, temperature settings, or other options. Watch the winds, watch your power consumption, and verify that all systems are firing on all cylinders, to use an expression from the fossil-fuel era.

FIGURE 11-6   Make sure the yaw bearing (upper part) and slip rings (lower) are tight. Note the grease dripping from the former to the latter. *Jeroen van Dam/DOE/NREL.*

INTERCONNECT

Two Brothers-in-Law in the Turbine Business

On Thanksgiving Day a few years ago, two brothers-in-law in rural Kansas got to talking about small wind power, and they soon decided to pool their talents and start a cottage industry, which they called Prairie Turbines. According to their website (www.prairieturbines .com), Timothy D. McCall and Alan E. Plunkett wanted to "create a simple, cost-effective homebuilt wind turbine fully capable of completely offsetting the average electric bill. It had to be simple, reliable, low cost, and easy to build and maintain."

The result was the Breezy 5.5, a batteryless, grid-connected design that the designers say produces 5.5 kW with high wind speeds of 21 to 23 mph (Figure 11-7). "With Alan's unique rotor design we've had no problems handling winds of 70 mph," McCall wrote. Each of the four wooden blades on the Breezy 5.5 is nine feet long. On the larger 10 kW model, each blade is 11.5 feet long.

In a phone interview, McCall told us he and his brother-in-law have sold about 3,000 plans for their turbine design, as well as 250 to 300 microcontrollers. "This is a design a person can build entirely from scratch, as long as they have a controller," McCall said. "It is assembled from readily available, off-the-shelf parts," he added. "We are a niche market in what I like to refer to as the 'sensible price range,' way less expensive than package turbines."

McCall and Plunkett provide phone support to those who buy their plans and kits, and often end up spending time explaining the electronics or doing some of the wiring. McCall estimates that 60 to 70 Breezy turbines have been put up in several states and

Figure 11-7   Through their cottage company Prairie Turbines, two brothers-in-law from Kansas sell unique small wind systems, like this Breezy 5.5 installed in Maine. *Prairie Turbines.*

Canadian provinces, Europe, Australia, Brazil, Nicaragua, and elsewhere. "They are mostly on ranches and farms; these are not recommended close to developed communities, because it's too big," McCall told us. He said he recommends them only for people with at least five acres, and only for those with class 3 or higher winds.

McCall told us his Breezy 5.5 produces 1,000 kWh a month in his class 3 wind, on a 60-foot tower in open terrain. Interestingly, these turbines don't use inverters. The controller and connection to the grid keeps the power on frequency and in phase with line level. Generated power can be used locally or fed back into the grid for credit.

According to McCall, customers who order plans can build their own Breezy 5.5 with locally obtainable parts for $3,500 to $4,500, a significant chunk of which goes toward the motor, a part that is normally used in overhead cranes for factories. McCall claims a reasonable payback period of two to three years.

McCall claims installation is a lot of fun, although he admits some folks sometimes need a little help. "We've had guys with towers that fell over, and a guy in central Texas who put the blades on backwards, and loosely coupled the wires, but he learned a lot along the way, and he ended up having a complete success story," he said.

Before starting Prairie Turbines, Plunkett had some experience with 25 kW turbines, and McCall had long been working with electronics for consumer and industrial applications. McCall still works for Spirit AeroSystems, although Plunkett recently quit his day job to spend more time on their business. "I knew this is what I wanted because it's so simple," McCall told us.

Prairie Turbines proves that low-cost kit and DIY turbines can be successful projects, in the right settings, and with the right know-how and support. They're definitely more challenging than putting together IKEA furniture, but they aren't exactly rocket science, either.

Maintenance

Currently, there are no maintenance-free wind turbines (Figure 11-8). The rotor spins away in a pretty severe environment, which means it encounters dirt, moisture, high winds, and wear and tear. And there are no quick-lube franchises for

FIGURE 11-8   The safest way to work on a turbine is with a bucket truck, if you can get the height. For a wind turbine, annual maintenance, at a minimum, is required. *Warren Gretz/DOE/NREL.*

wind turbines just yet (here's your chance to start one). So you will either have to strap in and do some maintenance work yourself or hire out help, or you will soon end up with an expensive lawn ornament.

It's also true that certain parts need more scheduled maintenance than others. Off-grid battery banks typically require more attention than battery banks with grid-tied systems. Inverters and other electronics are fairly maintenance-free. If the blades show wear and tear (e.g., flaking paint), their maintenance would need to be more frequent.

The following is a general guide to small wind system maintenance. Note that you should also be keeping a close eye on your energy production at all times. If that tapers off while the winds stay strong, that should be your first indication that something is amiss.

- **Frequency of check-ups** Every 6 to 12 months. Select a windless day before storm season.
- **General guidelines**:
  - Follow maintenance instructions in your user's manual.
  - Check with your manufacturer for any maintenance updates.
  - Attend a maintenance and repair workshop, or during certification training pose maintenance questions.
  - Check wind community websites.
- **Initial assessment**:
  - Look for unusual behavior, such as sticky, chipped, or discolored blades or unusual vibrations.
  - Listen for unusual sounds, such as banging, grinding (bad alternator or bearings), or clunking.

A good visual and auditory inspection can be like kicking the tires of a car. If you notice unusual vibrations, it might indicate that the rotor blades have become unbalanced. A minor increase in blade noise, or chipped paint, could suggest that it's time to take them down, clean them up, and refinish them. If you hear whining or grinding, it could be worn bearings, a frozen break, or failing alternator.

Now that you have a general idea of maintenance, let's take a look at a detailed checklist.

Tower and Turbine Maintenance Checklist

(Print 2 copies and place in your [1] hardware bag and [2] pants pocket)

TURN POWER OFF AND LOCK BLADES

1. Verify Tools and Supplies

- ❏ closeable tool bag
- ❏ torque wrench
- ❏ sandpaper
- ❏ cable cutter
- ❏ Loctite
- ❏ replacement bolts, clamps, and other connectors

- ❏ load-bearing carabiners
- ❏ protective gloves
- ❏ naval jelly
- ❏ assorted brushes
- ❏ grease gun

- ❏ wire brush
- ❏ safety helmet
- ❏ liquid wrench
- ❏ galvanizing paint can
- ❏ highly absorbent towel
- ❏ measuring tape

Climbable tower:
- ❏ full-body safety harness with lanyards
- ❏ pulley rope
- ❏ cell phone

- ❏ treestand (optional)

- ❏ pulley
- ❏ step bolts

Electrical tools:
- ❏ electrical tape
- ❏ wire caps/unions
- ❏ rechargeable drill

- ❏ voltmeter
- ❏ wire cutters
- ❏ marking pen

- ❏ screwdrivers
- ❏ electrical wire

2. Write torque specs of all bolts from manual (or best practices)

tower stub bolts _____

tower bolts _____

tail vane bolts _____

rotor bolts _____

3. Ground Assessment

- ❏ Listen for unusual sounds (clicking, grinding, scraping, vibrating, rubbing, loud whining) before, during, and after system shutdown.
- ❏ Inspect for rust, cracks, stress on ALL bolts, turnbuckles, guy grips, and cable clamps to anchors, guy wires, and tilt hinge.
- ❏ Inspect guy wires and attachments to anchors for wear and fraying.
- ❏ Tighten all bolts in ground assembly with torque wrench.
- ❏ Check to see ground rods are securely attached to guy cables.
- ❏ Inspect guy wire tension for uniform tension, free of slack, not overly taut.
- ❏ Inspect anchors for signs of movement (lifting, tilting).
- ❏ (Tilt-up) Inspect gin pole to tower for rust, cracks, or distortion.

4. Tower Inspection

- ❏ Inspect for rust, cracks, and stress of ALL bolts, welds, wires, turnbuckles, guy grips, and cable clamps.

❏ Tighten all bolts with torque wrench.

❏ All cracked, frayed, and severely rusted hardware needs to be replaced immediately.

❏ Any crack or kink in structural steel requires that the guyed tower be lowered and repaired on the ground.

❏ (Climbable tower) Consider installing one or more rust-resistant treestands so you can rest on your way up.

❏ Any anchor movement suggests soil problem and inadequate reinforcement.

5. Check the Wind Generator

If you have access to a crane or bucket truck, it is the safest route. Otherwise, you will have to climb.

Read through the entire inspection and maintenance checklist before ascending the tower to be certain you have all tools and parts you need. Add to the list as needed.

Carefully study diagrams before you ascend the tower.

Make sure all replacements are completed before descending or tilting up the tower.

❏ Check all bolts and tighten with torque wrench.

❏ Check tail vane bolts and any other fasteners.

❏ Tighten all hardware to factory specifications with torque wrench.

❏ Grease any fittings as specified in the manual.

❏ Inspect blades for flaking paint, wear, cracks, and crud.

❏ Check condition of leading-edge tape (if any), and reapply if necessary.

❏ Check tips and trailing edge for damage.

❏ Measure the distance from the stub tower to the tip of each blade. Contact manufacturer or check manual for tolerance and shim blades at the root if necessary.

❏ Check blade balance and tracking.

❏ Inspect all wiring connections and strain relief.

❏ Inspect mechanical parts of the turbine, such as the furling system, yaw bearings, and main bearings.

❏ Inspect anemometer, if present.

❏ Open the nacelle, and see if there is any moisture, rust, burned-out sections, or spider webs.

❏ Check to see that no cables are chaffing.

❏ Check alternator for cracks or damage.

❏ Check for obvious signs of corrosion.

❏ Check the slip rings for damage or for oil leaks from yaw bearings.

❏ Check the shutdown mechanisms.

❏ Inspect wire running down the tower.

❏ Worn blades may need to be removed, resanded, and repainted/refinished. Exposed blade material may absorb moisture and result in an unbalanced blade, severely stressing the wind generator.

❏ Any cracked or severely rusted metal requires replacement before you can turn the system on again.

6. Check the Batteries

Frequency: Every month or as needed

Equipment: Baking soda, distilled water, hydrometer (for measuring battery liquid), protective gloves, protective eyeglasses, battery nob brush, screwdriver/wrenches

Most important: Keep them clean and tight; ensure that the disconnect is turned to OFF.

Distilled water can be used to wash your eyes and hands if you get splashed with battery acid.

❏ Check battery fluid and top off with distilled water as necessary.
❏ Check battery connections for tightness and corrosion.
❏ Check specific gravity (battery fluid density) with hydrometer. See battery manual.
❏ Clean battery tops (with water only, no solvents).
❏ Ideally, you also may want to fully recharge batteries by disengaging their use (particularly during a windy day and when you don't need the energy).
❏ Verify the voltage range is within acceptable levels.
❏ Perform an equalization charge.

7. Perform Final Inspections

❏ Verify monitoring equipment.
❏ When finished, turn power back ON and remove brakes.

Troubleshooting

Many things in life can go wrong, and wind systems are not immune. There are a number of things that could explain your system's flagging performance or that noise pollution violation notice in your mailbox.

As we mentioned in Chapter 6, you are better off dealing with service contracts and warranties first and insurance last. Remember that dealers and manufacturers are there to help.

Don't do anything dangerous or that you aren't sure about. It's much better that you live to fight another day (Figure 11-9).

We put together this brief guide to help you troubleshoot some common problems.

FIGURE 11-9   When climbing a tower, take it slow and steady, and always have a backup crew on the ground. *Jerry Bianchi/DOE/NREL.*

Troubleshooting for a Small Wind System

⚎ system design   ◁ aerodynamic   🔧 mechanical   ⚡ electrical   🕐 controller

Symptom	Problems		Solution
No Output While Turbine Runs Slowly or Stopped	⚎	Insufficient wind	Relocate
	◁	Blades fitted poorly, worn or damaged; ice buildup	Inspect blades; verify that convex side of blade is on downwind side of turbine
	🔧	Main bearing failure; dislodged magnet jamming alternator; alternator iced up; yaw bearing failure; tail stuck in furled position	Inspect and replace failed main or yaw bearing, magnet, or tail; de-ice alternator and reseal nacelle
	⚡	Electrical short-circuit	Inspect brake switch; if okay, inspect for short-circuit and repair
	🕐	Faulty or blown controller dumping load	Inspect and replace controller

(continued)

No Output While Turbine Runs Fast		Insufficient wind	Relocate
		Disconnected from the load	Apply the short-circuit brake at base of tower. If brakes work, investigate wiring and make sure fuses, circuit breakers, and connections are conducting properly
Low Energy Production		Incompatible with wind conditions	Relocate
		Blades fitted poorly, worn or damaged, or ice build-up	Inspect blades; verify that convex side of blade is on downwind side of turbine
		Worn bearings, jammed yaw, rusted or broken bushings or springs, shaking machine	Inspect parts and replace
		Blown diode in rectifier; bad connection in wire	Inspect and replace
		Turbine furls too early	Refer to mechanical problems

Summary

Installing a small wind energy system can be a lot of fun, but it is hard work (Figure 11-10). There are a number of inherent hazards, including risk of falls, falling objects, and electrical shock. We strongly recommend that you get help with installation from a dealer, installer, or experienced friends or mentors.

It is possible to do all or most of the work yourself, but it requires diverse skills, including pouring concrete, wiring, and working with a crane or gin pole. If you have to climb a tower, make sure you have the right equipment and enough practice to stay safe.

Your system will also require regular maintenance, at least once a year, and likely more often. You will need to inspect for rust and wear, tighten bolts, and apply grease.

If you run into other trouble, we provided a checklist of possible solutions. Good luck!

Figure 11-10   WindAid S.A.C. leads service projects in Peru that help bring renewable energy to those in need. *WindAid S.A.C.*

CHAPTER **12**

Future Developments in Wind Power

There is nothing new under the sun, but the sun begets the wind, whose unyielding, unpredictable force unearths what was once hidden.

—KEVIN SHEA

Overview

- Can buildings accelerate wind in helpful ways?
- Can we design better turbines?
- Will super-small turbines work?
- What future wind energy technology can we expect?

Wind technology is constantly evolving, with new frontiers of research being explored in multimillion-dollar industrial and university labs as well as in the garages of tinkerers. This chapter takes a look at some of these emerging designs, with an eye toward what may be powering the future, whether that's rotating blimps or carouseling kites flying thousands of feet up in the air.

Although research is ongoing and field tests are limited, we'll try to give some insight into the feasibility and limitations of such cutting-edge systems when it comes to small-scale applications, the subject of this book. Some of these new designs show definite potential, while others call for the invention of smart wallets that automatically lock up when sensing the presence of unfeasible technology.

One word of caution: although we enjoy reading about novel designs as much as anyone, we would advise against getting out your credit card and ordering your very own experimental wind turbine. Most of the technology in this chapter is at best unproven, if not totally speculative. If you have the time, money, and desire to serve as a test case, we salute you, and may the winds of invention blow great success in your direction. But if you are like the rest of us, on a budget and interested in producing some clean kilowatts today, we strongly recommend that you stick with a tried and true design that's backed up with years of real performance data, manufacturer warranties, and a support network.

That exotic machine with 28 scissoring, helium-filled blades may look awesome when you first nail it to your roof, but what happens when a storm breaks it in half, or the complicated electronics go on the fritz?

Architectural Wind: Can Buildings *Improve* Wind Conditions?

We placed this section on so-called *architectural wind* in the future tech chapter, because although a few examples do exist "in the field" and the idea has received considerable media attention, the concept is unproven, and likely needs more refinement if it is ever going to contribute to meaningful amounts of energy. The term architectural wind refers to the fact that moving air often accelerates as it rushes past large buildings, leading enterprising people to wonder if they could harvest this energy with well-placed small wind turbines.

Up until recently, the dominant player in architectural wind was the California-based company AeroVironment, which was founded in 1971 by famed inventor and aeronautical engineer Paul B. MacCready, Jr. MacCready is perhaps best known for co-creating the *Gossamer Condor*, a pioneering human-powered aircraft. His follow-up, the *Gossamer Albatross*, became the first human-powered airplane to cross the English Channel. MacCready helped NASA develop solar-powered airplanes, and he helped General Motors build the prize-winning Sunraycer solar car and the ill-fated EV-1 electric car (star of the documentary *Who Killed the Electric Car?*). MacCready also built a working replica of the pterosaur Quetzalcoatlus for the Smithsonian, which received considerable media attention.

Today, AeroVironment is a public company that makes unmanned surveillance aircraft (aka drones) and various advanced power systems, including chargers for electric cars. As of this writing, they are installing home chargers for the Nissan Leaf, with the first customer having been ubercyclist Lance Armstrong. AeroVironment also recently flew a prototype of the first airplane powered by hydrogen fuel cells, the *Global Observer*. According to Steve Gitlin, AeroVironment's vice president of marketing strategies and communications, the company's name refers to the goal of taking aerospace technologies and applying them to the environment.

In recent years, AeroVironment has also been in the business of measuring air quality. In addition to monitoring pollution, the company has helped wind farms evaluate their potential resource. Given this pedigree, it's perhaps not surprising that the company made the move into small wind turbines in 2004.

AeroVironment's sleek, five-bladed AVX 1000 (1 kW) turbines were specifically designed to take advantage of architectural wind (Figure 12-1), which the company

Figure 12-1 AeroVironment AVX 1000 (1 kW) wind turbines installed on the roof of a "green" building at the Brooklyn Navy Yard. These "architectural wind" turbines were discontinued after buyers reported low energy production. *Brian Clark Howard.*

claimed would "increase the turbines' electrical power generation by more than 50 percent compared to the power generation that would result from systems situated outside of the acceleration zone."

The small, compact turbines were also intended to provide a "visible, compelling, and architecturally enhancing statement of the building's commitment to renewable energy." Twenty high-profile installations were made, including at Boston's Logan Airport, on an office building at the revitalized Brooklyn Navy Yard, on top of The School of Sustainability at Arizona State University, and on the roof of a Kettle Foods potato chip factory in Beloit, Wisconsin. The small, aviation-inspired turbines are extremely quiet and relatively easy to install, thanks to their modular design.

Perhaps not surprisingly, the AVX turbines earned several design awards, as well as thousands of inches of coverage in print and online. "They were very positively received," Steve Gitlin told us in an in-person interview in New York, at the LEED Gold–certified Hearst Tower (home of The Daily Green). "People loved them," said Gitlin.

However, despite their buzz and sexy profile, there was a problem with the AVX 1000. It turned out that the installed turbines were producing only small amounts of energy for their owners. In a scathing presentation, wind installer, author, and consultant Mick Sagrillo estimated that the 20-turbine Logan installation would take thousands of years to pay for itself. Paul Gipe calculated that the dimensions of the turbines—a "width" of 72 inches (1.83 meters) and swept area of 2.6 square meters—should be expected to produce a rating of 500 watts, half the manufacturer's claim.

For his part, Gitlin told us, "They produce electricity, but not at a rate that is economically viable." He added, "People want to go green, but they also want to make electricity." Could that be a motto for this book?

Gitlin told us his company would have little trouble selling more of the turbines if they wanted to, but he said they recently made the conscious decision to put the product "on the back burner." Asked if there was a problem with the architectural wind effect itself, Gitlin responded, "There is acceleration, but not enough when you take into account the cost."

That cost was pretty substantial for the businesses that bought AVX 1000 turbines. An apartment building in the Bronx spent about $100,000 for a ten-turbine AVX 1000 installation on its roof. The idea was to generate power that could be used for the building's hallways, elevators, and common areas. However, the results have been disappointing, according to Les Bluestone, principal for the developer, Blue Sea Development.

"If the wind was blowing consistently enough it could feasibly do that, but 'if' is a big question," Bluestone told us via phone. "We've found that winds have not been as consistent as we would have liked them to be to make payback reasonable." Bluestone admitted that the decision to install wind turbines had been "an experiment," and he added, "It's a little too early to tell, but I would guess we would not rush into it again." Bluestone added that they had known their wind resource was "a bit iffy," from looking at NOAA data, although he said they didn't do any long-term data collection themselves.

"We're in a very urban setting, with only a little bit of clearance, and there are all sorts of things that work against us," Bluestone said. "This particular product takes advantage of updrafts up buildings, so we thought this was our best shot rather than trying to use a vertical axis or conventional turbine. But for our application there's still not enough wind."

When asked what the community thinks of the turbines, Bluestone said they're very popular. "When they're spinning everyone is in awe, and when they're not some people ask if they're fans to cool the neighborhood. People like them, and I haven't heard any negatives," said Bluestone.

"If we were going to [install the turbines] again we would have to have much more data before we would make that decision," Bluestone concluded. He added that he was frustrated that AeroVironment is no longer manufacturing or supporting the turbines, and said he didn't know what they'd do if they have a problem.

The disappointing results of the AVX 1000 in the field don't necessarily prove that architectural wind will never work, but they do underscore the challenges of trying to harvest wind energy in an urban environment. As you should know now, wind speed is critical to the success of any wind turbine, regardless of what it looks like or where it is placed. Although it's true that there can be an acceleration of wind around obstacles, like buildings or cliffs, it also tends to be true that the less flat the surface, the more turbulence you have (which is bad), and the more likely winds coming from other directions will get deflected.

A number of designers have recently proposed other turbine systems that purport to channel wind around buildings, sometimes with tunnels, flues, or other structures. It's possible some configuration could work, but the site's wind resource would have to be really strong to compensate for the expense of additional materials. Most likely, the owner would be better served by going higher, into unimpeded winds, rather than trying to wrangle fickle winds around buildings.

Others have proposed adding small turbines between large generators on wind farms, or lower on their massive towers. Although the idea of wringing more energy out of a utility turbine may sound appealing, it's also true that wind farms take major capital investments, and few owners are going to want to risk damaging or hampering their main rotors by tinkering with unproven schemes. Lots of testing will need to be done.

Building Integrated Wind Turbines

In the aftermath of September 11, a number of bold designs were suggested for a new building at Ground Zero, for what was then going to be called the Freedom Tower. One of the more memorable proposals included several wind turbines, hung many stories above the city in a space between adjacent parts of the skyscraper. To many bloggers, the symbolism wasn't subtle. This was a bold statement against our fossil fuel–reliant past and the dirty dealings with unstable regimes that it required. This was a symbol of hope for a cleaner, freer future.

Of course, it didn't take long for skeptics to point out that the turbines would provide only a fraction of the massive tower's energy needs. There were also concerns about cost, safety, and maintenance, especially since the guy who walked between the Twin Towers on a tightrope in the 1970s is getting on in years.

The plan for so-called *building-integrated wind* at Ground Zero was scrapped, but the concept has lived on, starting at the site of another world trade center. In 2008, the 50-story twin towers of the Bahrain World Trade Center went up in the Middle East. The towers are linked via three skybridges, each of which supports a 95-foot-diameter, 225 kW wind turbine (Figure 12-2). The turbines face north, where prevailing winds blow from the Persian Gulf. In an example of architectural wind, the sail-shaped towers are designed to funnel and accelerate the wind.

The building's designers hope the turbines will provide 11 to 15 percent of the towers' total power consumption, or approximately 1.1 to 1.3 GWh a year. That's roughly enough power to light 300 homes for a year. As of this writing, actual performance data is hard to come by.

FIGURE 12-2   The 50-story twin towers of the Bahrain World Trade Center boast three 95-foot-diameter, 225-kW wind turbines, oriented to catch breezes off the Persian Gulf. *Tim Miller/Wikimedia Commons.*

In London, a 43-story residential high-rise was recently built with three building-integrated turbines. The building is officially known as Strata SE1, though it is often called the Razor, since it looks like a giant Norelco, with the turbines on top resembling shaving blades (Figure 12-3). The nine-meter turbines are rated at 19 kW each and are hoped to produce 50 MWh of electricity per year, enough for the common areas (8 percent of total building energy). Though the tower was completed in 2010, the turbines are undergoing ongoing testing, so actual performance data isn't available.

Figure 12-3   London's Strata SE1, often called "the Razor," hosts three 19 kW integrated wind turbines, which are hoped to produce about 8 percent of building energy. *Skyscrapercity User: SE9/Wikimedia Commons.*

In the past year, a number of other building-integrated wind projects have been announced. The appeal is obvious, although the expense is considerable, and the results unproven. It's possible that we may one day see a turbine as part of every building, but that seems pretty unlikely, given the limitations of current technology and the poor wind quality of most neighborhoods.

Long-time green building writer Alex Wilson recently wrote, "I want to like building-integrated wind. There's a wonderful synergy in the idea of combining form and function by generating electricity with turbines that reach into the sky on the buildings they will help to power. But in most cases, at least with today's technology, it just doesn't make sense …"

Biomimicry: Wind Systems Inspired by Animals

Some new experimental wind turbines borrow directly from nature, in a design process now called *biomimicry*—literally mimicking life. Examples of biomimicry in action include buildings that use passive cooling based on the architecture of termite mounds, adhesives based on the natural secretions of mollusks, and coatings that naturally repel dirt based on the texture of the lotus flower.

When it comes to wind turbines, the Toronto-based company WhalePower is marketing blades modeled after the flippers of humpback whales (Figure 12-4). The idea is based on research done at Harvard and elsewhere on the bumps, called *tubercles*, on the leading edge of humpback flippers. The tubercles are thought to help reduce stall when gliding through the water, and therefore increase speed and

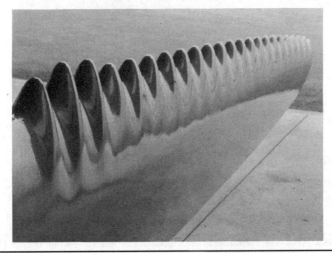

FIGURE 12-4A Toronto-based WhalePower is working on wind turbine blades with bumps, called tubercles, modeled after the flippers of humpback whales. *Joseph Subirana/ WhalePower.*

FIGURE 12-4B Tubercles are thought to increase speed and decrease stall. *Whit Welles/ Wikimedia Commons.*

make the giant mammals more agile. This discovery led researchers to suggest that tubercles could make airplanes easier to control and fans more efficient.

WhalePower claims adding tubercles to wind turbine blades would reduce noise, increase stability, and improve energy performance up to 20 percent overall. The company says the bumps work by channeling the airflow and reducing stalling, just like a whale flipper in water. Theoretically, this means turbine blades can have steeper attack angles, and therefore draw more power out of the wind. WhalePower says tests show the bumps double performance at wind speeds of 17 miles per hour, and also help turbines capture more energy out of lower-speed winds.

WhalePower says existing wind turbine blades can be retrofitted with tubercles, in a process the company says makes the hardware even sturdier. We asked a representative of the company on when we might expect the technology.

Danielle Dewar told us WhalePower is currently working on small fans (Figure 12-5), computer fans, and large and small wind turbines. "These products are not yet commercialized, but we hope that they will be over the next 18 months to two years," Dewar explained over e-mail.

So, in the near future, you may be able to buy a turbine inspired by a whale … or maybe even something smaller.

FIGURE 12-5   WhalePower's first commercial product, the Envira-North Altra Air Fan, also borrows from whale flippers. *Joseph Subirana/WhalePower.*

A recent paper from researchers at the California Institute of Technology argued that planners of future vertical axis wind farms might do well to mimic the schooling patterns of fish. Robert W. Whittlesey and colleagues pointed out that spacing horizontal axis wind turbines too close together results in decreased performance, while that effect is diminished with VAWTs.

Publishing in the peer-reviewed journal *Biosinspiration and Biomimetics*, Whittlesey's team concluded, "A geometric arrangement based on the configuration of shed vortices in the wake of schooling fish is shown to significantly increase the array performance coefficient based upon an array of 16 × 16 wind turbines. The results suggest increases in power output of over one order of magnitude for a given area of land as compared to HAWTs."

There aren't really any vertical axis wind farms around, for a number of reasons we covered earlier in this book. But the scientist's research suggests they may one day play a role, possibly even at a small scale.

Let's Go Fly a Wind Turbine

Since you already know that wind tends to get stronger and more consistent the higher up you go from the Earth's surface, it would seem like an obvious invitation to loft a turbine into the heavens. In fact, a lot of designers have taken a crack at the idea and Google has invested $10 million into research, though the practical challenges remain formidable, and there are few working examples.

Still, so-called *airborne wind energy* has its boosters. Joe Faust, editor of UpperWindPower.com, told us, "Trees and towers sit in the boundary zone where little wind occurs; above the boundary zone is a vast resource that is thick, steady, predictable, fast, accessible, cleaner, and minable without tall expensive towers; the power affected by the velocity-cube law combined with those aspects opens a game-changer for wind energy via traction, task, and direct-electricity generation. A new aviation is being born."

Of course, to harvest airborne wind energy, you probably have to get the power down to the ground, where it is useful. As you may know by now, wire isn't that cheap, especially when you are talking hundreds of feet. Sure, you could try to beam the power down via microwaves or lasers, but that technology is currently expensive. That may be an option in the future, however. There are obvious safety concerns with lifting electricity generators above our heads, and no one wants to see their investment come crashing down. Pilots may also have a few concerns. So there are some issues to work out, but supporters of airborne wind energy argue that the challenges are solvable, and that the benefits of reaching strong winds will eventually outweigh the challenges.

Cristina L. Archer has studied the potential of high winds in her role as an Assistant Professor in the Department of Geological and Environmental Sciences at California State University, Chico. She told us via email, "Wind speeds can easily double between 100 and 500 meters above the ground, which causes a growth in the available power by up to 8 times. I found that between 500 m and 2000 m there is not much growth (on average), but actually above 2,000 m you get rapid gradients of winds, with the jet streams being La Mecca of winds."

Archer told us she envisions a huge benefit in reaching the 500-meter height. "The same turbine would generate up to 8 times more there (the exact amount depends on the actual wind and air density, turbine design, and efficiencies, etc.)," she said. "Not to mention that it would be cheaper, lighter, and generally have a higher capacity factor due to the lower (although not insignificant) intermittency of winds." Archer is currently working on mapping jet stream currents around the world that appear between 300 and 2,000 meters above the ground.

It's certainly possible that new designs and advances in electronics and materials science will one day have us harvesting energy from floating turbines. Bryan Roberts, a professor of engineering at the University of Technology in Sydney, Australia, has proposed a helicopter-like turbine craft that would fly to 15,000 feet (4,600 m) altitude, where it would allegedly hover in place, buoyed by lift generated by action of the constant wind. It would be anchored to the ground by a very long cable.

Dutch ex-astronaut and physicist Wubbo Ockels has proposed a "Laddermill," a big loop of kites that generates power as one end of the kites lifts itself up, as if climbing an endless ladder in the sky. In Italy, inventor Massimo Ippolito is building a prototype of his Kite Gen idea, which he says was inspired by watching kitesurfing. Big airfoils will be lofted into the sky, and tethered to a vertical shaft, which will generate energy as it rotates (Figure 12-6). A computer will control the airfoils to keep them optimally turned into the high winds.

Another Italian company, Twind Technology, promotes a concept of two tethered balloons, each with an inflatable sail. They would take turns being pulled by the wind, and thereby move a shaft or other mechanical device on the ground, to make power or do work like sawing.

One of the more well-known attempts in the space is by Magenn Power, which has offices in California and Ontario, Canada. The company is currently taking pre-orders for its 100 kW Magenn Air Rotor System (MARS), a helium-filled mini blimp that is designed to float up to 1,000 feet (Figure 12-7). When wind hits its fins, the

FIGURE 12-6 Inspired by kitesurfing, Kite Gen's airfoils will be tethered to a vertical shaft, which will generate energy as it rotates. *Kite Gen Research.*

cylindrical body rotates around a horizontal axis, generating electricity that it sends down the tether wire.

The MARS system is said to stay especially stable in the air, and is said to meet FAA guidelines for airspace safety. The company is promoting it for use on oil rigs, in wilderness areas and other remote applications. As of this writing, pricing has not been announced.

Rival turbine designer Doug Selsam (see more below) has criticized the MARS design, and told us their turbine looks inherently inefficient. Magenn did not respond to repeated requests for comment.

Closer to the utility scale, California's Makani Power claims to be developing a 1 MW airborne wind energy system, thanks in part to a $10 million investment from nearby Google. Makani's 10 kW prototype is an airplane-shaped kite that flies in circles, while wing-mounted rotors extract energy and send it down a tether (Figure 12-8).

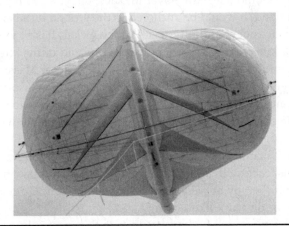

FIGURE 12-7   The 100 kW Magenn Air Rotor System (MARS) is a helium-filled mini blimp designed to float up to 1,000 feet. When wind hits its fins, the cylindrical body rotates, generating electricity that goes down the tether. *Magenn Power.*

FIGURE 12-8   Makani Power's 10 kW prototype is an airplane-shaped kite that flies in circles, while wing-mounted rotors extract energy and send it down a tether. *Makani Power.*

Makani also declined to comment for this book.

Supporters say airborne wind holds great potential, so in a few years, you may be able to order a floating turbine from a catalog, and tether it to your cabin for hours of free juice. But until they are more widely tested, you'll likely have to keep your dreams more firmly anchored on the ground.

INTERCONNECT

WindMade Label Makes Supporting Clean Energy Easy as Shopping

Consumers will soon have another option to support wind power, without having to give any thought to towers, inverters, or wind patterns. They can simply do what consumers already do best: shop.

Just as shoppers can now support sustainable agriculture by choosing certified organic foods, or support just labor practices by looking for the Fair Trade seal, people will soon be able to buy products marked with a WindMade logo (Figure 12-9). As of this writing, the precise details of the program are still being worked out, but the founders hope WindMade will achieve global adoption, and that it will help support new wind energy capacity.

WindMade was the brainchild of Morten Albaek, senior vice president for global marketing of Denmark's Vestas, the world's largest wind turbine manufacturer. Vestas provided startup funding to WindMade.org, the independent nonprofit that will actually administer the program. Certifications will be audited by PricewaterhouseCoopers, another founding partner. Other partners include the World Wildlife Fund, the UN Global Compact, Lego (yes, the toy maker), Bloomberg, and the Global Wind Energy Council.

When we met with Albaek in New York in winter 2011, he told us the WindMade working group was considering launching with a minimum requirement that 12.5 percent of a product's energy footprint must come from wind, at least for initial certification. Albaek said they were hoping to be able to increase the requirement over time, and to eventually

FIGURE 12-9 An ambitious new global label hopes consumers will choose products marked as "made with wind power." The idea is to raise funds for new wind farms. *WindMade*.

list the amount used for each product. Albaek also noted that they hoped to be able to certify whole companies, starting with a baseline of 5 percent of their operations powered by wind energy, "with a commitment to a fairly ambitious target to increase the amount of renewables used," he said.

According to Albaek, WindMade would require a significant percentage of wind energy to come from new installations (probably turbines placed in service within two years). "This ensures we have more going into the grid," he said. Albaek also added that he was hoping fees collected by WindMade could be put toward new wind farms in the developing world, "so they can avoid the same mistakes as us."

It's not hard to see why Vestas and others in the wind industry would be behind WindMade, since it has the potential to stimulate growth, or at least drive awareness of their business. For his part, Albaek hopes the program can become a model for other types of renewable energy. He added, "There's also a reason why wind is first; it's the most mature, developed, and organized clean energy."

It's an open question whether consumers will embrace the WindMade label, and if it will become powerful enough to actually influence purchasing decisions. Thousands of progressive labeling schemes have been launched around the world over the years, for everything from green hotels to less toxic dry cleaners. Most of these labels remain locally focused and niche players, although a few have broken into the global consciousness. But even if WindMade doesn't become as big as certified organic food, it can still become a viable way for consumers to support the wind industry.

Ultra-Micro Wind Turbines

If a tiny solar cell can readily power a watch or a calculator, why can't a tiny turbine charge up your cell phone? Actually, such products already exist, which we'll get to shortly. Still, current ultra-small turbines are essentially novelties, or at best prototypes. The small WePOWER turbines on the Ricoh ad in Times Square didn't last long, if they ever were installed in the first place. Not many sites on buildings are likely to have very good wind.

Still, if tiny turbines could be scaled up massively, could they make an impact?

The Hip HYmini

Billed as a portable charger for the active lifestyle, the HYmini is a sleek, handheld device that can charge cell phones or other small 5V electronics (Figure 12-10). Made by MINIWIZ Sustainable Energy Dev, the HYmini has an onboard lithium-ion battery pack that can hold a charge from a wall outlet (ho-hum) or from renewable energy (yay!).

Figure 12-10 The sleek, handheld HYmini is designed to power portable devices with the power of the wind or sun. *MINIWIZ Sustainable Energy Dev.*

The product comes with a small external solar cell; or you can blow away your friends by charging with the built-in micro-turbine, which cuts in with 9 mph breezes. It glows green when charging; how cool is that?

It's difficult to gauge how much power can actually be produced by the tiny turbine, though the manufacturer's claims that it is "supplemental" are perhaps telling. Some consumers have complained that it takes a lot of wind to be able to charge a phone, and they question the practicality of the product. (Some people have taken to driving around town with a HYmini strapped to their roof, though it's unclear if the power produced offsets the fuel cost resulting from additional drag.)

To skeptics, products like this discredit the renewable energy industry and fail to even pay back their embodied energy of production and shipping. To boosters, they educate the public about the possibilities of going greener. Funny, some say the same thing about small wind systems.

Tiny Piezoelectric Wind Turbines

You may have heard a news report about dance floors that harness energy through the stomping of feet, or speed bumps at drive-through restaurants that power lights by the passing of cars. Such motion-based energy harvesting works by taking advantage of the *piezoelectric effect*, which occurs when certain solids, typically crystals, emit a small amount of electricity after being placed under strain.

The precise electromagnetic properties that result in this phenomenon are beyond the scope of this book, but suffice it to say that inventers have envisioned tiny piezoelectric wind turbines, on the scale of centimeters, that produce electricity by flexing piezoelectric crystals as they rotate. According to *Nature News*, such tiny turbines would be 18 percent more efficient than if they harvested energy through a standard electrical and mechanical generator.

One potential application would be powering small electronic devices as stand-alones or perhaps as a finely distributed network.

The Windbelt

At the 2007 Popular Mechanics Breakthrough Awards ceremony, young inventor Shawn Frayne caused a sensation with his simple, ultra-small wind device, which he dubbed the Windbelt (Figure 12-11). Frayne's device is built from off-the-shelf parts that cost only a few dollars, and it is said to generate 40 mW in 10 mph winds. Power is generated by the fluttering of the belt, which causes a pair of magnets fitted on a membrane to oscillate between two wire coils.

Frayne envisioned the Windbelt as more than a toy, although it is a pretty nifty one. He hopes that it can be used in isolated communities in developing countries as an affordable source of power for lights, charging cell phones, or other small loads. He is currently working on larger and more efficient designs.

Traffic Turbines

Some clever folks have wondered if we could get energy out of the "wind" produced by moving cars, trucks, and trains (Figure 12-12). It's a nice idea, but you should know by now that such wind is likely to be highly turbulent and intermittent. Current technology is unlikely to be able to harvest enough energy over time to recoup an investment in equipment, but it's possible that such a scheme could work in the future, perhaps if turbines get cheap enough.

That would be good news for Tokyo drifters.

FIGURE 12-11   Young inventor Shawn Frayne examines his inexpensive Windbelt, which generates 40 mW in 10 mph winds. Frayne hopes the technology will be useful in the developing world. *Humdinger Wind Energy.*

FIGURE 12-12   Some designers wonder if we can harvest the wind created by moving traffic. It seems like it would be highly turbulent and intermittent, but if turbines can become cheap enough one day.... *Quietrevolution*.

Alternative Turbine Designs

It sometimes seems that every engineer and their brother has an idea for an "alternative" wind turbine that promises to be "more efficient," quieter, more bird-friendly, or some other superlative. As we have pointed out, there is no way to cheat nature, and it seems unlikely that any technology will beat the Betz limit of harvesting 59.3 percent of the kinetic energy from the wind.

But that hasn't stopped inventors from trying, and the Internet is full of images of oddly shaped turbines. If you want to try your hand at inventing the next design, by all means go ahead. But be aware that few have succeeded.

In the next section, we take a look at a couple of the more interesting recent alternative designs, both of which have garnered significant press.

RidgeBlade

A startup company in North Yorkshire, England, called The Power Collective won the Netherlands' 500,000-euro Postcode Lottery Green Challenge in 2009 for its RidgeBlade design, a low-slung wind generator intended to hug the peaks of rooftops, even in urban areas. With its wind-channeling louvers, the RidgeBlade resembles a hotel air conditioning unit or a bat house.

We hope this book has taught you to be skeptical of design awards, which often reward novelty, not real-world energy production. The Power Collective isn't releasing any energy or price data at the time of this writing, but the group claims it will be going into production of a small run of products in the near future.

On their website, The Power Collective writes:

> In the past there has been a lack of confidence in small wind turbines, because some of the systems sold simply did not deliver the sort of output that was promised—we want to be absolutely sure that the RidgeBlade performs well, and we will be working closely with a well known UK university to provide external verification of all our figures. Secondly, in the UK, we have a new requirement for all small wind turbines to be accredited under the MicroGeneration Certification Scheme (MCS) before they can be eligible for the generous feed-in tariffs, and this certification takes at least 12 months (and an awful lot of money)… What we can say is that we are quite confident that the RidgeBlade will pay for itself more quickly than any comparative technologies.

Will the RidgeBlade actually produce a meaningful amount of energy once installed on your roof? One potential red flag is that the company boasts it will start making energy in winds of 4 mph. As you should know by now, there is so little energy available in winds of that speed that it's irrelevant. Further, the company admits that its device can't yaw to face the winds, which are likely to be changing and turbulent on a roof.

As far as the underlying science, The Power Collective claims the RidgeBlade "uses the existing roof area to collect and focus the prevailing wind using the Aeolian wind focus effect. This is where the wind is forced to travel over the roof surface and forms a pinch point at the roof ridge, accelerating the airflow through the turbine." The company claims this so-called effect boosts wind speed around the ridge two to three times.

We were unable to find any other references to the Aeolian wind focus effect online, and a representative from the company, Bernie Cook, declined to comment on it. The so-called effect sounds a lot like AeroVironment's hopes for architectural wind, which clearly didn't pan out, despite a significant investment by a company with decades of experience in the aerospace sector.

We asked Cook what he thought of AeroVironment's failed project, and he claimed, "The RidgeBlade is a very different technology in terms of both concept and design, and we are currently engaged in a process of testing and assessment." Cook said he couldn't get into any more details of the RidgeBlade at this time, but indicated the company hoped to be ready to more openly discuss their technology in nine to twelve months.

He added, "Because micro-wind, as a sector, has suffered from controversy in recent times, we are committed to building a robust, third-party-verified set of real-world performance data before we start to communicate any outputs or commercial claims (other than that which is already in the public domain)."

Cook said, "We are working with a number of organizations that are keen to incorporate our technology into products, which is very much our preferred way of working— given our skill set and company strategy."

Given AeroVironment's experience and the fact that rooftops are usually high-turbulence zones—and at heights that normally do not give good wind speeds anyway—we're highly skeptical of the RidgeBlade. However, we'd like to see independent field trials—and get our specific questions on turbulence, energy production, and so on answered—before passing further judgment.

The Honeywell/WindTronics "Blade Tip Power System" Turbine

Another unusual small wind generator that has received a lot of press, including a spot on the *Today* show and a Breakthrough Award from *Popular Mechanics* magazine, is the Honeywell WT6500 2.8 kW turbine (Figure 12-13). Made by Michigan-based WindTronics, the turbine is marketed under the Honeywell brand in North America, due to a licensing agreement. Calls to Honeywell International were not returned, but insiders say the Fortune 100 conglomerate had little to do with the design or marketing of the turbine that bears its name.

Outside of North America, the unique turbine is sold under the brand WindTronics. Note that the company is distinct from EarthTronics, which is also based in Muskegon, Michigan, and was founded by the same person, Imad Mahawili. EarthTronics also advertises the Honeywell turbine, and listed it as the "Earthtronics 760" at one point, but today focuses on lighting, medical equipment, and small motors. Mahawili, a chemical engineering PhD and wind energy consultant who grew up in Baghdad, invented the Honeywell turbine after being inspired

FIGURE 12-13   The Honeywell/WindTronics WT6500 2.8 kW turbine is a bold new design, but can it produce meaningful amounts of energy in the real world? *Brian Clark Howard.*

by the 2004 Asian tsunami. His vision was an efficient, affordable generator that could one day help power the developing world.

WindTronics claims that its design "turns a wind turbine inside out." The WT6500 uses many nylon blades, with a magnet attached to each blade tip. The stator is the ring around the outside of the generator, which has coils of copper inside. Mahawili's idea was that the configuration would be inherently more efficient because it produces electricity in a single step, without a separate alternator.

At first glance, the WT6500 resembles the Éoliennes Bollée wind turbine patented by Ernest Sylvain Bollée in France in 1868. That design, also a HAWT with many blades arranged inside an outer ring, was the first to have wind pass through the blades of the stator before hitting the rotor. The Éoliennes Bollée was successful in France before widespread grid power, and numerous historic examples survive. Like the WT6500 today, the Éoliennes Bollée could yaw to face the wind or tip out of the wind under extreme conditions for protection.

As of this writing, about 50 Honeywell turbines have been installed, including one on the Solarium "green" apartment building in Queens (Figure 12-14). In a phone interview, Brian Levine, a co-founder of WindTronics and the vice president of business development and marketing, told us his company has an order backlog "in excess of $15 million." Besides the head office in Michigan, the company has an R&D facility in Napa Valley and a new assembly plant in Windsor, Canada. Levine said the company has 50 to 60 employees, but has ambitious plans to set up more "modular" factories around the world.

According to a March 2010 report prepared for the Tulane University Climate Committee, a Honeywell turbine costs approximately $9,000 up front, with annual maintenance costs of $250. The six-foot-diameter machine weighs 170 pounds, and is "significantly smaller than classic horizontal-axis wind turbines," according to the report. The estimated carbon reduction for each Honeywell turbine is 0.995 metric tons of carbon dioxide per year.

The Tulane report notes that Honeywell turbines are "designed to start generating electricity at 2 mph while high resistance traditional horizontal-axis turbines begin at 7 mph." The report authors went on to explain that this was a "crucial feature," given relatively low winds on campus. The authors failed to point out that there is actually very little energy in low winds, so the point is essentially moot. The authors did not question WindTronics' claim that the turbine "need only be five feet above a roofline."

The report indicated an average roof could support three WT6500 turbines, and that the suggested retail price was $5,995 each, plus an estimated $3,000 installation cost for each unit. The turbines are covered by a five-year warranty and are said to be designed to last 20 years.

Levine told us Mahawili's design philosophy had been to break with what he called the problems of conventional turbines, which he said, "have to be really high, need a hell of a lot of wind, have impacts on animals, vibrate, and make noise." Levine added, "What we've learned from users is that they're concerned about noise, animals, vibration, and height. Those constraints have kept people from adopting this technology. I thought people would ask about ROI first, but they ask about that last. The first thing that drives them is whether it is feasible where they live, based on permits, neighbors, noise, and other constraints. Then they start looking at the economics."

Levin said people like the fact that the blades seem shrouded inside the stator. "People tell us they're scared of blades swirling in the background," he said of more traditional wind turbines. "Our goal is to get the resistance and friction out of the turbine so it engages sooner and longer. That addressed these other issues: noise, height, vibration, etc."

When pressed exactly how much resistance and friction were reduced, Levin said, "It has to do with the fact that there's no gearbox." When we pointed out most small wind turbines don't have a gearbox, he said, "If you want to bring turbines closer to where we live and work, they're going to have to be a lot more agile and durable, and take the volatility out of the wind. It's in our interest to get small wind successful for everybody, not just for us."

Perhaps not surprisingly, some in the small wind industry are critical of the WindTronics concept.

Green-trust.org wrote, "I keep telling folks that the new 'Honeywell' wind turbine is a case of marketing trumping science." Dan Fink told us, "The claims on the WindTronics website defy the laws of physics, exceed the Betz limit, and overstate the amount of power that's in the wind. I can't believe Honeywell lent their name to something like that." Rejecting their claim of "reducing resistance and friction," Fink said energy "boils down to size of collector and speed of the wind."

Fink added, "Honeywell is a high solidity rotor, so there's no way they could get a coefficient of performance of even 30 percent, which is what most small wind turbines do, like the ones we build and the models from Southwest Windpower."

Ian Woofenden told us, "I'm flabbergasted Honeywell would put their name on a product like that. It's very unfortunate. It's more of the same, marketed as rooftop. It's another black eye for the industry. It marginalizes real machines. No one buys wind turbines to see something that spins. They want renewable kWh and cheaper kWh, and a rooftop turbine will not deliver either of those things."

Paul Gipe wrote an essay that's highly critical of the WindTronics. Among other issues, he pointed out that the WT6500 has a diameter of only 6 feet (1.8 m), or a swept area of 2.6 m². "Using the standard power rating, a turbine of this size should be rated at 500 W"—although the company lists it at 2.8 kW. Gipe also stressed that most small wind turbines don't have a gearbox, and wrote, "Mechanical resistance is not a significant factor in wind energy generation."

We asked Levine to respond to these experts. "Not one of those individuals has ever picked up the phone and called us or come to our labs. This is not a company cloaked in secrecy," Levine told us. "I've never seen an industry more resistant." Levin added, "I bet in a year we can double the output, but I need to go to market now, and be honest with people."

Levin said he did not know "where we rate on the Betz scale." He offered to put us in touch with the inventor, but repeated attempts to follow up were unsuccessful. Levin added, "You can't only look at swept area, because when you move the magnets and stator to the perimeter you create a different generator. Rather than turning a generator with a wind turbine, we built a generator. There is no traditional gear- or generator-based turbine that is as efficient in class 4 winds."

We asked Levin the most important question: How much energy are people actually seeing with the turbine in the real world? His answer: "We're not old enough to do that yet."

Levine added, "Our goal is to drive cost per kWh down to the lowest in the world. There's no company less interested in putting product in the wrong place." Levine went on to explain that although his company is known for targeting rooftops, he hopes the majority of buyers will use "poles" instead, on "open spaces."

"We all subscribe to the exact same laws, but by changing friction and resistance, getting noise and vibration out, and getting size to a proper level we're producing a mainstream product. This is the most flexible, lightest, quietest, least vibration-prone turbine, with the least constraints."

Power Up! *Our conclusion: Prove it. We wouldn't think of buying a Honeywell turbine (Figure 12-14) until we see a few years of hard performance energy data, preferably from multiple sites at different wind speeds. Whether the company's stated design advantages pan out remains to be seen. We think the burden of proof is on them.*

Figure 12-14   A look through the back of the Honeywell/WindTronics turbine installed on the Solarium, a "green" apartment building in Long Island City, Queens. *Brian Clark Howard.*

Advantages of the Honeywell/WindTronics Turbine

- Looks novel
- Relatively inexpensive ($5,995, plus $3,000 installation)
- Quiet
- Five-year warranty

Disadvantages of the Honeywell/WindTronics Turbine

- Unproven technology
- Unlikely to produce much power on rooftops
- No available real-world data
- No long-term track record

Optimizing Existing Designs

Now that we've had some fun looking at kites, blimps, and pocket-sized turbines, let's take a look at some of the advancements being developed for more conventional equipment. Many inventors claim to have made improvements on the standard HAWT, but not so many examples have caught on.

Here we take a look at some of the more promising areas of research.

Ducted Rotors

Also called diffuser-augmented wind turbines (DAWTs), this concept places the turbine inside a duct that flares out at the back. The idea is to enable the turbine to work in a wider range of winds and to get higher power per unit of rotor area. However, the ducts can also be bulky and heavy, which is a bummer.

Multirotor Turbines

Increasing the number of rotors would increase the total swept area, but it can also lead to turbulence problems that rob you of precious energy, not to mention maintenance headaches. Still, California-based inventor (and former heavy metal rocker) Doug Selsam has been working to overcome the challenges, and has built hundreds of prototypes. He has also sold several dual-rotor, 2 kW systems to home-owners (Figure 12-15).

Selsam charges $3,000 for the 10-foot diameter SuperTwin and $3,200 for the 12-foot diameter model. In both versions, both rotors combine their power to turn the same alternator.

Selsam worked on some of the fluid dynamics for his multirotor concepts while a student at the University of California Irvine, before he dropped out. He then began building generators and electronics for the wind industry, and eventually

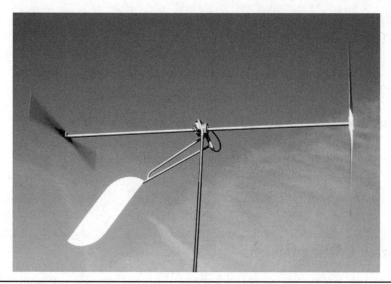

FIGURE 12-15   Former heavy metal rocker-turned wind turbine designer Doug Selsam has sold a handful of his SuperTwin dual-rotor systems. *Selsam Innovations.*

landed a grant from the California Energy Commission to support his inventions. He has tested numerous designs, including a "5-star superturbine" and a 7-rotor model (Figure 12-16). Selsam claims the rotors are carefully spaced and offset so that each one receives fresh wind, avoiding the wake of the previous. He says they all drive the same shaft, reducing the potential number of moving parts, and that their small size helps them easily achieve high RPMs.

Selsam captured a lot of press with his prototype 25-rotor turbine, which he says produces three kilowatts. One end of this "Sky Serpent" is tethered to a small

FIGURE 12-16   Selsam says his multirotor creations offer more swept area for less blade material, plus higher RPMs. *Selsam Innovations.*

tower while balloons hold the other aloft. With the blades spinning, the whole thing looks like a giant party streamer or prayer flag.

Over email, Selsam told us that the Sky Serpent was constructed specifically at the request of Popular Science Magazine and that it won "Popular Science Invention of the Year 2008" and was the centerfold of the June 2008 Issue. He said he was also invited to speak about the project at the U.S. Patent Office.

"The Sky Serpent is the leading candidate for airborne wind energy and was the only working prototype at the first world Airborne Wind Energy Conference at Cal State Chico and Oroville in 2009," Selsam explained. "This in a field against Boeing, Honeywell, etc."

Selsam envisions the skies full of long strands of multirotor turbines, with some floating or anchored to the sea floor (Figure 12-17). It's possible that a string of small turbines would be less likely to inspire NIMBY resistance than fewer large ones.

Selsam says his designs get the same amount of power as a large turbine with less blade material and less weight. "The power multiplication by far outperforms any previous method of power augmentation for a wind turbine," he told us. "This is a fact and not anyone's opinion." Selsam says his seven-rotor Superturbine was independently tested by Windtesting.com and verified to multiply output 6-7 times over a single-rotor turbine using the same blades (Whisper H-40 or Whisper 100 from Southwest Windpower). These promising results were published in a paper prepared for the California Energy Commission, which sponsored the project.

In 2008, Paul Gipe installed one of Selsam's SuperTwin dual-blade turbines at his test facility. He tried to run a test for a year, noting, "The unit is more powerful

FIGURE 12-17   Selsam envisions millions of multirotor turbines placed out at sea and lofted into the air. *Selsam Innovations.*

than turbines I've tested before and this has required several time consuming and expensive modifications of our test equipment. These upgrades were never fully completed. Because of other commitments, testing the turbine has not been a priority. Consequently, the tests remain incomplete."

Gipe concluded, "The Super Twin is experimental, but it is an option for someone who likes to tinker with their wind turbines."

Over email, Selsam told us, "Paul also tested a Bergey of the same eight-foot diameter, but mine was more powerful, again proving beyond a shadow of a doubt that the extra rotor makes more power. Additionally, the Bergey uses the best NREL airfoils, whereas my SuperTwin used 2×4's shaped on a belt sander, and still outperformed the Bergey."

Selsam argued, "The dirty secret of all this clean energy stuff is that despite millions and millions in funding, the big labs categorically refuse to build or test anything truly new. They'd rather waste millions on a few select dubious ideas that passed through the gauntlet of bureaucrats than adopt a broader, high-throughput approach to research that could identify and leapfrog promising new technologies by building and testing many new ideas."

Selsam did concede that some of that resistance to "new designs" is well founded. "Most 'new' designs are not new at all and most are over 1,000 years old," he told us. "They almost invariably repeat past mistakes. Imagine an airplane sporting a riverboat paddle wheel for propulsion, and you can see how ridiculous most 'alternative' wind turbines seem to those of us with 101+ burnt-out turbines under our belts." Still, he said, "But even a whole bushel of rotten apples can have one good apple in there somewhere!"

Selsam told us that rather than try to reinvent the wheel, his designs build on the time-tested "propeller" shape, but sweep more area with less material. "Imagine if the flashlight industry had never discovered stacking multiple batteries." Selsam said. "Imagine if all cars had only single-cylinder engines! Imagine if you had to cover your roof with a single large tile (Fred Flintstone?) Sweeping a huge area with a single large propeller is the Fred Flintstone approach to wind energy."

Doug Selsam has big ideas for wind power at large and small scales, and he hopes to get more support for his multirotor concepts. Early tests suggest that it's something that certainly deserves more study. One thing to note is that, as expected, the power curves on Selsam's website show that his multirotor concepts harvest very little energy out of low winds. There are no designs that cheat nature.

Counter-Rotating Dual Turbines

A typical wind turbine creates a significant amount of tangential or rotational air flow, the energy of which is wasted. However, clever engineers have suggested that this flow could be harvested by a second rotor spinning in the opposite direction.

In fact, a friend of German wind pioneer Albert Betz, Hans Honneff, wrote a book on counter-rotating dual turbines in the 1930s. According to an article in the e-magazine *Alternative Energy,* Honneff described an array of large turbines, set three to a tower, with each rotor revolving separately from the others. Honneff called for rotor diameters of 150 meters and a massive output of 21 MW. The Nazis gave his idea a whirl, and the Third Reich Wind Energy Ministry tested ten-meter-diameter models.

In the late 1970s, Trimble Windmills produced some small, 5 kW counter-rotating dual turbines, which were sold for farms and isolated heating operations. According to *Alternative Energy,* these machines had a contra-rotating permanent magnet alternator and sail wing blades attached to each half. They achieved a "respectable Cp (coefficient of performance) of 0.37," although the magazine suggested that the two sets of blades were actually spaced too close together for optimal results.

In the early 2000s, Kari Appa tested counter-rotating turbines with funding from the California Energy Commission. Appa reported extracting 30 to 40 percent more power out of the winds versus a comparable single-rotor system.

So why aren't more turbines dual-action? For one thing, they would cost more up front, although some supporters have argued that they are likely to be less expensive than choosing a larger single-rotor system. However, dual turbines are significantly more complicated machines. Although they can help minimize gyroscopic forces on equipment, dual-rotor systems must be configured just so, or they can quickly run into trouble. With more moving parts comes greater chance of failure and increased maintenance time and expense.

Still, if dual turbines can be developed that are easy to install and maintain, and that show consistently higher energy returns, they may become economically advantageous in the near future.

We may one day look back with amusement on the quaint old days of single rotors.

Telescopic Blades

Some designers have proposed making turbine blades telescopic so they can extend their swept area when winds are optimal, but retract during high winds to avoid damage. They would also be easier to ship and store.

However, making blades telescopic is also likely to make them less strong. If one of the blades gets stuck, it could be a real hassle to loosen it, especially if you don't have a climbable or tilt-up tower.

Telescopic blades could also increase the price of the rotor in a way that could take a long time to pay off financially.

Laser-Guided Turbines

Danish researchers have recently been experimenting with laser-guided active controls for utility-scale wind turbines. The idea is to pair a laser with an anemometer to actively measure the winds.

Torben Mikkelsen, a professor at Denmark's Risø DTU, told *Science Daily*, "The results show that this system can predict wind direction, gusts of wind, and turbulence. So we estimate that future wind turbines can increase energy production while reducing extreme loads by using this laser system, which we call wind LIDAR."

So far, tests suggest lasers can boost energy production of wind turbines by around 5 percent. Although the expense and relatively low payoff is bound to keep such technology out of the hands of small wind farmers for years, it is conceivable that lasers could one day trickle down. Of course, most small wind turbines don't currently use computer-assisted active controls, but that could change as technology improves and gets cheaper.

You may one day be able to boast to your neighbor about your turbine's "frickin laser."

Advanced Active Controls

As stated previously, most small wind turbines don't have active controls, as utility-scale machines do. However, that might not always be the case. Active controls, in general, keep blades positioned to maximize energy harvested at every moment. They allow turbines to extract more energy at higher wind speeds, and help prevent damage during storm events.

A number of researchers are working on improving active controls through advanced software programming, algorithmic modeling, and improving sensory outputs, through the use of lasers and other devices. Some researchers are looking at ways to improve the evenness and "quality" of power generated to make it less spiky according to changes in the wind. One idea is to keep some of the energy from a strong gust in the form of kinetic energy—in the form of fast spinning blades—instead of converting as much as possible directly to electricity. Then, when the wind dies down a moment later, the blades will still be spinning fast, and the electricity can be produced at that point. The overall effect could be more even power production.

Other analysis seeks to better optimize the amount of power harvested at each wind speed by actively adjusting blade rotation speed as needed.

Given high costs of implementation and the relatively small size of most home systems, we don't expect to see that many active controls on the market very soon. But they could become standard one day. True, some active controls are already available on Quietrevolution VAWTs (Chapter 8), though those machines tend to run a bit expensive.

Summary

We hope this book has been a useful introduction to the exciting world of small wind power (Figure 12-18). As we have shown, the industry is far from mature, and

Figure 12-18 This Bergey Excel 10 kW turbine still serves as a teaching tool at the Liberty Science Center in New Jersey, which has gorgeous views of Manhattan. An early plan for the new World Trade Center included building-integrated wind turbines, but that idea was scrapped. *AWS Scientific, Inc./DOE/NREL.*

it faces a number of challenges, including relatively high cost of entry, a scarcity of qualified technicians and installers, and fairly demanding maintenance schedules. Oh, and you need a great wind resource, or you're paying a lot of money for a fancy lawn ornament.

In this chapter, we presented a brief survey of cutting-edge wind technology, some of which may result in outstanding products in the near future. As we've mentioned, there tends to be a great deal of hype around novel designs. The mainstream media tends to breathlessly report on every press release from a rising number of secretive startups, which often claim to have invented "revolutionary" small wind turbines that can fit neatly on your roof, power a small village, and even start working in the faintest breeze. By the way folks, it is totally bird- and bat-friendly and can make you a nice hot cappuccino.

By now, you should be armed with the tools to evaluate initial claims of any wind power purveyor. You will ask to see real-world energy data, preferably from independent sources, and you will go to great pains to verify the quality of your wind resource, not to mention the local regulatory environment. You'll know to run away from anything that claims to be a "Betz beater," and you'll insist on finding out the real coefficient of performance.

Of course, if you are an engineer or a hobby tinkerer, you may read this book and decide that you are ready to try out your design for an eight-rotor, kite-lofted piezoelectric power plant. If so, more power to you. Just stay safe, and let us know how she turns out. Don't forget to send a picture.

For his part, Paul Gipe told us, "If you want a windmill as a statement, to be green, I'm fine with that. You want it as a recreational thing that also produces electricity, that's better than a Jacuzzi, snowmobile, or wine bar, it will do something useful. But I'm talking about putting up a wind turbine to make a profit. The only way we're going to get enough windmills to save our butts is that they have to be profitable."

Dan Fink told us, "The general public understands solar panels pretty well, but the wind resource is totally confusing to people. If someone gave you a solar panel the size of a playing card, and said you could stick it on your roof and power an electric car, no one would believe you. Or if they said a one-ton pickup could get 50 mpg, no one would believe you. But people don't understand wind energy."

Fink added, "The small wind industry has a high proportion of dissatisfied customers, because people have unrealistic expectations."

Ian Woofenden told us, "If you misunderstand how much energy is in a gallon of gas, who cares about your car design? If you don't know your wind resource, you're sunk."

When we asked Hugh Piggott to look at the big picture, he said, "Most people have assumptions that they can put something on their roof that will generate most of their electricity in an urban environment. That's a really, really long way from the truth."

We hope this book has brought you closer to understanding wind energy.

Morten Albaek, senior vice president for global marketing of Vestas, told us he sees "a fairly bright future for microturbines." The Dane added, "They're tangible to consumers, just like solar panels on rooftops. We think most wind energy will [continue] to be developed by classic large wind farms, but we think the next developments in wind energy technology will be in micro-turbines. The amount of energy we will be able to get from micro-turbines is going to be limited, but important, compared to what we will get from large classical wind turbines, but I really hope the micro-turbine industry sees success."

Albaek added, "The stupidest thing we can do is put a turbine, large or small, in a place that has no wind. It's also stupid to drill in Saudi Arabia when we have such great wind here."

May the wind always be at your blades!

Glossary

Electrical Terms

Alternating current (AC): Electron (charge) flow in two directions, usually alternating back and forth numerous times per second. The electrical grid provides alternating current.

Alternator: A device that generates AC electricity by passing magnetism past coils of wire.

Ammeter: A device that measures amperage (current).

Ampacity: The maximum amount of current that a wire or device can carry safely, without getting damaged.

Amperage: Also called current, the rate of charge flow in circuits, measured in amperes, or amps (A).

Ampere-hour: Commonly called amp-hour (Ah), a unit for the amount of charge that is transferred by a steady current of one amp for one hour; used to measure battery capacity.

Balance of system (BOS): Electrical components of a wind system besides the turbine and tower, typically including the wiring, voltage clamp, charge controller, batteries, inverter, disconnects and grounding.

Battery: An electrochemical device that stores electrical energy via chemical reactions.

Batteryless grid-tied system: A wind electric system that connects to the grid without batteries, meaning it provides no backup for utility outages. If the grid is down, the system stops working.

Capacity credit: The output of a power source that may be statistically relied upon, expressed as a percentage. At low levels of penetration, the capacity credit of wind is about the same as the capacity factor. As the concentration of wind power on the grid rises, the capacity credit percentage drops.

Capacity factor: Also called load factor, the average expected output of a generator, usually over an annual period. Expressed as a percentage of the nameplate capacity or in decimal form (e.g., 40 percent or 0.40).

Charge: The moving electrons that create electricity.

Charge controller: An electronic device that regulates battery charging and prevents overcharge.

Circuit: An electrical pathway for charges; a complete loop must be present.

Conduit: Tubing that carries electrical wires, often made out of metal or plastic.

Current: Rate of electron (charge) flow, also known as amperage.

Density: Mass per unit of volume.

Depth of discharge (DOD): The level of battery discharge; it's the inverse of state of charge.

Digital multimeter (DMM): An electronic device that measures voltage, amperage, resistance, and sometimes other electrical parameters.

Direct current (DC): Flow of charge in one direction. Direct current is provided by batteries, solar panels and some small wind turbines, and it is typically used in recreational vehicles, boats, and sometimes small cabins.

Direct drive: When wind turbine blades connect directly to the generator, without going through a gearbox.

Disconnects: Commonly known as breakers, these switches disconnect electrical elements from each other.

Dispatchability: Also called maneuverability, the ability of a given power source to increase and/or decrease output quickly on demand. Wind power is often said to be "highly nondispatchable."

Dump load: A load designed to dissipate excess energy from a power source, usually controlled by the charge controller. The most common examples are air or water heaters.

Efficiency: The ratio of energy out (or work done) to energy in, expressed as a percentage.

Energy: Work done over time; power (watts) × time (hours) = energy (watt-hours).

Fall-arrest system: Safety features that prevent climbers from falling, even in the case of fainting.

Fixed guyed towers: A tower supported with guy wires.

Frequency: The number of times per second that alternating current (AC) changes direction, measured in hertz (Hz).

Grid: The network that connects electricity generators to electricity users via utility lines and other infrastructure.

Grid-tied: A system that is wired into the grid. It may or may not have batteries for outage protection.

Grounding: Wiring structures to the ground, in order to safely dissipate excess electrical energy (such as from lightning strikes).

Hertz (Hz): The unit of frequency of alternating current (AC) electricity. Must grids are at 50 or 60 Hz.

Intermittency: The extent to which a power source is unavailable; a single wind turbine is highly intermittent, but a large wind farm spread over a wide area will be less so.

Inverter: A device that converts direct current (DC) electricity to alternating current (AC). A device called a rectifier converts AC to DC.

Kilowatt (kW): Rate of power or use of electrical current; also 1,000 watts.

Kilowatt-hour (kWh): A measure of electrical energy, equal to the use of one kilowatt in one hour.

Load: Something that uses electricity, such as an appliance or charging battery, or a whole building.

Maximum power point tracking (MPPT): A system that adjusts the operating voltage of a wind turbine at different speeds to maximize the output and capture as much energy as possible.

Megawatt: A measure of power; also 1,000,000 watts.

Nameplate capacity: The normal maximum output of a generating source.

Ohm (K): The unit of electrical resistance.

Ohm's Law: A law that states that amperage is equal to voltage divided by resistance $(A = V/K)$.

Overcurrent protection: Protection against high current that may otherwise melt wires or cause fires; typically consists of circuit breakers or fuses.

Parallel: Connecting electrical devices together in parallel paths, with all positives and all negatives wired together; it increases amperage in batteries while voltage remains the same.

Payback: The amount of time it takes energy-generating or energy-saving equipment to pay back its original investment.

Peak power: Maximum wattage a device can generate.

Photovoltaic (PV): Solar-electric panels, commonly called solar panels, that generate electricity directly from sunlight.

Power: The rate of energy generation, transmission, or use; also called wattage.

Power conditioning equipment: Devices like inverters, rectifiers, and controllers that change incoming energy so it's compatible with the grid or equipment.

Power formula: In electrical terms, power = voltage × amperage × power factor; in wind terms, power = ½ air density × swept area × wind speed cubed.

Rated power: The manufacturer's claim of energy output, often at the energy peak (and usually at high wind speeds).

Rectifier: A device that converts alternating current (AC) electricity to direct current (DC).

Resistance: A measure of how much an electrical device opposes flow of charge.

Series: Connecting devices together in a daisy-chain fashion, with each positive wired to a negative; it increases voltage while maintaining amperage in batteries.

Setback: The distance between a structure and the property line or a road, often mandated by zoning rules.

State of charge (SOC): Measure of the capacity of battery remaining, as a percentage; it's the inverse of depth of discharge (DOD).

Step-down: Going from higher to lower voltage in an electrical device.

Tag line: An additional rope used to control items being raised or lowered on a tower (not the line actually doing the lifting).

Volt (V): The unit of electrical force.

Watt (W): A unit of power (wattage), the rate of energy generation, transmission, or use.

Watt-hour (Wh): The unit of electrical energy; one watt-hour is the energy used to create one watt of power for one hour.

Wind Energy–Related Terms

30/500 rule: A rough guideline suggesting that a wind generator's lowest blade tip should be a minimum of 30 feet above anything within 500 feet to get into a quality wind resource and minimize destructive turbulence (higher still is almost always better).

Air density: Mass per unit volume of air; air is less dense by roughly 3 percent for every 1,000 feet of elevation above sea level. It also varies with temperature and humidity.

Airfoil: The shape of a bird's feather or an airplane wing, and commonly used for wind turbine blades to optimize the lift/drag ratio and maximize energy production.

Anemometer: A device that measures wind speed, often with cups on a rotating shaft.

Annual energy output (AEO): The amount of energy in kilowatt-hours that a specific wind generator will produce at a specific average wind speed over a year.

Average wind speed: The average recorded wind speed over a set period, usually one year.

Betz limit: The maximum percentage of wind energy (about 60 percent) that a perfect wind generator could capture. Note that a perfect turbine would have an infinite number of infinitely thin blades. However, large turbines are approaching the Betz limit, thanks to advanced materials and active controls.

Blade pitch control: Twisting the blades' angle of attack into or out of the wind, to either increase energy production or govern by slowing down rotation in high winds, to avoid damage.

Brake (wind): Various systems used to stop the rotor from turning.

Coefficient of performance: Also called power coefficient, the ratio of the power extracted by a wind turbine to the power available in the wind stream. Can never be higher than the Betz limit.

Cube law: Wind power is proportional to the cube of the wind speed (V^3).

Cut-in: The wind speed at which a wind generator starts to generate, typically 5 to 9 mph.

Cut-out: The wind speed at which a wind generator stops producing, in order to protect itself from damage in high winds. *See also* governing.

Downwind: A wind turbine with the rotor on the opposite side of incoming wind (lee side).

Energy curve: A graphic presentation of the energy produced by a wind generator in a range of specific average wind speeds.

Freestanding tower: A self-supporting tower, without guy wires.

Furling: A form of governing in which the rotor folds up or around the tail vane, to reduce exposure of the airfoils to high winds.

Gin pole: A pole used to raise tilt-up towers, or a vertical pole temporarily attached on a nontiltable tower to lift items up.

Governing: Slowing down the rotor in high winds, in order to protect it from damage. This is done through techniques such as furling, blade pitch, or electrical braking.

Guy anchor: The support for guy wires, anchored into the ground.

Guy wire: Steel cable used to hold up guyed towers.

Horizontal axis wind turbine (HAWT): A wind generator that spins on a horizontal shaft, like most available wind turbines do today.

Hub: The center of a wind turbine's rotor, where the blades are connected to the shaft.

Hub height: The height above the ground at the hub.

Instantaneous wind speed: Wind speed at a given moment.

Nacelle: The body of a propeller-type wind turbine, usually containing the gearbox (if present), alternator, and other parts.

Net metering: A system of utility interconnection that credits a user for renewable energy produced, often at the same rate the utility charges and up to the level of the building's usage.

Nose cone: Fiberglass or metal cowling in front of a wind generator's rotor (optional).

Off-grid system: A wind electric system not connected to the utility grid.

Peak wind speed: Maximum wind speed that a device experiences.

Power coefficient: *See* coefficient of performance.

Power curve: A plot of the instantaneous wattage of a wind generator in various wind speeds. Manufacturers like to show these off, but they aren't very useful for customers.

Power law method: A way to calculate the increase in wind speed with height. The formula is $V/V_o = (H/H_o)$, where V_o is the wind speed at the original height, V is the wind speed at the new height, H_o is the original height, H is the new height, and μ is the surface roughness exponent.

Power (wind): In the wind industry, power is a function of ½ air density (ρ) × swept area (A) × wind speed cubed (V^3) or $P = \frac{1}{2}\,\rho A V^3$.

Rated output capacity: The output power of a wind generator operating at the rated wind speed.

Rated wind speed: The wind speed at which a wind generator's rated power output is measured.

Renewable energy: Energy generated from "clean," naturally reoccurring sources like the wind, sun, falling water, and biomass (plant and animal materials).

Rotor: The part of a wind turbine (or generator or alternator) that rotates. In the case of a wind turbine, this includes the blades and hub.

Slip rings: Rings that form an electrical connection between the generator and the tower, while allowing the turbine to rotate around 360 degrees, without breaking anything. Slip rings are usually made of brass and graphite brushes, which make a sliding contact.

Start-up: The wind speed at which a turbine starts spinning, typically before cut-in (when the generator starts producing electricity).

Survival wind speed: The maximum wind speed equipment will "survive" without permanent damage.

Swept-area: The area (in square feet or square meters) that a wind generator's blades sweep. For HAWTs, this is equal to pi (π) times the blade rotor's radius (r) squared, or πr^2.

Tail boom: A rod that extends behind the wind turbine to support the tail vane (on upwind designs).

Tail vane: On upwind designs, the sheet of metal, wood, or plastic on a tail boom that keeps the rotor facing into the wind.

Tilt-up tower: A tower that's typically not climbable but can be lowered to the ground using cables, lifting gear, and a gin pole.

Tip speed ratio (TSR): The speed at the tip of the rotor blade as it moves through the air divided by the wind velocity.

Tower adapter: The hardware that connects the wind generator to its tower.

Turbulence: Randomized movement of air, frequently in the form of eddies, caused by obstacles. Turbulence decreases the amount of energy available to a turbine.

Upwind: A wind turbine with the rotor on the same side as the incoming wind (windward).

Vertical axis wind turbine (VAWT): A wind turbine that spins on a vertical shaft. The earliest recorded windmills were vertical, though horizontal designs have been dominant for more than a century. New VAWTs have enjoyed considerable interest recently, although they remain controversial.

Wind farm: A group of wind turbines, also known as a wind power plant.

Wind generator: A mechanical and electrical device that produces electricity from moving air, also called a wind turbine.

Wind rose: A graphical presentation of which directions a site's wind comes from.

Wind shear: A sudden change in the speed or direction of wind. Many in the small wind industry use this term to refer to the rate of increase in wind speed as you move away from the ground, though that can also be described as the wind gradient.

Yaw: Rotation of the turbine around the tower, which allows the turbine to stay into the wind.

Yaw bearing: The bearing (typically a ball bearing) that allows a wind generator to turn and face the wind.

Conversions, Abbreviations, and Acronyms

If you have a smart phone, conversions can now be done with free conversion tools:

- For an android: www.appbrain.com/search?q=conversion
- For an iPhone: www.apple.com/webapps/calculate

Size Conversions	
1 inch = 2.54 cm	1 cm = 0.39 inch
1 foot = 0.305 meter	1 meter = 3.28 feet
1 square foot = 0.093 square meter	1 square meter = 10.9 square feet

Speed Conversions
1 mile per hour = 1 cm = 0.39 inch = 1.61 kilometers per hour
1 meter per second = 2.24 miles per hour = 3.60 kilometers per hour
1 kilometer per hour = 0.62 mile per hour = 0.28 meter per second

If you have a smart phone, conversions can now be done with free abbreviations tools:

- For an android: www.appbrain.com/search?q=abbreviations
- For an iPhone: www.abbreviations.com/iphone.aspx

Abbreviations and Acronyms: Energy			
A	Amps	CFL	Compact fluorescent lighting
AC	Alternating current	DC	Direct current
AEO	Annual energy output	DOD	Depth of discharge
Ah	Amp-hours	HAWT	Horizontal axis wind turbine
AWG	American wire gauge	PURPA	Public Utility Regulatory Policies Act (1978), 16 USC § 2601.18 CFR §292 that refers to small generator utility connection rules
BOS	Balance of system	VAWT	Vertical-axis wind turbine
O&M	Operation and maintenance		

Abbreviations and Acronyms: Agencies	
AWEA	American Wind Energy Association
BWEA	British Wind Energy Association
CanWEA	Canadian Wind Energy Association
EWEA	European Union Wind Energy Association
DOE	Department of Energy
DSIRE	Database of State Incentives for Renewable Energy
EERE	The DOE's Office of Energy Efficiency and Renewable Energy
NOAA	National Oceanographic and Atmospheric Administration

Abbreviations and Acronyms: Agencies	
NREL	National Renewable Energy Laboratory
NWS	National Weather Service
PUC	Public Utility Commission, a state agency that regulates utilities. In some areas known as Public Service Commission (PSC).

Surface Roughness and the Wind Shear Exponent (μ)	
Terrain	Wind Shear Exponent (μ)
Ice	0.07
Snow on the flat ground	0.09
Calm sea	0.09
Coast with on-shore winds	0.11
Snow-covered crop stubble	0.12
Cut grass	0.14
Short-grass prairie	0.16
Crops, tail-grass prairie	0.19
Hedges	0.21
Scattered trees and hedges	0.24
Trees, hedges, a few buildings	0.29
Suburbs	0.31
Woodlands	0.43

Relative to a reference height of 10 m (33 ft).

Adapted from *Characteristics of the Wind* by Walter Frost and Carl Aspliden in *Wind Turbine Technology*, and *Windenergie: Theorie, Anwendung Messung* by Jens-Peter Molly.

Notes

Chapter 1

1. Mount Washington Observatory. "The Story of the World Record Wind." www.mountwashington.org/about/visitor/recordwind.php.
2. Leslie, Brewer. "Canadian Market for Small Wind Energy Systems Growing Rapidly." *The Epoch Times*, Oct. 31, 2010. www.theepochtimes.com/n2/content/view/45110.
3. CanWEA. "Small Wind Market Survey." 2010. www.canwea.ca/pdf/Small Wind/canwea-smallwindmarketsurvey-e-web.pdf.
4. Nao Nakanishi. "UK Small Wind Blows Strong Despite Recession." Reuters, November 20, 2009.
5. Number presented at 1st World Wind Energy Association, World Summit for Small Wind, Husum, Germany, March 18–19, 2010, provided by REN21.
6. BWEA. "Small Wind Systems UK Market Report 2009." www.bwea.com/pdf/small/BWEA%20SWS%20UK%20Market%20Report%202009.pdf.
7. Guidi, Daniele, Stephanie Cunningham. Ecosoluzione via REN21, March 2010.
8. Jayakumar, P.B. "Lighting Up Lives: Tatas Plan to Do a Nano with Power." *Business Standard*, July 2, 2010. www.business-standard.com/india/news/lightinglives-tatas-plan-to-donanopower/400128.
9. China data presented at 1st World Wind Energy Association, World Summit for Small Wind.
10. World Wind Energy Association. "Small Wind Systems and Their Applications: An Overview." www.wwindea.org/technology/ch05/en/5_1_1.html.
11. GWEC. "Wind Is a Global Power Source." www.gwec.net/index.php?id=13.
12. EcoGeneration. "Growing Small Wind." July/August 2009. http://ecogeneration.com.au/news/growing_small_wind/001331.
13. Bombourg, Nicolas. "Reportlinker Adds Global Small Wind Turbine Market 2009–2013." PRNewswire, July 22, 2010.

14. www.gwec.net/index.php?id=139.

15. EWEA. "Wind Energy and the Benefits." www.ewea.org/fileadmin/ewea_documents/documents/press_releases/factsheet_environment2.pdf.

16. Harsh, Stephen B., Lynn Hamilton, Eric Wittenberg. "Small Wind on the Farm: A Capital Budgeting Case Study." *Agricultural Finance Review*, Vol. 70 Issue: 2, pp. 201–213. www.emeraldinsight.com/journals.htm?articleid=1876533&show=abstract.

17. Bharat Book Bureau. "Financial Incentives and Government Policies Are Driving Wind Technology." My Efficient Planet, March 28, 2010. www.myefficientplanet.com/72307/financial-incentives-and-government-policies-are-driving-wind-technology.

18. Jacobson, D., D.J. Virginia, C. High. "Wind Energy and Air Emission Reduction Benefits: A Primer." Subcontract Report NREL/SR-500-42616, February 2008.

19. Ibid.

20. Energy Savers. "Small 'Hybrid' Solar and Wind Electric Systems." www.energysavers.gov/your_home/electricity/index.cfm/mytopic=11130.

Chapter 2

1. Colby, W., M. Det. David, et. al. "American Wind Turbine Sound and Health Effects: An Expert Panel Review." Wind Energy Association and Canadian Wind Energy Association, December 2009. www.awea.org/newsroom/releases/AWEA_CanWEA_SoundWhitePaper_12-11-09.pdf.

2. DeGagne, David C., Anita Lewis, Chris May. "Evaluation of Wind Turbine Noise." www.noisesolutions.com/uploads/images/pages/resources/pdfs/Wind%20Turbine%20Noise.pdf.

3. Dang, W.M., et. al. "Characteristics of Infrasound and Its Influence on Workers in Working Environment of Certain Thermoelectricity Works and Department." *Zhonghua Lao Dong Wei Sheng Zhi Ye Bing Za Zhi*. 2008 Dec. 26(12): 711–4.

4. Oerlemans, Stefan. "Detection of Aeroacoustic Sound Sources on Aircraft and Wind Turbines." University of Twente, September 3, 2009. www.utwente .nl/news/quieter-wind-turbines.

5. Hanning, Christopher M.D. "Sleep Disturbance and Wind Turbine Noise." June 2009. www.wind-watch.org/documents/wp-content/uploads/Hanning-sleep-disturbance-wind-turbine-noise.pdf.

6. Pedersen E., K.P. Waye. "Perception and Annoyance Due to Wind Turbine Noise—A Dose-Response Relationship." *J Acoust Soc Am*. 2004 Dec;116(6): 3460–70. www.ncbi.nlm.nih.gov/pubmed/15658697.

7. Pedersen E., F. van den Berg, R. Bakker, J.J. Bouma. "Response to Noise from Modern Wind Farms in The Netherlands." *Acoust Soc Am*. 2009 Aug;126(2): 634–43: www.ncbi.nlm.nih.gov/pubmed/19640029.

8. Pedersen E., Waye K. Persson. "Wind Turbine Noise, Annoyance and Self-Reported Health and Well-Being in Different Living Environments." *Occup Environ Med*. 2007 Jul;64(7):480–6. Epub Mar 1, 2007. www.ncbi.nlm.nih.gov/pubmed/17332136.

9. Ibid.

10. Smedley A.R., A.R. Webb, A.J. Wilkins. "Potential of Wind Turbines to Elicit Seizures Under Various Meteorological Conditions." *Epilepsia*. 2010 Jul;51(7): 1146–51. Epub Nov 16, 2009. www.ncbi.nlm.nih.gov/pubmed/19919663.

11. Harding G., P. Harding, A. Wilkins. "Wind Turbines, Flicker, and Photosensitive Epilepsy: Characterizing the Flashing that May Precipitate Seizures and Optimizing Guidelines to Prevent Them." *Epilepsia*. 2008 Jun; 49(6):1095–8. Epub Apr 4, 2008.

12. Wikipedia. *Viewshed*. http://en.wikipedia.org/wiki/Viewshed.

13. EWEA. "Visual Influence." www.wind-energy-the-facts.org/en/part-i-technology/chapter-4-wind-farm-design/factors-affecting-turbine-location/visual-influence.html.

14. AWEA. "How Popular Is Wind Energy." www.awea.org/faq/wwt_environment.html.

15. Chiras, Dan. *Power from the Wind*. New Society Publishers, 2009.

16. Department of Energy's Lawrence Berkeley National Laboratory. "The Impact of Wind Power Projects on Residential Property Values in the United States." eetd.lbl.gov/ea/EMS/reports/lbnl-2829e.pdf.

17. Renewable Energy Policy Project (REPP). "The Effect of Wind Development on Local Property Values." www.repp.org/articles/static/1/binaries/wind_online_final.pdf.

18. Dent, Peter, Dr. Sally Sims. "What Is the Impact of Wind Farms on House Prices?" by Oxford Brookes University, was published as part of the FiBRE (Findings in Built and Rural Environments) series. www.st-andrews.ac.uk/media/RICS%20Property%20report.pdf.

19. Edinburgh Solicitors' Property Centre. "Impact of Wind Farms on Residential Property Prices—Crystal Rig Case Study." February 2007, available from ESPC.

20. www.beaconhill.org/BHIStudies/Windmills2004/WindFarmArmyCorps.pdf.

21. Zeller, Tom. "For Those Near, the Miserable Hum of Clean Energy" *New York Times*. Retrieved October 5, 2010.

22. Promotion of Renewable Energy Act No. 1392 of 27 December 2008. www.windaction.org/?module=uploads&func=download&fileId=1910.

23. Yes2Wind. "Debunking the Myths." www.yes2wind.com/explore/debunking-the-myths/.

24. Wilson, Whitney. "Interference of Wind Turbines with Wide Area Communications." www.masstech.org/Project%20Deliverables/Comm_Wind/Eastham/Eastham_Cell_Tower_Analysis.pdf.

25. CanWEA. "Benefits and Issues with Small Wind." www.smallwindenergy.ca/en/Overview/BenefitsIssues/Issues.html.

26. Sengupta, Dipak L. "Electrogmagnetic Interference Effects of Wind Turbines." Department of Electrical Engineering and Computer Science, The University of Michigan, 1982.

27. Schwartz, Ariel. "Wind Turbine Projects Run Into Resistance." *New York Times*, Aug 27, 2010.

28. BWEA. "Feasibility of Mitigating the Effects of Wind Farms on Primary Radar." www.bwea.com/aviation/ams_report.html.

29. Schwartz, Ariel. "Wind Turbine Projects Run Into Resistance." *New York Times*. Aug 27, 2010.

30. Energy Insurance Brokers/South Bay Risk Management and Insurance Services/Garrad Hassan.

31. Erickson, Wallace P., Gregory D. Johnson, David P. Young, Jr. "A Summary and Comparison of Bird Mortality from Anthropogenic Causes with an Emphasis on Collisions." www.fs.fed.us/psw/publications/documents/psw_gtr191/Asilomar/pdfs/1029-1042.

32. "Wind Turbines Continue to Kill Birds." www.wind-watch.org/news/2010/ 02/05/wind-turbines-continue-to-kill-birds.

33. Drewitt A.L., R.H. Langston. "Collision Effects of Wind-Power Generators and Other Obstacles on Birds." *Ann N Y Acad Sci*. 2008;1134:233–66.

34. Baisner A.J., J.L. Andersen, A. Findsen, et. al. "Minimizing Collision Risk Between Migrating Raptors and Marine Wind Farms." Development of a Spatial Planning Tool. *Environ Manage*. Aug 15, 2010.

35. Fielder, J.K. "Assessment of Bat Mortality and Activity at Buffalo Mountain Windfarm, Eastern Tennessee." Thesis, University of Tennessee, Knoxville. 2004.

36. Kerns, J., P. Kerlinger. "A Study of Bird and Bat Collision Fatalities at the Mountaineer Wind Energy Center, Tucker County, West Virginia." FPL Energy and Mountaineer Wind Energy Center Technical Review Committee. Curry and Kerlinger, LLC, Cape May, NJ, 2004. www.responsible wind.org/docs/MountaineerFinalAvianRpt3-15-04PKJK.pdf.

37. Arnett, E.B., W.P. Erickson, J. Kerns, J.W. Horn. "Relationships between Bats and Wind Turbines in Pennsylvania and West Virginia." Bats and Wind Energy Cooperative, 2005. www.batcon.org/wind/BWEC2004finalreport .pdf.

38. Kerns, J., W.P. Erickson, E.B. Arnett. "Bat and Bird Fatality at Wind Energy Facilities in Pennsylvania and West Virginia." pp. 24–95. In E.B. Arnett, editor, 2005. Relationships between bats and wind turbines in Pennsylvania

and West Virginia. Final Report. Bats and Wind Energy Cooperative. February 15, 2007: www.batcon.org/wind/BWEC2004finalreport.pdf.

39. Desholm M., J. Kahlert. "Avian Collision Risk at an Offshore Wind Farm." *Biol Lett*. 2005 Sep 22;1(3):296–8.

40. Baerwald, Erin F., Genevieve H. D'Amours, Brandon J. Klug, Robert M.R. Barclay. "Barotrauma Is a Significant Cause of Bat Fatalities at Wind Turbines." *Current Biology*, Vol. 18, Issue 16, R695-R696, August 26, 2008.

41. Kunz, T.H., L.F. Lumsden. 2003. "Ecology of Cavity and Foliage Roosting Bats." In T.H. Kunz and M.B. Fenton, editors. *Bat Ecology*. University of Chicago Press, Chicago, Illinois. pp. 3–89.

42. Vonhof, M.J., R.M.R. Barclay. 1996. "Roost Site Selection and Roosting Ecology of Forest Dwelling Bats in Southern British Columbia." *Canadian Journal of Zoology*. 74:1797–1805.

43. Kunz, T.H., L.F. Lumsden. 2003. "Ecology of Cavity and Foliage Roosting Bats." In T.H. Kunz and M.B. Fenton, editors. *Bat Ecology*. University of Chicago Press, Chicago, Illinois. pp. 3–89.

44. Long, C.V., J.A. Flint, P.A. Lepper. "Insect Attraction to Wind Turbines: Does Colour Play a Role?" *European Journal of Wildlife Research*. DOI: 10.1007/s10344-010-0432-7.

45. Nicholls B., P.A. Racey. "The Aversive Effect of Electromagnetic Radiation On Foraging Bats: A Possible Means of Discouraging Bats from Approaching Wind Turbines." *PLoS One*. Jul 16;4(7):e6246, 2009.

46. Horn, et al. "Behavioral Responses of Bats to Operating Wind Turbines." *Journal of Wildlife Management* 72:1 123–132, 2008: http://docs.google.com/viewer?url=http://www.bu.edu/cecb/files/2009/08/Horn_et_al_2008.pdf.

47. Boyle, Rebecca. "Increasing Wind Turbine Turn-On Speeds Could Help Reduce Bat Deaths." www.popsci.com/science/article/2010-10/increasing-wind-turbine-turn-speeds-could-help-reduce-bat-deaths-new-study-says.

Chapter 3

1. Beaty, William. "How Are Watts, Ohms, Amps, and Volts Related?" Amasci.com, April 2, 2000: http://amasci.com/elect/vwatt1.html.

2. Ibid.

3. upload.wikimedia.org/wikipedia/commons/1/1b/Lee_Ranch_Wind_Speed_Frequency.png.

4. Lee Ranch Data 2002. www.sandia.gov/wind/other/LeeRanchData-2002.pdf.

5. Wikipedia. "Wind Power." Wikipedia http://en.wikipedia.org/wiki/Wind_power#Distribution_of_wind_speed.

6. Realcalc scientific calculator. www.appbrain.com/app/realcalc-scientific-calculator/uk.co.nickfines.RealCalc.

7. GE Power's H Series Turbine. www.ge-energy.com/prod_serv/products/gas_turbines_cc/en/h_system/index.htm.

8. en.wikipedia.org/wiki/Thermal_efficiency.

9. U.S. Energy Information Administration. "The Impact of Increased Use of Hydrogen on Petroleum Consumption and Carbon Dioxide Emissions." 2008-04. www.eia.doe.gov/oiaf/servicerpt/hydro/hfct.html.

10. National Fuel Cell Research Center. "NFCRC Tutorial: Energy Conversion." www.nfcrc.uci.edu/EnergyTutorial/energyconversion.html.

11. Electrical Generation Efficiency—Working Document of the NPC Global Oil & Gas Study, July 18, 2007.

12. Ibid.

13. Sandia National Laboratories. "Sandia, Stirling Energy Systems Set New World Record for Solar-to-Grid Conversion Efficiency." Feb. 12, 2008.

14. Northern Arizona University. "Geothermal: Electricity Generation." geothermal.nau.edu/about/generation.shtml.

15. U.S. Department of Energy. "Solar Energy Technologies Program." apps1.eere.energy.gov/solar/cfm/faqs/third_level.cfm/name=Photovoltaics/cat=The%20Basics#Q43.

16. www.alternative-energy-guide.com/articles/bio-hydro/Ennercy-effiecincy.htm.

17. "Cree Raises Near-UV LED Efficiency Record." LED News. Apr 1, 2001: www.compoundsemiconductor.net/csc/features-details.php?cat=features&id=11513.

18. Examples of conversion efficiency of lamps. After Meyer & Nienhuis, 1988 and Anon, 1987.

19. Miyamoto, K. "Renewable Biological Systems for Alternative Sustainable Energy Production." FAO Agricultural Services Bulletin – 128. Chapter 1: Biological Energy Production. Food and Agriculture Organization of the United Nations.

20. www.alternative-energy-guide.com/articles/bio-hydro/Ennercy-effiecincy.htm.

Chapter 4

1. In 2008, the average annual electricity consumption for a U.S. residential utility customer was 11,040 kWh, an average of 920 kilowatt-hours (kWh) per month. Tennessee had the highest annual consumption at 15,624 kWh and Maine the lowest at 6,252 kWh. www.eia.doe.gov/cneaf/electricity/esr/table5_a.xls.

2. EIA. "U.S. Residential Electricity Consumption by End Use." 2008. www.eia.doe.gov/ask/electricity_faqs.asp#electricity_lighting.

3. Federal Energy Regulatory Commission Assessment of Demand Response & Advanced Metering. www.ferc.gov/legal/staff-reports/12-08-demand-response.pdf.

4. Kshemkalyani-Tavkar, Amruta. "Energy Efficiency Standards for Appliances." GreenLiving.com: lead-green-life.blogspot.com/2009/06/energy-efficiency-standards-for.html.

5. McCade, Dallas. "The Do-It-Yourself Energy Audit, Part I." www.youtube.com/watch?v=ptYMHoPzVgk and from Green Homes America, "Home Energy Audit Overview." www.youtube.com/watch?v=oZ_kO1M-1rY.

6. Shea, Kevin. "Energy Usage Spreadsheet." http://bit.ly/myEnergy.

7. itunes.apple.com/app/wattulator/id297265823?mt=8. Compatible with iPhone, iPod touch, and iPad.

8. *Consumer Reports Magazine* issue, March 2009.

9. The manufacturer's spec sheet for the energy meter reader E2: www.energymonitor.com/E2-Data-Sheet.pdf.

10. National Trust for Historic Preservation. "The Cost of Going Green." *Preservation Magazine Online.* www.preservationnation.org/magazine/2010/march-april/green-guide.html.

11. Software Product. "Power Stoplight Mobile: Application for iPhone, iPad, iPod Touch, and Android." www.powerstoplight.com/?page_id=8.

12. "Cut Energy Costs." *Consumer Reports.* www.consumerreports.org/cro/money/real-estate/plug-your-homes-money-leaks/cut-energy-costs/plug-your-homes-money-leaks-cut-energy-costs.htm.

13. Kanellos, Michael. "Sensors Drafted to Turn Off Lights Nationwide." CNET News, February 4, 2005 10:57 AM PST. news.cnet.com/Sensors-drafted-to-turn-off-lights-nationwide/2100-1008_3-5563775.html#ixzz 1ICA8zTqD.

14. Energy Star specifications. http://energystar.com.

Chapter 5

1. A worldwide-based free online classified service at craigslist.org.

2. National Renewable Energy Lab. "Wind Maps." www.nrel.gov/gis/wind.html.

3. European Commission. "Urban Wind Resource Assessment in the UK: An Introduction to Wind Resource Assessment in the Urban Environment." WINEUR Deliverable 5.1, February 2007.

4. Wind Shear and Wind Speed Variation Analysis for Wind Farm Projects for Texas.

5. WindTurbines.net. "Wind Shear." www.windturbines.net/wiki/Wind_Shear_Exponent.

6. Heier, Siegfried. *Grid Integration of Wind Energy Conversion Systems.* Chichester: John Wiley & Sons. p. 45, 2005.

7. Harrison, Robert. *Large Wind Turbines*. Chichester: John Wiley & Sons. p. 30, 2001.

8. "What Is Wind Shear?" www.weatherquestions.com/What_is_wind_shear.htm.

9. Zbigniew, Lubosny. "Wind Turbine Operation in Electric Power Systems: Advanced Modeling." Berlin: Springer. p. 17, 2003.

10. For more recent updated list on surface roughness coefficients, check the following article: Oliver Probst and Diego Cárdenas "State of the Art and Trends in Wind Resource Assessment." Energies 2010, 3, 1087-1141; doi:10.3390/en3061087, ISSN 1996-1073. www.mdpi.com/1996-1073/3/6/1087/pdf.

11. NASA. "Catalog of NASA Images and Animations of Our Home Planet." visibleearth.nasa.gov/view_rec.php?id=2154.

12. Natural Resources Conservation Service. "Wind Rose Data." www.wcc.nrcs.usda.gov/climate/windrose.html.

13. Australian Government, Bureau of Meteorology. "Wind Roses." www.bom.gov.au/climate/averages/wind/wind_rose.shtml

14. Sagrillo, Mick. "Warwick Urban Wind Trial Project Sheds Much Needed Light on Rooftop Wind Resources." urbanturbines.org.

15. news.cnet.com/8301-11128_3-10157474-54.html#ixzz11VJJ7GjE.

16. Sklar, Scott. "Can the Mounting of Small Wind Turbines Directly on Roof Structures Be Effective?" Renewable Energy World website, August 15, 2006. www.renewableenergyworld.com/rea/news/article/2006/08/can-the-mounting-of-small-wind-turbines-directly-on-roof-structures-be-effective-45725.

Chapter 6

1. CNET. "The Chart: HDTV Power Consumption Compared." reviews.cnet.com/green-tech/tv-consumption-chart/.

2. Blanchard, Olivier. *Macroeconomics* (2nd ed.). Englewood Cliffs, N.J: Prentice Hall, 2000.

3. Visual Economics. "Inflation by Country." www.visualeconomics.com/inflation-by-country/.

4. AWEA, "Small Wind Turbine Global Market Study." AWEA, 2009. e360.yale.edu/images/digest/2010_AWEA_Small_Wind_Turbine_Global_Market_Study-1.pdf.

5. Sagrillo, Mick, "Wind System Operation and Maintenance Costs." *Factsheet from AWEA Windletter,* December 2002. renewwisconsin.org/ wind/Toolbox-homeowners/Operation%20and%20maintenance%20costs.pdf.

6. Fischer, Mike. "Betting the Farm." *Home Power Magazine,* Aug–Sept 2003, issue 96, p.34-41, homepower.com/view/?file=HP96_pg34_Fischer.

7. Andy Black. andy@ongrid.net.

8. AWEA. "Financial Incentives." archive.awea.org/smallwind/toolbox/IMPROVE incentives.asp.

9. Center for Resource Solutions. "International Tax Incentives for Renewable Energy: Lessons for Public Policy." www.resource-solutions.org/lib/librarypdfs/IntPolicy-Renewable_Tax_Incentives.pdf.

10. Fischer, Mike. "Betting the Farm." *Home Power Magazine*, Aug–Sept 2003, Issue #96. homepower.com/view/?file=HP96_pg34_Fischer.

11. Gipe, Paul. "Renewable Energy Policy Mechanisms." windworks.org/FeedLaws/RenewableEnergyPolicyMechanismsbyPaulGipe.pdf.

12. Lipp, J. "Lessons for Effective Renewable Electricity Policy from Denmark, Germany and the United Kingdom," *Energy Policy*, Volume 35, Issue 11, pp.5481–5495, 2007.

13. Europa. "Support for Electricity from Renewable Energy Sources." europa.eu/legislation_summaries/energy/renewable_energy/l24452_en.htm.

14. REN21. "Renewables Global Status Report: 2009 Update," Paris: REN21 Secretariat, 2009. www.unep.fr/shared/docs/publications/RE_GSR_2009_Update.pdf.

15. en.wikipedia.org/wiki/Renewable_Energy_Certificates#cite_note-usdoe_gpn-0.

16. SREC Trade. "SREC Auction History." www.srectrade.com/auctionhistory.php.

17. www.ofgem.gov.uk/Sustainability/Environment/RenewablObl/Pages/RenewablObl.aspx.

18. World Future Council. "Unleashing Renewable Energy Power in Developing Countries: Proposal for a Global Renewable Energy Policy Fund." Hamburg. November 2009.

19. Center for Resource Solutions. "International Tax Incentives for Renewable Energy: Lessons for Public Policy." www.resource-solutions.org/lib/librarypdfs/IntPolicy-Renewable_Tax_Incentives.pdf.

20. Union of Concerned Scientists. "Production Tax Credit for Renewable Energy." www.ucsusa.org/clean_energy/solutions/big_picture_solutions/production-tax-credit-for.html.

21. United Kingdom provides 150% tax credits for small businesses; extrapolated from Table 8.1. Tax Credits for RD&D in Center for Resource Solutions. "International Tax Incentives for Renewable Energy: Lessons for Public Policy." www.resource-solutions.org/lib/librarypdfs/IntPolicy-Renewable_Tax_Incentives.pdf.

22. Center for Resource Solutions. "International Tax Incentives for Renewable Energy: Lessons for Public Policy." www.resource-solutions.org/lib/library pdfs/IntPolicy-Renewable_Tax_Incentives.pdf.

23. Ibid.

24. AccionUSA. "Green Business Resources." www.accionusa.org/home/small-business-loans/green-business-resources.aspx.

25. Black, Andy, Erin Moore Bean. "Cashing in on Renewable Energy." homepower.com/view/?file=HP121_pg48_RE_Sources.

26. Gipe, Paul. *Wind Energy, Renewable Energy for Home, Farm, and Business* (Chelsea Green Publishing, 2004).

27. Cohn, Lisa. "Safeguard Your RE Investment." *Home Power* 128, December 2008 & January 2009. www.realwriters.net/rew/HP128_pg72_Cohn.pdf.

28. Sagrillo, Mick. "Insuring Your Wind System Evaluating Policy Options." Sagrillo Power & Light Co. www.awea.org/smallwind/sagrillo/ms_insur1.html.

29. Ibid.

30. Cohn, Lisa. "Safeguard Your RE Investment: Finding a Policy That Works for Your System." *HomePower* Jan-Feb 2009, Issue #128. homepower.com/view/?file=HP128_pg72_Cohn.

31. McMullen, Mike. "Twisting in the Warranty Winds." windsystemsmag.com/media/pdfs/Articles/2010_Dec/1210_Powerguard.pdf.

Chapter 7

1. thinkexist.com/quotation/given-the-fact-that-you-can-count-on-one-hand-the/1479580.html.

Chapter 8

1. Wikipedia. ISO 9001. en.wikipedia.org/wiki/ISO_9000#Contents_of_ISO_9001.

2. European Commission. "CE Marking." ec.europa.eu/enterprise/policies/single-market-goods/cemarking/faq/index_en.htm#product Group.

3. www.rohs.eu/english/index.html.

4. IEEE. "IEEE 1547 Standard for Interconnecting Distributed Resources with Electric Power Systems." grouper.ieee.org/groups/scc21/1547/1547_index.html.

5. IEC. IEC61400-2. webstore.iec.ch/preview/info_iec61400-2%7Bed2.0%7Den_d.pdf.

6. NEMA. ANSI C84.1. www.nema.org/stds/complimentary-docs/upload/ANSI%20C84-1.pdf.

7. Underwriters Laboratory, UL 6140-1. ul.com/global/documents/offerings/perspectives/regulators/technical/ul_WindTurbines.pdf.

8. "Wind Certification Labels Expected in Late 2010." *Homepower Magazine.* Jun/Jul, #137, pp. 14–15, 2010.

Chapter 9

1. Gipe, Paul. *Wind Power: Renewable Energy for Home, Farm, and Business.* Chelsea Green Publishing Company: 2004.

2. Adapted from the June/July 2009 issue of *Home Power Magazine*, www .homepower.com.

3. "Acoustic Characteristics of the Bergey Excel-S 10 kW Wind Turbine." July 14, 2010. Per the AWEA standard, the tests were conducted in accordance with IEC 61400-11, "Wind Turbine Generator Systems, Part 11 – Acoustic Noise Measurement Techniques."

4. "Eoltec Scirocco 5.6 m/6 kW Wind Turbine Source Noise Level Measurement." HM: 1820/R1 Version 1.3, April 10, 2007. www.dpea .scotland.gov.uk/Documents/qJ8750/J70632.pdf.

5. Manufacturer's rating. www.skystreamenergy.com/documents/datasheets/ skystrea_%203.7t_datasheet.pdf.

6. Ibid.

7. Power-talk.net. "Wind Turbine Weight." www.power-talk.net/wind-turbine -weight.html.

8. Summerville, Brent. "Small Wind Turbine Performance in Western North Carolina." Appalachian State University Fall 2005. www.at.appstate.edu/ documents/researcharticlesmallwindperformanceBJS.pdf.

9. Sagrillo, Mick, Ian Woofenden. "Wind Energy Predictions." *Homepower Magazine*, Aug/Sep 2009 (#132) pp. 24–28. homepower.com/article/ ?file=HP132_pg24_Mail_2.

10. www.engr.colostate.edu/ALP/ALP_51_Stonington.htm.

11. "Wind Power: Capacity Factor, Intermittency, and What Happens When the Wind Doesn't Blow?" Renewable Energy Research Laboratory, University of Massachusetts at Amherst.

12. The British Wind Energy Association. "Blowing Away the Myths." February 2005. www.bwea.com/pdf/ref_three.pdf.

13. Power Talk.net. "Wind Turbine Speed." www.power-talk.net/wind-turbine -speed.html.

14. Woffenden, Ian, Mick Sagrillo. "2010 Wind Generator Buyer's Guide." *Homepower Magazine*, #137, pg. 44. homepower.com/view/?file= HP137 _ pg44 _Woofenden.

15. www.reuk.co.uk/Wind-Turbine-Tip-Speed-Ratio.htm.

16. Gipe, Paul. *Wind Power: Renewable Energy For Home, Farm, and Business*, Chelsea Green Publishing, 2004.

17. netfiles.uiuc.edu/mragheb/www/NPRE%20475%20Wind%20Power%20 Systems/Optimal%20Rotor%20Tip%20Speed%20Ratio.pdf.

18. Wind Turbine Overspeed Protection: Blade Pitch or Furl. Power-talk.net. www.power-talk.net/blade-pitch.html.

19. "Acoustic Characteristics of the Bergey Excel-S 10 kW Wind Turbine." July 14, 2010. Per the AWEA standard, the tests were conducted in accordance with IEC 61400-11, "Wind Turbine Generator Systems, Part 11 – Acoustic Noise Measurement Techniques."

Chapter 11

1. Chiras, Dans. *Power of the Wind* (New Society Publishers, 2009):168.

Index

Note: Page numbers followed by "f." and "t." indicate figures and tables, respectively.

A Wind Turbine
Recipe Book

Workshop plans and courses

www.scoraigwind.com

Hugh Piggott

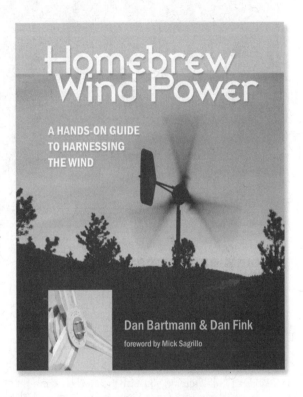